Molecular and Genetic Analysis of Human Traits

Molecular and Genetic Analysis of Human Traits

Gustavo Maroni

Department of Biology
University of North Carolina
Chapel Hill, North Carolina

b

Blackwell Science

© 2001 by Blackwell Science, Inc.

Editorial Offices:
Commerce Place, 350 Main Street, Malden, Massachusetts
02148, USA
Osney Mead, Oxford OX2 0EL, England
25 John Street, London WC1N 2BL, England
23 Ainslie Place, Edinburgh EH3 6AJ, Scotland
54 University Street, Carlton, Victoria 3053, Australia

Other Editorial Offices:
Blackwell Wissenschafts-Verlag GmbH, Kurfürstendamm 57,
10707 Berlin, Germany
Blackwell Science KK, MG Kodenmacho Building, 7-10 Kodenmacho Nihombashi, Chuo-ku, Tokyo 104, Japan

Distributors:

USA
　Blackwell Science, Inc.
　Commerce Place
　350 Main Street
　Malden, Massachusetts 02148
　(Telephone orders: 800-215-1000 or 781-388-8250; fax or-
　ders: 781-388-8270)

Canada
　Login Brothers Book Company
　324 Saulteaux Crescent
　Winnipeg, Manitoba, R3J 3T2
　(Telephone orders: 204-837-2987)

Australia
　Blackwell Science Pty, Ltd.
　54 University Street
　Carlton, Victoria 3053
　(Telephone orders: 03-9347-0300;
　fax orders: 03-9349-3016)

Outside North America and Australia
　Blackwell Science, Ltd.
　c/o Marston Book Services, Ltd.
　P.O. Box 269
　Abingdon
　Oxon OX14 4YN
　England
　(Telephone orders: 44-01235-465500;
　fax orders: 44-01235-465555)

Acquisitions: Nancy Hill-Whilton
Development: Jill Connor
Production: Irene Herlihy
Manufacturing: Lisa Flanagan
Marketing Manager: Carla Daves
Marketing Director: Lisa Larsen
Cover and interior design by Boynton Hue Studio
Typeset by Modern Graphics, Inc.
Printed and bound by Courier Companies / Westford

Printed in the United States of America

00 01 02 03 5 4 3 2 1

The Blackwell Science logo is a trade mark of Blackwell Science
Ltd., registered at the United Kingdom Trade Marks Registry

　Library of Congress Cataloging-in Publication Data
　Maroni, Gustavo.
　　Molecular and genetic analysis of human traits / by Gustavo
　Maroni.
　　　p. cm.
　　Includes bibliographical references.
　　ISBN 0-632-04369-5
　　1. Human genetics. 2. Medical genetics. I. Title.

　QH431.M3226 2000
　599.93′5—dc21

　　　　　　　　　　　　　　　　　　　　　　　　00-023792

Brief Contents

Table of Contents

CHAPTER **4**

Genome Organization II　86

CHAPTER **5**

Chromosomes and Karyotypes　125

How Mutant Alleles Affect the Phenotype 150

Mutations: Damage and Repair of DNA 175

Cancer: A Genetic Disease   208

Preface

The goal of this book is to present human genetics as a distinct and coherent field of study. Human genetics has become a discipline not just because the subject of analysis is special, but also—and more importantly—because methods have been developed that are either unique or very specially applied to the study of humans. We could define the broad aim of this discipline, at its current stage, as the characterization of the molecular basis of what we are, of our traits. A precondition of this characterization is, naturally, establishing which traits are genetically determined and which are not. This is one of the primary specific objectives of human genetics.

To appreciate the uniqueness of the subject matter it is necessary to get past the elementary aspects of the study of heredity, those that are better illustrated by examples from model organisms. Thus, this book is aimed at undergraduates and graduate students who have had the equivalent of a one-semester introduction to general genetics. It assumes familiarity with meiosis and Mendelian genetics as well as the basic principles of transmission of information from DNA to RNA to proteins. For those topics deemed most likely in need of reviewing, a series of summary diagrams is provided in the appendix.

The foregoing description applies to the first seven chapters of the book. The last two chapters could be considered applications of human genetics to two very diverse situations: the understanding of cancer, and the use of genetic predictions to counsel families at risk of genetic disease.

The selection of examples throughout the text was driven by two principles. The first is that advances in the field are such that we are long past the point where it would be either possible or desirable to present all the "interesting" cases in which genetics plays a role in human affairs. The second guiding principle is that the organization of this book follows the logic of concepts, rather than that of case studies. Examples, therefore, were chosen to illustrate specific methods or principles under discussion, and for this reason certain genes or genetic diseases may appear in various places in the book. No effort was made to tell "the whole story" about any given trait because this was thought to be distracting from the main effort of conveying the idea of human genetics as more than a collection of genetic diseases.

Illustrations are, to a large extent, taken from or based on the original literature. In these cases figures are visual pedagogical devices, and are used to present data that support contentions made in the text. The end-of-chapter questions reinforce this use of figures by asking the reader to extract information that was intentionally left out of the text.

Summary of Pedagogical Features

- Applications of human genetics: how genetics is used in understanding cancer and the use of genetic predictions to counsel families at risk of genetic disease.
- Examples throughout the text to illustrate specific methods or principles under discussion.
- Over 150 illustrations taken from or based on the original literature.
- End-of-chapter questions.

- General and specific references to primary literature.
- Internet references, with pertinent site addresses of distinction in the field of human genetics.

Acknowledgments

I would like to gratefully acknowledge my colleagues who read various portions of the manuscript and made suggestions: Rita Calvo, David Carson, Rosann Farber, Elaine Hiller, Ellen Kittler, Joan Rutila, Robert Sheehy, and Georgia Smith. I am also indebted to those scientists who so graciously made available published and unpublished figures for our use. A particular note of gratitude is owed to Ms. Susan Whitfield for her skill and patience in developing the artwork. I would also like to thank Irene Herlihy and Jill Connor of Blackwell Science for their invaluable contributions to the development of the final manuscript.

Gustavo Maroni
gmaroni@unc.edu

The Inheritance of Simple Mendelian Traits in Humans

1

DETECTING MENDELIAN INHERITANCE IN HUMANS

For supplemental information on this chapter, see Appendix Figures 1, 2, 3, and 7

The first example of simple Mendelian inheritance in humans was alkaptonuria, a metabolic abnormality documented by Sir Archibald E. Garrod soon after the rediscovery of Gregor Mendel's work in 1900. Alkaptonuria leads to the darkening of urine in newborns; it is a very rare and relatively mild condition caused by improper metabolism of the amino acid tyrosine. A metabolic intermediate, homogentisic acid, accumulates in tissues and is excreted in the urine where, upon exposure to air, it oxidizes to dark alkaptans. Because 1) the trait tended to occur within certain families, 2) the affected children were usually born from normal parents, and 3) parents of these children had often some degree of consanguinity, Garrod deduced that alkaptonuria followed the pattern of inheritance of a simple, autosomal, recessive trait.

1

Test-Crosses and Pure Lines versus Pedigrees and Inferred Genotypes

If one were dealing with experimental plants or animals, demonstrating the validity of a hypothesis of recessive inheritance might involve preparing homozygous strains of each phenotype and mating them to obtain F1 and F2 progenies in sufficient numbers to demonstrate a statistically significant 3:1 ratio of phenotypes in the F2. Alternatively, one might carry out a test-cross of presumptive heterozygotes to the homozygous recessive type with the expectation of obtaining a phenotypic ratio of 1:1. Of course, neither of these approaches is possible in humans. The human geneticist must instead rely on identifying informative matings a posteriori and then pooling results from many families to obtain statistically significant numbers. Identification of those informative matings is through the analyses of pedigrees.

Pedigree analysis is useful both in formulating a preliminary hypothesis regarding the mode of inheritance of a trait (as in the case of alkaptonuria) and in gathering the data needed to test that hypothesis. We will first introduce pedigrees for several traits for which the mode of inheritance is known, and we will return later to the question of selecting families for statistical tests. It will be useful to distinguish between the terms "genetic cross" and "pedigree."

Genetic cross is often used to denote a symbolic notation showing the proportions with which various offspring can be expected to occur given a certain mode of Mendelian inheritance (assuming a large enough progeny). Genetic cross can also refer to actual matings carried out in the laboratory whereby defined numbers of males and females produce sufficient progeny to test a particular prediction. In either case, all parents of the same sex have the same genotype and the genotype of at least one of the parents is known either from previous crosses or because they are members of a homozygous, pure strain.

Box 1.1	*Early observations of Mendelian inheritance in humans*

As we have mentioned, the earliest publication that attempted to explain a human trait in Mendelian terms was by Garrod in 1902. This study on alkaptonuria was several decades ahead of its time in that it linked the effects of genes to cellular chemistry. It would not be until the 1930s and 1940s that studies in *Neurospora* and *Drosophila* finally established the correspondence between genes and enzymes.

Only one year later, in 1903, W.C. Farabee submitted a doctoral thesis at Harvard University in which he interpreted brachydactyly in terms of dominant Mendelian inheritance. Farabee reached this conclusion studying a five generation family with over 100 members.

Even earlier, a report published in the *Philosophical Transactions of the Royal Society* (London) in the year 1779 was so precise that the pedigree of a sex-linked trait has been reconstructed from it (Fig. 1-2). A Mr. J. Scott reported that he did "not know any green in the world," that full red and full green appeared the same although he could tell yellows and most blues. Mr. Scott indicated,

> It is a family failing; my father has exactly the same impediment; my mother and one of my sisters were perfect in all colors; my other sister and myself, alike imperfect; my last mentioned sister has two sons, both imperfect; I have a son and a daughter, who both know all colors without exception; and so did their mother; my mother's own brother had the like impediment with me, though my mother, as mentioned above, knew all colors very well.

Interpretation of this report in Mendelian terms had to wait until the 20th century.

Pedigree, on the other hand, refers to a form of shorthand representation of a particular family; a pedigree notes the presence or absence of a particular trait in each member of the family as well as information about the birth order of siblings and their sexes (Fig. 1-1 and Table 1-1). As a rule, genotypes are deduced, but not known with certainty; however, with the advent of molecular tools for identifying hereditary traits, inclusion of the actual genotype of family members has become more common (we will come back to this later). Except in rare cases of very extended families, human families are usually too small to reveal Mendelian ratios from a single pedigree.

PATTERNS OF INHERITANCE AND EXAMPLES

Five basic modes of inheritance can be deduced from the analysis of pedigrees: *autosomal dominant, autosomal recessive, X-linked recessive, X-linked dominant,* and *Y-linked*. We will discuss the first three in some detail. The latter two are much rarer, and the characteristics of their pedigrees can be deduced by extension from the ensuing discussion.

To these five modes of *nuclear* inheritance should be added *mitochondrial inheritance*. The mitochondria contain a small amount of genetic material, and some mutations may result in distinguishable phenotypes. Because the oocyte, but not the spermatozoon, contributes mitochondria to the zygote, any trait determined by mitochondrial genes is transmitted from the mother and not the father, and it is transmitted to all her progeny. For this reason, mitochondrial inheritance is also called *maternal inheritance*. A caveat to be kept in mind with respect to mitochondrial in-

FIGURE 1-1. Partial pedigree of a large family affected with brachydactyly, an autosomal dominant disorder that causes shortening of the index finger. There are a number of conventions that are usually followed when representing pedigrees (see Table 1-1). Generations are indicated by roman numerals (on the left), and within each generation each individual receives an Arabic number ordered from left to right. Within each sibship the offspring are arranged by age with the oldest to the left. A bar across III-8 indicates that this woman was normal or with a very mild form of the trait, but may have been a carrier nonetheless. (Based on a pedigree by Mohr and Wriedt, 1919.)

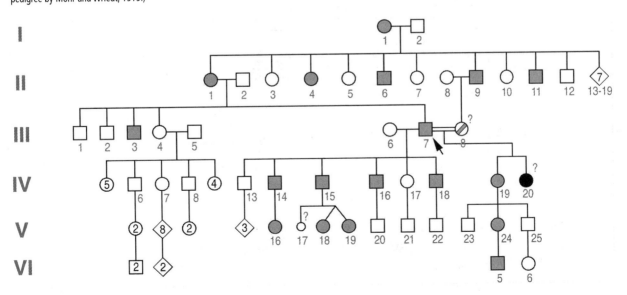

TABLE 1-1. Conventional use of symbols in the representation of pedigrees. Often various modifications of the symbols are used for specific purposes, and they are defined in each case.

◯	Female.
☐	Male.
◇	Individual whose sex is unknown or not disclosed.
② ◇③	Number of consecutive offspring.
○ ●	Spontaneous abortion, often with no specific information about genotype or phenotype.
☐ ◯	Normal individuals.
● ◆	Affected individuals.
◐ ◪	Usually indicate heterozygotes; the legend needs to describe, for example, how it is known that the individual is heterozygous.
⧄ ⊘	Deceased individuals.
◯─☐	Mating line.
☐ ◯	Offspring line.
● ☐	Dizygotic or fraternal twins.
■ ■	Monozygotic or identical twins.
◯═☐	Consanguineous matings, not always indicated.
➤◯	Propositus or proband, the individual at whose request the study was initiated.

heritance is that the oocyte contains thousands of mitochondria, and that when mutations occur, a mixture of both mutant and nonmutant mitochondria are passed on to the progeny. In this way many generations may pass before an individual acquires only mutant mitochondria.

Autosomal Dominant Traits

The pedigree in Figure 1-1 is that of a family in which many members are affected with a form of the autosomal dominant trait brachydactyly, a condition that involves

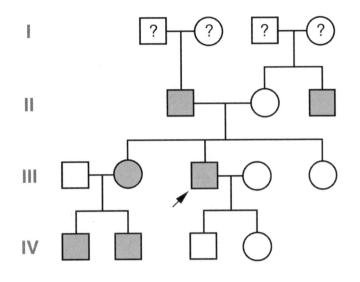

FIGURE 1-2. Pedigree of a family with color blindness reconstructed from an 18th century report. (Based on Cole, 1919.)

the shortening of the bones in certain digits. Figure 1-1 illustrates the distinctive characteristic of pedigrees representing families in which a dominant trait is segregating: all affected individuals are offspring of affected parents; as a consequence, the trait tends to appear in every generation. Sibships usually have affected and unaffected members, and the branches of the family initiated by unaffected members can be expected to be free of the trait from that point on (offspring of female III-4, for example). Dominant traits with simple Mendelian inheritance are rare in the population, and, except for special cases, the mating of two unrelated, affected individuals is unlikely. Hence, as Figure 1-1 shows, matings involving affected individuals in these families are usually between one individual heterozygous for the dominant trait and one who is homozygous for the normal, recessive, allele. The expected proportions among the progeny of such matings is ½ normal and ½ affected children.

The definition of dominance in experimental species is provided by the following relationships:

$$\text{Phenotype } (AA) = \text{Phenotype } (Aa) \neq \text{Phenotype } (aa) \qquad [1]$$

where A represents a mutant allele and a the normal allele. A consequence of the mating pattern described in the previous paragraph is that individuals homozygous for the dominant trait are extremely rare; therefore, only the second of the relationships shown in [1] is usually tested in human genetics.

The pedigree in Figure 1-1 is exceptional because it includes a marriage of first cousins (III-7 and III-8), both of whom may have been affected. From this marriage, a seriously deformed child (IV-20) was born; if as suspected, this child was homozygous AA, the phenotype of the homozygous AA is much more severe than that of heterozygotes. This pattern seems to be true for most "dominant" alleles affecting morphogenetic processes (in humans as well as in other species). It is likely that truly dominant mutations (as defined by relationship [1]) are represented by a few rare cases of genes affecting visible but nonvital functions, such as the production of certain pigments in flowers and animals. The rest could be considered partially dominant, such that phenotype (*Aa*) is different from the normal phenotype (*aa*), while phenotype (*AA*) is an even more extreme expression of the same trait and often inviable. Because in humans the genotype *AA* is so rare, the distinction between dominant and partially dominant is not made unless necessary; and any mutant allele that manifests itself in the heterozygote is said to be dominant.

An exception to the general rule described above is supplied by Huntington's disease, which appears to be a truly dominant trait, with homozygotes showing the same phenotype as heterozygotes (Fig. 1-3).

Extensive pedigrees are not always available for dominant traits. In some cases, the disabilities caused by dominant alleles are severe, even in the heterozygote, and affected individuals tend to leave few progeny, if any. Thus some dominant traits

FIGURE 1-3. Transmission of Huntington's disease, an autosomal dominant disease, in a five-generation family. The symbols below each individual represent their genotype for genetic markers located very close to the *HD* gene. In this family, HD is associated with allele *C1* of the marker locus. (The genotypes of deceased individuals were inferred from those of their progeny and are shown in parentheses.) The marriage of individuals III-6 and IV-6 produced 14 children; in this figure their birth order, as well as their sex and phenotype, are hidden to protect confidentiality. HD is a neurological disorder caused by a dominant mutation, but carriers of the gene remain asymptomatic until middle age; thus, we would not expect all *C1* carriers to be affected. At the time of this study, 7 of the 14 sibs showed some signs of disease, but the neurological profile of *C1* homozygous individuals was not significantly different from their heterozygous sibs. Also, the proportion of the various genotypes fits with the expected Mendelian ratios and does not support the hypothesis of in utero lethality for homozygotes. (Based on Wexler et al., 1987.)

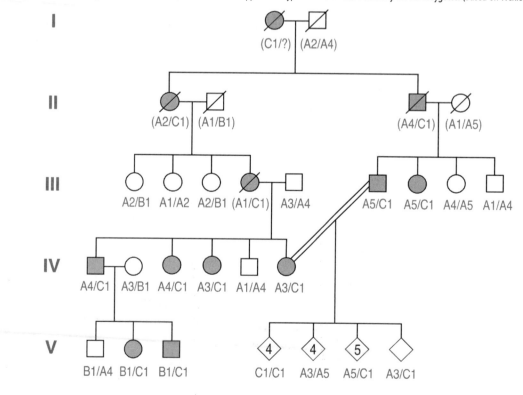

appear only sporadically and are maintained in the population mainly by new mutations.

Autosomal Recessive Traits

In most families in which an autosomal recessive trait is segregating, affected children are offspring of normal parents—when two affected individuals have children, all of them can be expected to be affected (see Fig. 2-11). These traits, then, appear most often when two heterozygous individuals bear one or more affected children. The unaffected offspring may or may not be carriers, but they usually marry unrelated people and the trait is very unlikely to reappear in subsequent generations. Figure 1-4 shows the pedigree of a family in which albinism, an autosomal recessive trait, is expressed. This family is an exception to the pattern of recessive traits as isolated cases, because it illustrates a trait that occurs at high frequency in that particular population, the Hopi people of Arizona. Another situation in which a recessive trait might occur with a greater probability than expected from the general population is among the children of parents with some degree of consanguinity.

In cases of recessive traits, relationship [1] applies fully, with *A* representing the normal allele and *a* the mutant one. Because, for the most part, recessive mutant alleles have negligible negative effect on the heterozygote, many of these alleles are perpetuated in the population at higher frequencies than are dominant mutations. Indeed, it is suspected that in some cases there may be *heterosis*, a situation where heterozygotes have a slight selective advantage over the homozygous normal genotype. It has been suggested, for example, that individuals heterozygous for the cystic fibrosis mutation may be better able to fight off some infections.

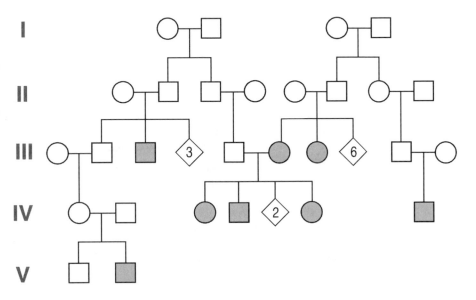

FIGURE 1-4. Transmission of albinism as an autosomal recessive trait in a Hopi family of Arizona. In this population the frequency of albinos in the population is approximately 1 in 200. (Redrawn with permission from Fig. 9 in Woolf, C.M. and Dukepoo, F.C. Hopi Indians, inbreeding, and albinism. Science 164:30–37 [1969]. Copyright 1969 American Association for the Advancement of Science.)

FIGURE 1-5. Transmission of the X-linked recessive form of retinitis pigmentosa, a degenerative disease of the retina that manifests itself in childhood and progresses to partial or complete loss of sight in 20 to 30 years. The cause is a mutation in the *RPGR* gene (*R*etinitis *P*igmentosa *G*TPase *R*egulator) at position Xp21.1. In this pedigree, half-filled symbols represent *obligatory heterozygous*, unaffected females who have transmitted the trait to their progeny. Notations under some symbols denote genotypes that were determined by direct analysis of DNA (*M*, mutant allele), and that agree in all respect with the phenotypic inferences. Some females are labeled *+/M*, indicating that they are heterozygous, but they are not represented by a half-filled circle because they have not yet transmitted the condition to progeny. (Based on Meindl et al., 1996.)

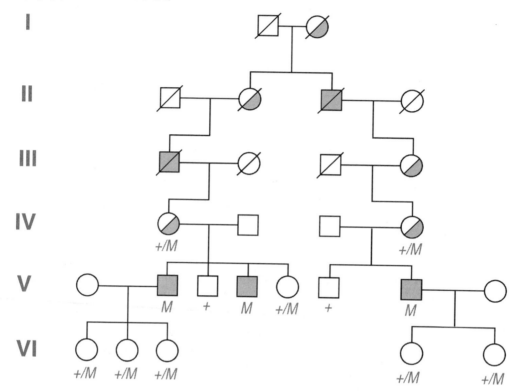

X-Linked Recessive Traits

Figure 1-5 shows the pedigree for a family transmitting retinitis pigmentosa, an X-linked progressive form of blindness caused by premature death of photoreceptor cells. Patients begin to show symptoms in their 20s and usually lose their sight completely in another 10 to 20 years. The pedigree shows the main characteristics of this type of inheritance: the trait is transmitted from affected males to daughters, who are usually heterozygotes and phenotypically normal, and from heterozygous mothers to half of their sons, thus producing a characteristic *skipping-of-generations* pattern, and to half their daughters.

Figure 1-2 illustrates a family with a very mild condition, color blindness, and shows a rare pedigree with affected females. In the case of other, more severe conditions, affected males never have offspring and the syndrome is transmitted only by heterozygous females. One such disease is Lesch-Nyhan syndrome (Fig. 1-6), which results in severe neurological and developmental abnormalities.

FIGURE 1-6. Pedigree of a family transmitting the X-linked, recessive, and fatal Lesch-Nyhan syndrome. This syndrome is caused by mutation of the structural gene for the enzyme hypoxanthine-guanine phosphoribosyl transferase, which leads to improper purine metabolism and excessive accumulation of uric acid. Characteristics of the syndrome are severe neurological disorders and self-mutilation (biting of lips and fingers); death usually ensues in infancy. (Based on Nyhan, 1968.)

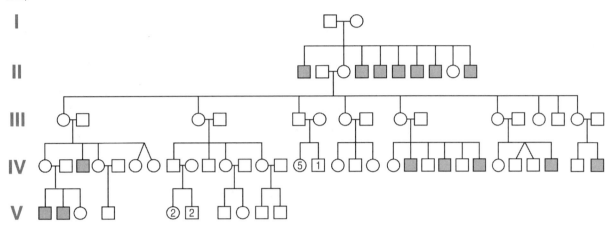

In a way, the situation with X-linked recessive inheritance is intermediate between those for autosomal dominant and recessive traits. The X-linked recessive mutation is masked in heterozygous females; thus, even fairly deleterious alleles are kept at relatively high frequencies in the population. At the same time, the trait is expressed in all hemizygous carrier males. These characteristics make X-linked traits the most common form of hereditary syndromes, and their tendency to reappear in a family, even in the absence of inbreeding, makes for unusually rich pedigrees.

THE USE OF PEDIGREES TO PREDICT THE RESULTS OF MATINGS

Probability of Individual Genotypes

One of the main uses of pedigree analysis is to predict whether children born by a particular set of parents will be affected by a given hereditary trait. In some cases, one can predict with almost complete certainty; more often, the best that can be achieved is an estimation of the probability of a given outcome. For example, from the pedigree in Figure 1-7, we can say that male III-1 is certain not to transmit the trait to his progeny (barring a new mutation); female II-4 can transmit either the mutant X or the normal X, so her sons have a 50% probability of being affected; her daughters will all be phenotypically normal but will have a 50% probability of being carriers. When probability estimates must be obtained, the calculation is always based on the Mendelian rules of transmission.

FIGURE 1-7. Hypothetical pedigree of a family in which a mild recessive X-linked trait, such as color blindness, is transmitted. The genotypes ascribed to various individuals are deduced from their phenotypes and those of their parents.

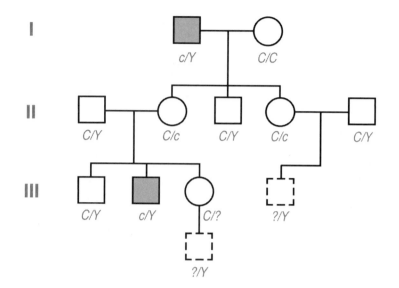

In many cases, one must predict the transmission of a trait by individuals whose genotype is not known with certainty. For example, female III-3 in Figure 1-7 may or may not be a carrier. For her to have a color blind son, two unrelated events are required: first, she must have received the mutant allele from her mother (probability = 0.5); second, that mutant allele must be carried in the oocyte that gives rise to her son (probability = 0.5). Thus, the probability that if female III-3 has a son he will be affected is the product of the probabilities of the two unrelated events, or 0.5 × 0.5 = 0.25. Notice, however, that if this individual were to have an affected son that would tell us that she is heterozygous (i.e., the probability that she carries the mutant allele becomes 1.0 rather than 0.5), so the probability of this woman bearing another affected son would be 0.5. If, on the other hand, she were to have three normal sons, this would give us some statistical information about her genotype. Because a woman homozygous for the normal allele is more likely to have three normal sons than one who is heterozygous, we could use conditional probabilities to estimate that the likelihood that individual III-3 is heterozygous—having had three normal sons—would be 0.11 rather than 0.5 (we will omit a discussion of conditional probabilities).

In practice, most calculations of probabilities in human genetics are carried out to assess risk to an individual or the progeny; the estimates are usually quite straightforward and come down to one of the Mendelian ratios, but see Chapter 9 for the difficulties presented by the *perception of risk* as compared to the actual objective value of that risk.

Probability of Group Outcomes

As we saw in the last section, it may sometimes be necessary to estimate the probability that a sibship of a particular phenotypic or genotypic composition will occur. The two basic rules of probability that are necessary to make these calculations are

1. If two independent events have probabilities of occurrence *p1* and *p2*, the probability that both events will occur is equal to the *product* of their individual probabilities (*p1* × *p2*). Thus, if we draw a card from a well-shuffled deck and throw a

die, the probability of drawing a 4 from the cards *and* throwing a 4 with the die can be obtained by multiplying the probability of drawing a 4 from the cards, 1/13, to the probability of throwing a 4 with the die, 1/6. The result is 1/78.

2. If two independent events have probabilities of occurrence *p1* and *p2*, the probability that either one of them will occur is equal to the *sum* of their individual probabilities (*p1* + *p2*). Thus, if we draw a card and throw a die, the probability of drawing a 4 from the cards *or* throwing a 4 with the die can be obtained by adding the probability of drawing a 4 from the cards, 1/13, to the probability of throwing a 4 with in the die, 1/6. The result being 19/78.

We can apply these rules to genetics cases. For example, when two heterozygotes for a recessive trait mate, the probability of any one of their offspring being normal is 0.75, and the probability of their having four normal children is the product of the probabilities for each individual or

$$0.75^4 = 0.32$$

The probability that the first three children will be normal and the last one affected is

$$0.75^3 \times 0.25 = 0.11$$

(Note that this "3:1" family, which matches the Mendelian expectation, is considerably less likely that than the "4:0" family.)

However, if we were to ask for the probability of a family with three normal and one affected children, we must take into account that there are four possible ways in which a family can have that genetic constitution. That is, the affected child can be the first, second, third, or last and each of those combinations has a probability of $0.75^3 \times 0.25$. Because any order satisfies the requirement of "three normal and one affected," we must add the probabilities for all possible orders. Thus, the probability of parents having three normal and one affected child, regardless of the order in which they are born, is

$$4 \times [0.75^3 \times 0.25] = 0.42$$

The first member of this equality consists of two parts: the factor within squared brackets indicates the probability that a family with the desired composition *in one particular order* will occur; the number 4 represents *all the possible orders* in which that composition can occur. In the general case, if we have a family with *n* children and *k* of them have one phenotype (*A*) and *n* – *k* have another phenotype (*B*), the number of possible orders in which that condition may be satisfied is given by the expression:

$$n!/k!(n - k)!$$

If *p* = probability of *A* and *q* = probability of *B*, the probability that *k* of them will have phenotype *A* and that *n* – *k* (the rest of the children) will have the alternative phenotype *B* is given by

Probability of [*k*(*A*) and *n* – *k*(*B*)] = $n!/k!(n - k)!p^k q^{(n - k)}$ [2]

For example, the probability that a family of five children (*n* = 5) will have three boys (*k* = 3) and two girls (*n* – *k* = 2) with *p* = *q* = 0.5 is

$$5 \times 4 \times 3 \times 2 \times 1/[(3 \times 2 \times 1) \times (2 \times 1)] \times [(0.5)^3(0.5)^2] = 10 \times 0.125 \times 0.25 = 0.3125$$

GATHERING DATA TO TEST THE MODE OF INHERITANCE OF A TRAIT

Pooling Data from Many Families and Ascertainment Errors

As was suggested earlier in this chapter, pedigrees can be used to identify families in which a particular trait occurs. Then, information from individual families can be pooled to obtain enough data for a statistical test of whether the predicted Mendelian ratios obtain. However, in selecting such families, some care must be taken to avoid systematic errors that can seriously skew the data; these are called *ascertainment errors* or *biases*.

Ascertainment errors in studies of dominant traits.   As was discussed earlier, most transmission of dominant traits in humans are through matings of heterozygotes with homozygotes for the recessive, normal allele. Let us assume that we are trying to confirm that a particular trait such as *polydactyly* is in fact dominant, as it appears from pedigree analysis. We could set up a program to look for the occurrence of the condition in kindergarten pupils; then, whenever we discovered an affected child, we would contact the parents to obtain permission to study the rest of the family. In this way, we would identify matings of interest, which as we said earlier would be, in most cases, of the type: *Aa × aa*. We would then tally the number of normal and affected phenotypes among their children. It is apparent that this approach will lead to an overestimate of affected children, because parents of genotype *Aa × aa* who by chance have only normal children will not be included in the count.

We can get an idea of the magnitude of the error generated by the approach described in the previous paragraph by considering the case of families with two children: there are four types of families with the constitutions and proportions shown in Table 1-2.

We can see that if we counted all the offspring in all the families, the proportion among the progeny would be four normal and four affected children, as would be expected from this type of mating. Note, however, that if we only counted families with

TABLE 1-2.   The four types of families of two children in which parents have the genotype *Aa × aa* (*a* is the normal allele). For each type of family is indicated the genotype of the children, the number of children with each genotype, and the frequency with which each family is expected to occur.

Family Type	Genotype of 1st Child	Genotype of 2nd Child	#Aa:#Aa	Frequency
1	Aa	Aa	2:0	1/4
2	Aa	aa	1:1	1/4
3	aa	Aa	1:1	1/4
4	aa	aa	0:2	1/4
Total	2Aa:2aa	2Aa:2aa	4:4	1

affected children, we would fail to count 25% of the families (Family Type 1). Also, among the families that are counted, there will be four affected to two normal children, or a proportion of 2:1, which would suggest other than simple dominant Mendelian inheritance. One easy way to obviate this problem is to select families for inclusion in the pool based on the identification of an affected parent, rather than on the existence of an affected child.

Ascertainment errors in studies of recessive traits. With recessive traits the problem of ascertainment bias is even more serious because parental couples in which both members are heterozygous can usually be identified only if they have an affected child. When both parents are heterozygotes, 9/16 of families with two children will be cases in which both children are normal.

Table 1-3 demonstrates the extent of relative undercounting of normal individuals that occurs among families with three children and the distorted phenotypic ratio derived as a consequence. Three-child families produced by parents heterozygous for a recessive trait can be classified into any one of eight possible sibships (see the sibship types column on the left in Table 1-3). The probability for each one of the sibships is given in the middle column. The right column gives the proportion of normal to affected children from each type of sibship. Thus, we would expect that a sample of 64 such three-children families will include 27 families with only normal children, 27 with one affected child, 9 with two affected children, and one with three affected children. In the 27 families having one affected child, the 81 children (27×3) will include 54 normal and 27 affected individuals, and this is the proportion shown in the right column.

An ascertainment error is introduced when we fail to include in the count the 27 families in which all three children are normal (the first row of data). Then, the 37

TABLE 1-3. The various types of families of three children in which parents have the genotype $Bb \times Bb$ (B is the normal allele).

Sibship Types			Probability of SibshipType		Ratio of Phenotypes	
Birth Order					**B_:bb**	
1st	**2nd**	**3rd**				
B_	B_	B_	$(3/4)^3 =$ 27/64		81:0	
B_	B_	bb	$(3/4)^2 \times 1/4 = 9/64$			
B_	bb	B_	$(3/4)^2 \times 1/4 = 9/64$	= 27/64	54:27	
bb	B_	B_	$(3/4)^2 \times 1/4 = 9/64$			
B_	bb	bb	$3/4 \times (1/4)^2 = 3/64$			
bb	B_	bb	$3/4 \times (1/4)^2 = 3/64$	= 9/64	9:18	
bb	bb	B_	$3/4 \times (1/4)^2 = 3/64$			
bb	bb	bb	$(1/4)^3 =$ 1/64		0:3	
Total	**Expected**		64/64		144:48 = 3:1	
	Observed		37/64		63:48 = 1.3:1	

TABLE 1-4. Inverse relationship between sibship size and ascertainment error.

Number of Sibs	Proportion of Families Uncounted	Biased Ratio B_:bb
1	3/4 = 0.75	0.00:1
2	$(3/4)^2 = 0.56$	0.75:1
3	$(3/4)^3 = 0.42$	1.31:1
5	$(3/4)^5 = 0.24$	2.05:1
10	$(3/4)^{10} = 0.06$	2.77:1

families that are observed include normal and affected phenotypes in the ratio 1.3:1.0 instead of the expected 3:1. The extent of the bias resulting from ascertainment errors depends on the size of the families being considered: the larger the family, the smaller the error. However, as is shown in Table 1-4, even for families of 10 children, if ascertainment errors are incurred, the ratio obtained deviates significantly from 3:1. There are several ways of correcting for this type of bias; one of them is to exclude the *proband* (the affected individual who brought this family into the study) from the count. (Note that each affected child in turn has to be considered the proband.) It is as if the proband is used to identify heterozygous parents, but the ratio of normal to affected children is obtained from among the sibs of the proband.

INHERITANCE OF MULTIPLE TRAITS: INDEPENDENT ASSORTMENT AND LINKAGE

Determination of Linkage from Human Pedigrees

Genes on different chromosomes assort independently of one another during meiosis. In other words, in a double heterozygote *A/a B/b*, the presence of one particular allele of gene *A* in a gamete gives no information about which of two alleles of gene *B* will occur in the same gamete. Thus, as shown in Figure 1-8A, gametes receiving *A* are equally likely to carry *B* or *b*. The converse is true when two genes are located close to each other on the same chromosome. This is illustrated by Figure 1-8B, the loci for genes *C* and *D* are close together, and inheritance of allele *C* will be associated with inheritance of *D* more often than with *d*. In such case, the genes are said to be *linked*. Establishing whether there is linkage between genes and the application of such knowledge for the creation of *linkage maps* is one of the most important tasks of human genetics. The practical implications of this will become apparent in later chapters.

In experimental animals, a *test-cross* or *back-cross* is carried out to establish whether there is linkage between genes of interest. One crosses a double heterozygote to the double homozygous recessive; and, if there is no linkage, as in Figure 1-8A, four phenotypic classes are recovered with equal frequency. If there is linkage, as in Figure 1-8B, the two parental classes predominate and the two recombinant classes are rarer, being that the latter two depend on the occurrence of crossing over, and that the frequency of crossing over depends on the distance between the genes. The specific phenotypes of the parental and recombinant classes depend on the *phase* of the alleles: if

FIGURE 1-8.   Proportion of parental-type and recombinant-type progeny in two-factor crosses of independently assorting and linked loci.

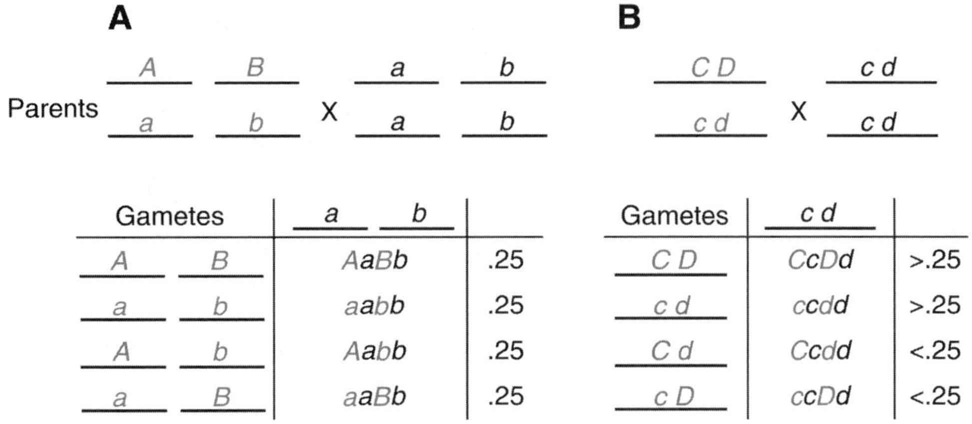

A

	A	*B*		*a*	*b*			
Parents	*a*	*b*	X	*a*	*b*			

Gametes		*a* *b*	
A	*B*	*AaBb*	.25
a	*b*	*aabb*	.25
A	*b*	*Aabb*	.25
a	*B*	*aaBb*	.25

B

C D		*c d*		
c d	X	*c d*		

Gametes	*c d*	
C D	*CcDd*	>.25
c d	*ccdd*	>.25
C d	*Ccdd*	<.25
c D	*ccDd*	<.25

C and *D* are on one member of the pair of homologues, and *c* and *d* are on the other, as in Figure 1-8B, the alleles are said to be *in coupling* and the parental classes are the ones that display either both dominant or both recessive traits. If *C* and *d* are on one chromosome and *c* and *D* on the other, the alleles are said to be *in repulsion* and the parental classes display one dominant and one recessive trait.

Using experimental test-crosses, it is possible to collect sufficient progeny to obtain statistically significant data and so ascertain with 95% or 99% probability whether the genes in question are linked or not. When genes are linked, the frequency of recombination can be used as an indirect measure of the physical distance between the loci occupied by these genes, and a genetic map can be constructed. Linkage studies of human genes are not so straightforward. Matings of double heterozygotes and homozygous recessive individuals do not occur frequently enough to be useful. This is especially true in the case of mutations that cause disease.

Human linkage information can, however, be obtained from other types of matings as is illustrated in Figure 1-9. In this example, both parents are heterozygous for the cystic fibrosis mutation (*c*), an autosomal recessive. The transmission of these alleles can only be followed with certainty in children who inherit both recessive alleles and who therefore suffer from cystic fibrosis. Unaffected children could be heterozygous or homozygous for the normal allele. With respect to the blood group alleles *M* and *N*, both alleles can be revealed in any combination because they can be typed with appropriate antisera (alleles such as these, are said to be *co-dominant*): the father is homozygous and the mother heterozygous for *M/N*. Notice that if the locus for cystic fibrosis (*C/c*) and the locus for *M/N* blood groups are on the same chromosome, we do not know their phase in the mother; the *c* allele could be in the same homologue as either *M* or *N*.

Despite those uncertainties, the pedigree in Figure 1-9 still yields some information on the linkage between *c* and *M/N*. The normal children 1 and 2 are noninformative because we do not know whether they carry the recessive *c* allele; children 3 and 4, on the other hand, developed from maternal gametes carrying both *N* and *c*. There are three possible explanations for this:

FIGURE 1-9. Transmission of cystic fibrosis and the *M/N* blood group in one of the families analyzed for possible linkage between the two loci. (Data from Steinberg and Morton, 1956.)

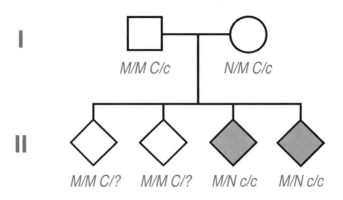

1. The loci are not linked, and *c* and *N* just happened to segregate to the same oocyte.
2. The loci are linked, *c* is in coupling with *N*, and the two alleles segregated together because there was no recombination.
3. The loci are linked, *c* and *N* are in repulsion, but recombination occurred during the meiotic divisions that produced the oocytes in question.

Although we cannot ascertain which of the three explanations is correct, we can assign a certain probability or likelihood that this combination of offsprings would occur given each of the three possibilities. Explanations 2 and 3 assume that the loci are linked, and thus the sum of the corresponding probabilities will be equal to the likelihood that this particular outcome would have occurred if there were linkage. We can then compare the likelihood of obtaining this particular family if the loci were linked (mechanisms 2 or 3) to the likelihood of obtaining it if they were not linked (mechanism 1) and in this fashion calculate the odds that the loci in question are linked.

FIGURE 1-10. Hypothetical family 1 used to estimate LOD scores.

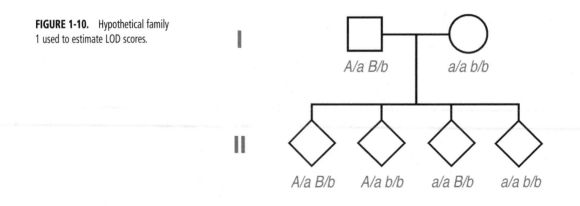

Genetic Recombination and LOD Scores

Although the method outlined above appears to be a very roundabout way of estimating linkage, it has the advantage of allowing for data from multiple pedigrees to be pooled, even when there is uncertainty about the phase of the alleles (as is usually the case), and even when the types of matings are different in the various families. In the study from which Figure 1-9 was derived, for example, data from the family described were combined with data from many other families, including one in which the mating was *C/c M/N* × *C/c M/N*. A more detailed example will make the concept clear. We will use the simple case of a hypothetical family in which the mother is homozygous for the recessive alleles (i.e., a test-cross), to illustrate how some of the calculations are carried out (Fig. 1-10).

The probability of each genotype among the offspring is simply the probability (p) of the corresponding type of paternal gamete; for example,

$$\text{Probability of genotype } A/a \; B/b = \text{Probability of paternal gamete } AB$$

If the genes were not linked, the probability of the four possible paternal gametes would be

$$p(AB) = p(Ab) = p(aB) = p(ab) = 0.25$$

and the probability of this particular sibship would be

$$p_1(0.5) = (0.25)^4 \approx 0.0039$$

where, in $p_1(0.5)$, the subscript 1 designates this particular family and 0.5 indicates 50% recombination, assuming independent assortment.

If the loci were linked, we would need to take into account two variables: the phase of the alleles and the genetic distance that separates them. Let us assume that the loci are 10 centimorgans apart (10% recombination). If the alleles were in coupling, the probability of the various gametes would be

gamete	AB	Ab	aB	ab
p	0.45	0.05	0.05	0.45

and the probability of this family,

$$p_1(0.1, \; coupling) = 0.45 \times 0.05 \times 0.05 \times 0.45 \approx 0.0005$$

If the alleles were in repulsion, the probability of the various gametes would be

gamete	AB	Ab	aB	ab
p	0.05	0.45	0.45	0.05

and the probability of this family,

$$p_1(0.1, \; repulsion) = 0.05 \times 0.45 \times 0.45 \times 0.05 \approx 0.0005$$

In this particular case, the probability estimated for the alleles in coupling happens to be the same as the probability for the alleles in repulsion. In the absence of specific

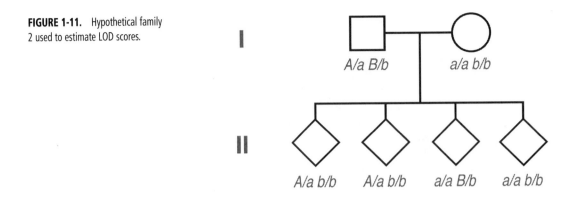

FIGURE 1-11. Hypothetical family 2 used to estimate LOD scores.

information, it is customary to assign equal likelihood to the two allelic arrangements, so the probability that this sibship would obtain if the loci were linked at a distance of 10% recombination is given by

$$p_1(0.1) = [0.5p_1(0.1, \text{repulsion})] + [0.5p_1(0.1, \text{coupling})] \approx 0.0005$$

A ratio of the likelihood of linkage to the likelihood of independent assortment gives the odds that the two loci are linked at a distance of 10 centimorgans,

$$p(0.1)/p(0.5) = 0.0005/0.0039 \approx 0.13$$

The odds are only 1 in 8 (0.13) that the loci are so close together. In other words, it is 8 times more likely that the loci assort independently than it is that they are linked 10 map units apart.

These calculations are repeated for several representative genetic distances (typically 0, 5%, 10%, 20%, 30%, and 40%) and if the odds indicate linkage, the distance between loci is approximated by the recombination fraction with highest odds.

Let us assume we have the opportunity to study a second family with the same traits, as shown in Figure 1-11. If the loci are not linked, the probability is the same as before, $p_2(0.5) = (0.25)^4 \approx 0.0039$. If we assume linkage with 10% recombination the probability of the various gametes are as before:

gamete	AB	Ab	aB	ab
p(coupling)	0.45	0.05	0.05	0.45
p(repulsion)	0.05	0.45	0.45	0.05

and the probability of this family if the alleles are in coupling is

$$p_2(0.1, \text{coupling}) = 0.05 \times 0.05 \times 0.05 \times 0.45 \approx 0.0001$$

The probability if the alleles are in repulsion is

$$p_2(0.1, \text{repulsion}) = 0.45 \times 0.45 \times 0.45 \times 0.05 \approx 0.0046$$

As before,

$$p_2(0.1) = [0.5p_2(0.1, repulsion)] + [0.5p_2(0.1, coupling)] \approx 0.0023$$

and the odds of linkage with 10% recombination are

$$p_2(0.1)/p_2(0.5) = 0.0023/0.0039 \approx 0.60$$

Again the odds are against the loci being linked that tightly, although in this case by less than 1.7:1. Because these two families are unrelated, the observations are independent of one another, and we can combine the two sets of observations simply by arguing that the likelihood of 10% linkage in both families is equal to the product of the individual probabilities. Thus,

$$p_T(0.1) = p_1(0.1) \times p_2(0.1) = 0.0005 \times 0.0023 \approx 1.1 \times 10^{-6}$$

Similarly, the probability of independent assortment in the two families is

$$p_T(0.5) = p_1(0.5) \times p_2(0.5) = 0.0039 \times 0.0039 \approx 1.5 \times 10^{-5}$$

and the odds of linkage relative to independent assortment considering both families is

$$p_T(0.1)/p_T(0.5) = (1.1 \times 10^{-6})/(1.5 \times 10^{-5}) \approx 0.076$$

Notice that

$$p_T(0.1) = p_1(0.1) \times p_2(0.1) \qquad \text{and} \qquad p_T(0.5) = p_1(0.5) \times p_2(0.5)$$

and the ratio

$$p_T(0.1)/p_T(0.5) = [p_1(0.1) \times p_2(0.1)]/[p_1(0.5) \times p_2(0.5)] = [p_1(0.1)/p_1(0.5)] \times [p_2(0.1)/p_2(0.5)]$$

This means that it is possible to pool data from different families without regard for the specific arrangement of alleles or the genotype of the parents simply by multiplying the odds calculated for each individual family ($0.13 \times 0.60 \approx 0.076$). By combining the results in this way, we calculate odds of 13:1 against linkage with 10% recombination as compared to odds of 8:1 and 1.7:1 calculated independently for each family. That is, the hypothesis that the loci are further apart (or even that they assort independently) becomes better supported by combining families.

When can we consider that there is enough data to reach a conclusion with confidence? It is customary to accept odds greater than 1000:1 as indicating that there is linkage at the specified distance and odds smaller than 0.01:1, as showing that the genes must be further apart. Such odds correspond to a probability of 0.99 that the conclusion reached is correct. For example, if a third family became available for study of the traits considered above, and it yielded a ratio

$$p_3(0.1)/p_3(0.5) \approx 0.1$$

the product of the three ratios would now be 0.0076 (0.076×0.1), and we could conclude with confidence that the loci are further apart than 10 centimorgans.

The Greek letter θ is used to indicate recombination frequency as a variable and L to indicate the ratio of probabilities, $L = p(\theta)/p(0.5)$. Thus, in general,

$$L = p(\theta)/p(0.5) = [0.5p(\theta, repulsion) + 0.5p(\theta, coupling)]/p(0.5)$$

$$L = [0.5 \times (\theta/2)^n \times ((1 - \theta)/2)^m + 0.5 \times (\theta/2)^m \times ((1 - \theta)/2)^n]/(0.25)^{m+n}$$

In this equation *m* and *n* represent the number of children of each genotype.

The formulations used in practice involve not the odds themselves, but their logarithm, known as *z* or *LOD* (for logodds) *score*. Pooling of data is done by adding LOD scores rather than multiplying ratios. Statistically significant LOD values are therefore smaller than −2 or greater than 3. As indicated above, the great value of this method is that it allows us to 1) obtain linkage information from various kinds of crosses and with no knowledge of the phase of alleles; and 2) combine results obtained by scientists working at different times and places.

It should be emphasized, however, that obtaining significant values is not always possible. The study of linkage between cystic fibrosis and the *M/N* blood group (a pioneering study involving 61 families from which the example discussed above was derived) could exclude the possibility of linkage closer than 5% recombination but proved to be inconclusive for longer distances. As Table 1-5 shows, although an assumption of 20% recombination yields the highest LOD score (0.72), we can draw no conclusions because the value is still well below 3, the lowest statistically significant value that is taken to demonstrate linkage. The hypothesis of independent assortment (i.e., that two loci are on separate chromosomes or very far apart on the same chromosome) is supported by LOD scores of less than −2 when assuming 40% recombination; because such low values are not reached for recombination frequencies of more than 5%, from these data we cannot conclude that there is no linkage. In other words, this study can reject neither independent assortment nor linkage at distances greater than 5 map units. Many years later it was demonstrated, by other methods, that the gene for cystic fibrosis is on chromosome 7 while that for the *M/N* blood group is on chromosome 4, and the two must therefore assort independently.

In practice, it is not necessary to carry out the calculations described above because tables and computer programs are available with all possible types of matings and for families with various numbers of children.

Informative and Noninformative Matings

An informative mating is one that can produce children whose phenotypes will help establish linkage relationships. In cases of complete dominance, the back-cross (see Fig. 1-8B) is the most informative type of mating because, for each child, the phenotype fully defines the genotype and one can ascertain whether any given child possesses one of the parental genotypes (*CDcd, cdcd*) or one of the recombinant ones (*Cdcd, cDcd*).

TABLE 1-5. Summary of data on the linkage analysis of the cystic fibrosis gene to the *M/N* blood group gene.

% recombination	0	5	10	20	30	40
z	−∞	−2.60	−0.27	0.72	0.53	0.18

SOURCE: Data from Steinberg and Morton, 1956.

FIGURE 1-12. Hypothetical family that may produce both informative and noninformative offspring, with respect to whether there was crossing over between the genes *c* and *D*.

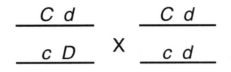

A somewhat less informative type of mating is the one in Figure 1-12. In this case, with respect to *C-c*, only *c/c* individuals reflect their genotype phenotypically and can be classified as parental (*cDcd*) or recombinant (*cdcd*). Individuals with the dominant allele *C* can be either *C/C* or *C/c* and are, therefore, uninformative.

Neutral Polymorphisms, Marker Loci, and Genetic Maps

Establishing linkage between genes that control major syndromes or genetic diseases is practically impossible because this would require data from individuals who are affected with two such traits. Such cases are so rare that information from multiple families is not usually available. Therefore, genetic studies usually involve the analysis of linkage between a disease-causing gene and some other more innocuous gene.

Notice that the matings discussed in the examples used to calculate LOD scores have the following characteristics: 1) one of the parents is heterozygous for the two traits being studied, and 2) the other parent's genotype allows recognition of an offspring's genotype in at least some cases. These conditions are most easily met if only one of the two genes includes a rare allele, as is usually the case for the gene of primary interest (e.g., cystic fibrosis or one of the hemophilias). The other gene is selected on the basis of the following criteria: 1) a high degree of polymorphism, that is, a large proportion of people are heterozygous at this locus, and 2) co-dominance, so the genotype is revealed by the phenotype, and the genotype of all progeny can be identified, regardless of the genotype of the parents.

---

www | Box 1.2 *Internet Sites*

A rich source of information on modern applications of LOD score analysis, including online and offline references and tutorials, is the European Molecular Biology Computing Network. CRI-MAP is the name of the program that generates a genetic map from raw family data. This site also provides numerous links to other sites of interest to geneticists, and can be reached at:

> **http:// biobase.dk /Embnetut/Crimap/**

A similarly useful resource is the web site of the Laboratory of Statistical Genetics at Rockefeller University:

> **http:// linkage.rockefeller.edu /**

Blood groups are among the traits that fit those requirements and were among the first to be used for such genetic analysis. For the study presented earlier—attempting to establish linkage between cystic fibrosis and the *M/N* blood groups determinants—the human geneticist was not entirely dependent on the serendipitous discovery of individuals carrying rare genetic combinations. Because a large fraction of the population is heterozygous for the *M/N* blood group, once a family is found in which both parents are heterozygous for the trait of primary interest, one can test whether one of them is also heterozygous for *M* and *N*. In the example presented earlier, both normal parents were known to be heterozygous for cystic fibrosis because of the appearance of the syndrome in their offspring; when the mother was also shown to be heterozygous for *M/N*, it was possible to proceed with the analysis.

Frequency of recombination between genes can be used as an approximate measure of physical distance between them and to create maps indicating the relative position of genes along each chromosome. Genes such as *M/N* are sometimes called *genetic markers, marker loci,* or simply *markers,* because they can be used as reference points against which we can determine the position of the genes of primary interest (sometimes called *trait loci*). It would be a great advantage to have a battery of biochemical tests that identify a large number of marker loci scattered throughout all the chromosomes. For any disease gene of interest a marker would be nearby, the distance to that marker could be measured, and the disease gene could be added to the genetic map. An overall map of all the chromosomes has long been a goal of human genetics, but only relatively recently have advances in molecular biology made the detection of a sufficiently large numbers of marker loci possible. In Chapter 4 we will discuss the question of a whole-genome genetic map.

The use of protein products to define marker loci.   Other than blood group determinants, early examples of genetic markers were the tissue typing genes responsible for the rejection of tissue transplants; in common with the blood groups, their various alleles are detected by immunological methods. In other words, the *phenotype* is the presence or absence of a particular antigen in the subject's tissues. That is, when we establish that an individual is of blood group AB, we are actually indirectly identifying the products of the two alleles of the *I* gene, I^A and I^B.

In population genetics studies, a *trait* is said to be *monomorphic* if one allele in the population is overwhelmingly prevalent, and no other allele has a frequency greater than 1%. A *polymorphic trait,* on the other hand, is one for which there is a second allele with a frequency of 1% or more. For example, in the major human populations, allele I^A varies in frequency between 20% and 30%, allele I^B is between 10% and 20%, and allele *i*, that produces no antigen, varies between 60% and 70%. The ABO blood group system is, thus, said to be a polymorphic trait and because there is no apparent adaptive difference in the various blood groups, it is referred to as a *neutral polymorphism.*

As a rule, the closer we inspect normal genes or gene products, the more likely we find them to be neutrally polymorphic. For example, other cases of neutral polymorphisms are detected when protein electrophoresis is used to analyze gene products. In this method, the protein is confined within a gel matrix under non-denaturing conditions and subjected to an electric field to observe its *mobility* relative to other proteins (the mobility of a protein under these conditions is determined largely by the balance of positively and negatively charged amino acids). Approximately 30% of proteins that have been tested display mobility polymorphism due to the allelic substitution of

FIGURE 1-13. Diagrammatic representation of the electrophoretic mobility of red blood cell acid phosphatase. To obtain this type of data, red blood cell extracts are run electrophoretically in a starch or polyacrylamide gel. The position of the enzyme is revealed when the gel is immersed in a buffered solution with a substrate that changes color by action of the enzyme. Homozygotes for ACP^A, ACP^B, and ACP^C each displays two bands (isozymes) due to post-translational modifications. Heterozygotes present four bands, two from each allele. (Based on Fig. 5-12 in Harris, 1975.)

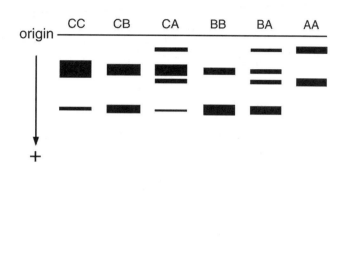

charged amino acids. These so-called electrophoretic variants are well-suited as genetic markers, and before the advent of DNA sequence analysis they frequently served as reference loci for the mapping of genes responsible for genetic disease. Electrophoretic variants are usually co-dominant traits; that is, in an individual heterozygous for a fast- and a slow-moving allele, both forms of the protein will be present and can be detected. Figure 1-13 is a schematic diagram of an electrophoretic gel stained for red blood cell acid phosphatase; it shows the distribution of the enzyme in individuals homozygous for alleles A, B, or C and in the heterozygotes.

The use of directly detected DNA differences as marker loci. Instead of depending on the characterization of gene products for detection of neutral polymorphisms, methods in molecular biology are used to identify differences in DNA molecules directly. One such method is based on *restriction endonucleases*; these are enzymes that hydrolyze the DNA double helix only where specific short sequences occur. The restriction enzyme *Bam*HI, for example, hydrolyses at all 5'GGATCC3' sequences and so digests human DNA into millions of fragments with an average length of a few thousand nucleotide pairs. Electrophoresis in a polyacrylamide gel matrix

FIGURE 1-14. Hypothetical distribution of restriction sites in a region of the genome for which there is a probe available.

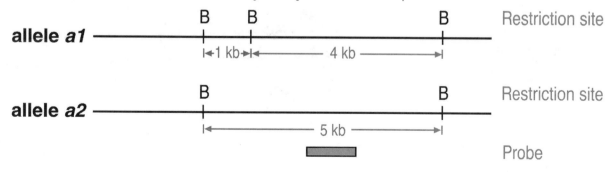

sorts these fragments according to size—the smaller ones run faster and so separate from the slower moving larger fragments. Molecular hybridization with a preselected short piece of DNA (the probe) that is labeled with radioactivity or a fluorescent dye allows the detection of pieces of DNA having a nucleotide sequence complementary to that of the probe (Southern analysis). Let us suppose that a probe corresponds to a region of the human genome with the distribution of *Bam*HI sites shown in Figure 1-14 (allele *a1*). When Southern analysis is performed on such DNA, a 4.0 kilobase (kb) band would be detected by the probe. Let us now assume that a fraction of the population is carrying a mutation in the middle restriction site such that a base substitution occurred within the *Bam*HI recognition site. The altered site would no longer be susceptible to cleavage by the enzyme and Southern analysis would highlight a 5.0 kb band instead of 4.0 kb one (Fig. 1-14, allele *a2*).

The presence or absence of the restriction site can be treated as alleles because they are inherited in a Mendelian fashion. The corresponding phenotypes are the differently sized fragments produced by the restriction enzyme, and are known as *restriction fragment length polymorphisms* (RFLPs). RFLPs are especially useful markers because they are co-dominant traits—that is, in heterozygotes, both alleles are detectable.

Notice that no function needs to be attributed to the segment of DNA we are studying. Often the DNA fragments of interest are not even associated with genes or they might be derived from stretches of DNA in the noncoding portion of genes; probes of this sort are useful simply because they detect a polymorphism. The higher the proportion of individuals in a population who are heterozygous for a particular site, the more useful is the probe. Figure 1-15 is an example of the kind of data provided by DNA markers; it reproduces the autoradiography of a Southern blot in which DNA was digested with the enzyme *Pvu*II and the hybridization probe detects a region in the neighborhood of the insulin gene. Aligned above each lane are the DNA donors from a three-generation family.

FIGURE 1-15. Inheritance of RFLPs in a nuclear family. The DNA of each individual is displayed in the lane directly below the symbol. DNA was digested with the enzyme *Pvu*II and hybridized with a probe from the insulin gene. Southern analysis of DNA shows bands of four possible sizes; dashes beside the autoradiograph serve as guides to the positions of the bands. (Redrawn with permission from Fig. 1a of White, R., Leppert, M., Bishop, D.T., et al. Construction of linkage maps with DNA markers for human chromosomes. Nature 313:101–105 [1985]. Copyright 1985 Macmillan Magazines Limited.)

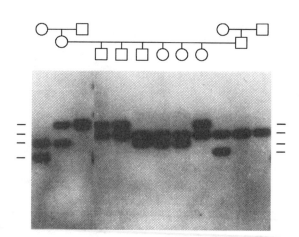

EXAMPLE 1.1	*Linkage of the Huntington's disease gene to an RFLP*

Huntington's disease (HD) is a progressive neurodegenerative disorder of late onset; the symptoms usually are not apparent until the patient is in his or her 30s or 40s. Usually the disease is fully penetrant and behaves as an autosomal dominant. Attempts to detect linkage between HD and classic polymorphic antigen and enzyme markers were negative, but only approximately 20% of the human genome was estimated to be covered by such traditional markers. With the advent of RFLP analysis, James Gusella and his collaborators started the systematic typing of two large affected families with anonymous human DNA fragments obtained from a genomic library, in search of a marker that would generate significant LOD scores with HD. Quite surprisingly, after very few such fragments had been tested, one of them, identified as G8, gave clearly positive results. The locus defined by the probe G8, designated *D4S10*, has seven *Hind*III sites; five of those sites are uniformly present in the population (they are monomorphic) and two are absent in some chromosomes (they are polymorphic) (Figs. 1-16 and 1-17). No recombination between *D4S10* and *HD* was detected in either family, and a maximum LOD score of 8.53 was obtained at 0% recombination; no linkage was observed with the *M/N* blood group marker, used as a control (Table 1-6). It should be noted, however, that the *HD* mutation was associated (in coupling) with the *A* haplotype in one family (American) and the *C* haplotype in the other (Venezuelan) (Fig. 1-18).

FIGURE 1-16. Top: Restriction map of *Hind* III sites in locus *D4S10*, and extent of the cloned G8 probe. Bottom: Various combinations of restriction sites (haplotypes) and the fragment sizes characteristic of each haplotype. H(1) and H(2) are the two polymorphic sites. (Redrawn with permission from Fig. 4 in Gusella, J.F., Wexler, N.S., Conneally, P.M., et al. A polymorphic DNA marker genetically linked to Huntington's disease. Nature 306:234–238 [1983]. Copyright 1985 Macmillan Magazines Limited.)

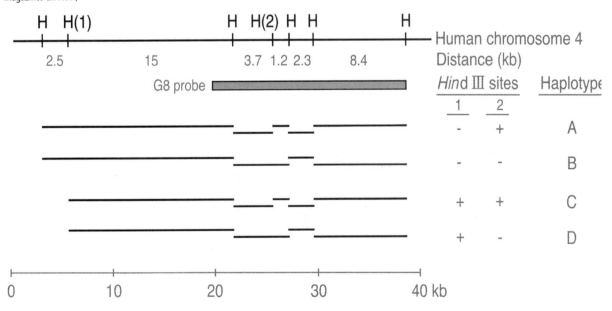

FIGURE 1-17. Southern blot autoradiography of DNA obtained from members of two families. The DNA was digested with *Hind* III and hybridized with the G8 probe shown in Figure 1-16. Haplotype designations are defined in Figure 1-16. (Redrawn with permission from Fig. 3 in Gusella, J.F., Wexler, N.S., Conneally, P.M., et al. A polymorphic DNA marker genetically linked to Huntington's disease. Nature 306:234–238 [1983]. Copyright 1985 Macmillan Magazines Limited.)

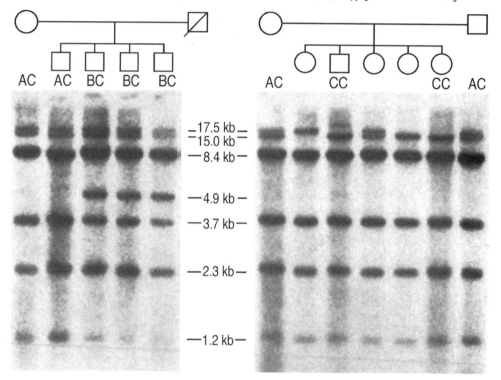

FIGURE 1-18. Branch of the Venezuelan family, indicating the haplotype and Huntington's disease status for each individual used for LOD score analysis. Haplotype designations are as in Figures 1-16 and 1-17. (Redrawn with permission from Fig. 2 in Gusella, J.F., Wexler, N.S., Conneally, P.M., et al. A polymorphic DNA marker genetically linked to Huntington's disease. Nature 306:234–238 [1983]. Copyright 1985 Macmillan Magazines Limited.)

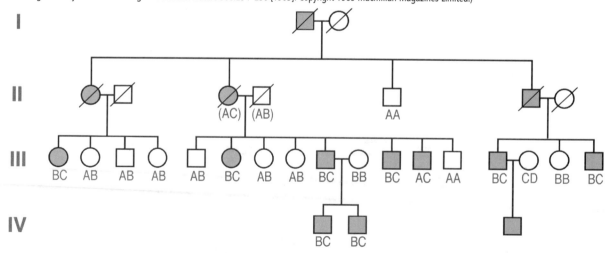

TABLE 1-6. Linkage of the Huntington's disease gene to the G8 probe and the *MN* blood group gene, summary of LOD scores.

% recombination	0	5	10	20	30	40
HD against G8	8.53	7.55	6.52	4.36	2.19	0.49
HD against MNS	$-\infty$	−3.22	−1.70	−0.43	−0.01	0.07

SOURCE: Data from Gusella et al., 1983.

In a later chapter, we will discuss other methods for detecting polymorphisms within the DNA sequence. These methods are very profitable because they identify more than just two alleles per locus. In some cases, more than 80% of the population can be heterozygous for the markers in question.

With the availability of many highly polymorphic DNA markers, it is now possible to select a set of marker loci such that all chromosomal regions are represented, then test the transmission of any genetic disease against the transmission of each DNA marker in turn. The existence of linkage between the disease gene and one or another of the marker loci determines the region where the gene is located.

CONCLUSION

The study of human genetics helps us to understand the biological principles that give rise to humans in all their diversity. From a more practical point of view, establishing whether complex traits and devastating diseases have genetic or only environmental causes is of great importance if we hope to improve the lot of those affected.

We started this chapter by introducing the approaches that can be used to establish that a given trait follows a Mendelian pattern of inheritance, the traditional way of demonstrating that a trait is hereditary. We ended the chapter by focusing on traits that are without a doubt hereditary and subject to simple Mendelian patterns of transmission, as these traits are the alternative forms of the DNA molecule itself. When faced with complex traits that do not exhibit simple Mendelian inheritance, it is often possible to demonstrate their hereditary nature indirectly, by establishing linkage of these traits to marker loci such as DNA polymorphisms. We will discuss this in more detail in Chapter 2.

EXERCISES

1-1. The pedigree in Figure 1-19 shows a family affected with amelogenesis imperfecta, redrawn from the study of the condition by Lageström et al. (1990). Black-filled symbols represent individuals in whom dental enamel is thinner and softer than normal; grey-filled symbols represent individuals with patches of normal and abnormal enamel. What form of inheritance appears to be followed by the trait in this family?

1-2. (A) What form of inheritance is displayed by the pedigrees in Figures 1-1 and 1-2? Define the symbols and assign genotypes to as many family members as possible. (B) Analyze Figure 1-3. Define the symbols and assign genotypes to as many family members as possible.

1-3. Analyze the pedigree in Figure 1-4. Indicate the genotype (*A* = normal, *a* = albino) of all individuals for which this is possible, starting with generations IV and V and proceeding upward. Notice in particular how many unrelated mates are heterozygous.

1-4. (A) Assume that the probability that an individual in the population at large will transmit a gamete with the autosomal mutant allele *a* for albinism is 0.01. What is the probability that two unrelated individuals will have an albino child? (B) The pedigree shown in Figure 1-20 represents a family in which albinism is transmitted as an autosomal recessive, and in which there are consanguineous matings. What is the relationship between individuals III-2 and III-3? What is the prior probability that they will have an albino child? What is the prior probability that individuals III-7 and III-8 will have an affected child? (C) Does the phenotype of the offspring that these two couples have had affect the probability that an unborn child will be affected? (D) What is the probability that III-4 and III-5 will have an albino child?

1-5. X-linked recessive inheritance, such as colorblindness, can be demonstrated by verifying that, among the offspring of carrier mothers and normal fathers, none of the daughters and half of the sons show the trait. If we were to pool families with affected sons to test that 1:1 ratio among them, what ascertainment error would we commit in families with three sons? In families with two sons? How could we avoid this ascertainment error?

1-6. What proportion of sibships of five children will have members of both sexes? What proportion will have two boys and three girls?

1-7. A customer is interested in purchasing two Siamese cats of different sex. At the store, the clerk says that he has two cats, and that he knows at least one is a male; he does not know the sex of the other. What is the probability that the customer will get her wish?

1-8. Lucy is an only child and her parents are expecting a baby. What is the probability that the new baby will be a boy? How does this situation differ from that in Exercise 1-7?

1-9. In Chapter 2, the pedigree in Figure 2-6 is of a family in which many members are affected with psoriasis (filled symbols). For each individual are also shown the alleles of a particular molecular marker on chromosome 17 (the numbers below each symbol); thus, individual II-1 has the allelic constitution L^2/L^3, and II-2 is L^6/L^5. Use this family to answer the following questions. (A) Propose a form of inheritance for psoriasis. (B) Suggest a genotype (relative to psoriasis) for each family member. (C) Does there appear to be linkage between a gene responsible for psoriasis and the molecular marker? Which is the marker allele most often associated with psoriasis? Could you say that marker allele causes psoriasis?

1-10. Using your answer to Exercise 1-2, calculate LOD scores between the Huntington's disease gene and the marker locus identified in Figure 1-3 for the nuclear family of individuals III-4 and III-5.

1-11. In Figure 1-19 each individual is characterized with respect to the trait amelogenesis imperfecta and also with respect to several RFLPs, whose alleles are indicated in the boxes below each symbol. (A) For the sake of this exercise, simplify the pedigree to assume that all the children in generation III are offspring of a single affected female in generation II. Then classify all offspring as being of parental or recombinant type for marker A/a and amelogenesis imperfecta, and calculate LOD scores for a distance of 10 map units between A/a and the trait. Repeat the exercise with marker E/e. (B) Is it possible to estimate a meaningful LOD score between generations I and II, using for example the children of I-2 and I-3?

1-12. Inspect Figure 1-16. Draw a gel with four lanes, one for each haplotype (combination of restriction sites A, B, C, and D) and indicate the positions at which you would expect radioactive bands for each, when using probe G8. Why does the 2.5 kb fragment fail to appear?

1-13. In Figure 1-15, assume that there are four discrete fragment sizes (in kb): 0.90, 0.95, 1.0, and 1,1. For each family member provide a genotype using the designation $I^{0.90}$, $I^{0.95}$, $I^{1.0}$, and $I^{1.1}$ for the four alleles. Notice their transmission as genetic markers.

1-14. Write down the genotype of individuals II-1, II-3, and II-4 in the second family of Figure 1-17, based on the bands present in the corresponding lanes. (The various haplotypes are defined in Figure 1-16.)

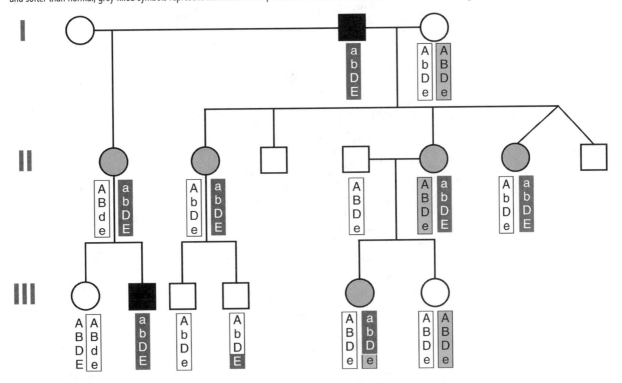

FIGURE 1-19. Pedigree of a family with amelogenesis imperfecta (OMIM301200). Black-filled symbols represent individuals in whom dental enamel is thinner and softer than normal; grey-filled symbols represent individuals with patches of normal and abnormal enamel. (Based on Lageström et al., 1990.)

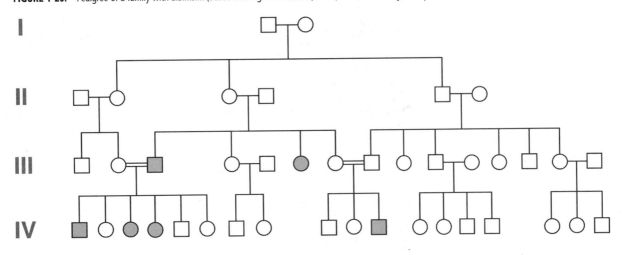

FIGURE 1-20. Pedigree of a family with albinism. (Based on a figure in Sinnott, Dunn, and Dobzhansky, *Principles of Genetics,* New York: McGraw Hill, 1950.)

REFERENCES

Cole LJ. (1919) An early family history of color blindness. J Hered 10:372–374.

Garrod AE. (1902) The incidence of alkaptonuria: a study in chemical individuality. Lancet 2:1616–1620.

Gusella JF, Wexler NS, Conneally PM, Naylor SL, Anderson MA, Tanzi RE, Watkins PC, Ottina K, Wallace MR, Sakaguchi AY, Young AB, Shoulson I, Bonilla E, and Martin JB. (1983) A polymorphic DNA marker genetically linked to Huntington's disease. Nature 306:234–238.

Harris H. (1975) The principles of human biochemical genetics. New York: American Elsevier.

Haws DV, and McKusick V. (1969) Farabee's brachydactylous kindred revisited. Bull Johns Hopkins Hosp 113:20–30.

Lageström M, Dahl N, Iselius L, Bäckman B, and Pettersson U. (1990) Mapping of the gene for amelogenesis imperfecta by linkage analysis. Am J Hum Genet 46:120–125.

Meindl A, et al. (1996) A gene (*RPGR*) with homology to the RCC1 guanine nucleotide exchange factor is mutated in X-linked retinitis pigmentosa (RP3). Nat Genet 13:35–42.

Mohr OL, and Wriedt C. (1919) A new type of hereditary brachyphalangy in man. Washington, DC: Carnegie Institute of Washington Publication No 295. Morton NE. (1995). LODs past and present. Genetics 140:7.

Nyhan WL. (1968) Seminars on Lesch-Nyhan syndrome. Discussion of epidemiology and genetic implications. Federation Proc 27:1091–1096.

Steinberg AG, and Morton NE. (1956) Sequential test for linkage between cystic fibrosis of the pancreas and the MNS locus. Am J Hum Genet 8:177–189.

Terwilliger JD. (1994) Handbook of human genetic linkage. Baltimore: Johns Hopkins University Press.

Wexler N, et al. (1987) Homozygotes for Huntington's disease. Nature 326:194–197.

White R, Leppert M, Bishop DT, Barker D, Berkowitz J, Brown C, Callahan P, Holm T, and Jerominski L. (1985) Construction of linkage maps with DNA markers for human chromosomes. Nature 313:101–105.

Woolf CM, and Dukepoo FC. (1969) Hopi Indians, inbreeding, and albinism. Science 164:30–37.

Hereditary Traits That Do Not Show a Simple Mendelian Pattern

2

For supplemental information on this chapter, see Appendix Figures 1, 2, 3, and 7

At the end of the previous chapter we said that one of the goals of human genetics is establishing which of the elements of diversity in the human phenotype are hereditary, of special interest being those phenotypes with medical implications. Most diversity in humans, however, does not follow a simple Mendelian pattern of inheritance. We should keep in mind that a good deal of the differences among humans are not hereditary at all, and many of those differences that are hereditary are not determined by single genes with two alleles, in what is called simple Mendelian inheritance. Even in those cases where a single gene is involved, the Mendelian ratios may not be realized for various reasons. For the sake of organizing the discussion, we will classify non-simple hereditary traits into *single-gene* traits, *quantitative* traits, and *complex* traits.

In this chapter we will also explore 1) how it is possible to test whether a given trait is hereditary, and 2) if a trait is hereditary, how scientists go about establishing the mode of inheritance and pinpointing the gene(s) responsible. In Chapters 3 and 4 we will see how the genetic approaches studied in these first two chapters are brought together with molecular methods to obtain a global understanding of human inheritance in the Human Genome Project.

SINGLE-GENE TRAITS LACKING A SIMPLE MENDELIAN PATTERN

Single-gene traits may fail to display simple Mendelian patterns because they show variable expressivity or incomplete penetrance, because they are of heterogenic origin, or because the gene in question has multiple alleles that differ from one another in their quantitative rather than qualitative effects. We will discuss this last case in the section on quantitative traits.

Variable Expressivity and Incomplete Penetrance

Variable expressivity refers to differences in severity of the condition observed in individuals having the same genotype, so the classification of phenotypes may be confusing because of the apparent existence of multiple classes. For example, one might propose that the mild cases are due to heterozygosis for a mutation and the more severe cases are due to homozygosis when both have, in fact, the same genotype.

The *penetrance* of a trait is defined as the fraction of individuals of a certain genotype who express a particular phenotype; *incomplete penetrance* refers to those cases where that fraction is less than one. In trying to obtain Mendelian ratios, this would lead to the counting of individuals in one category when by their genotype they belong in another.

The causes of variable expressivity and incomplete penetrance are diverse and specific to each gene. The two causes usually responsible for this failure of phenotypic homogeneity are environmental effects and differences in the genetic background— that is, the existence of *modifier genes* which control the expression of the primary gene. Depending on the alleles at these modifier genes, the expression of the mutation in the primary gene may be more or less severe.

The best understood cases of variable expressivity, however, are due neither to environmental differences nor genetic background. Huntington's disease and a number of other neuropathies present themselves with great variability within a family because of the inherent instability of the mutations, such that different members of the same family have alleles that become more and more severe from one generation to the next (see the discussion of trinucleotide repeat diseases in Chapter 7).

Complementation

Some traits may appear to lack simple Mendelian inheritance because they are of *heterogeneous origin* (as opposed to *monogenic*), meaning they represent clinically similar disorders that arise from mutations in different genes. For example, there are three genetically distinct entities that result in superficially similar blood clotting disorders— hemophilia A and hemophilia B are X-linked recessive mutations in the structural genes for clotting factors VIII and IX, respectively, while von Willebrand's disease is an autosomal dominant mutation in a large multimeric protein that acts as a carrier for factor VIII. Matings of people affected by any one variety of the disease will show

Mendelian inheritance, but if the existence of such multiplicity of varieties of a disease is unknown, matings in which one spouse carries a gene for one variety and the other spouse carries a mutation for another variety will have unexpected results.

Figure 2-1, for example, represents a family with hereditary hearing loss, a trait that is often determined by an autosomal recessive allele. Because they associate in educational institutions and other social activities, non-hearing people tend to marry one another more often than individuals homozygous for other traits. In this family we see that individuals II-7 and II-8 had, as expected, a family of all affected children. Individuals II-1 and II-2, on the other hand, both affected, had only hearing children. Given our understanding of this trait, this result is most likely due to the parents having mutations in two different genes (*AA bb* × *aaBB*) such that there is *complementation* (*AaBb*) in the children.

Phenocopies and Pleiotropy

The effect of a gene on some traits also may be hidden by phenocopies and/or pleiotropy. *Phenocopies* are cases in which nongenetic forces cause the production of a phenotype that is similar to that of a particular mutation. *Pleiotropy* is a term used to designate mutations that affect a number of different traits or organs. Breast cancer is a good example of both of these phenomena.

More than 90% of cases of breast cancer are sporadic, nonhereditary, and could be considered phenocopies of the few cases that are the result of a transmitted mutation in one of two genes that have been identified. One of those genes is *BRCA1*, which when mutant may cause breast cancer in some patients but ovarian cancer in others, thus showing pleiotropy. When we combine this capacity for pleiotropy with incomplete penetrance, the result is a pedigree such as the one illustrated later in Figure 2-7. We can see how this type of pedigree, combined with the overwhelming majority of sporadic cases of breast cancer, would make firm genetic predictions impossible if we were to rely only on the phenotype. It was the identification and characterizing the *BRCA1* gene by *molecular methods* that made possible our understanding of the extent to which this disease is hereditary and allowed us to tease apart, to some extent, the various factors that contribute to breast cancer. Breast cancer also provides an example of a heterogenic trait, because there are two genes with mutant alleles that can produce breast cancer, *BRCA1* and *BRCA2*. Note that in this case there is no complementation because both mutations are dominant.

FIGURE 2-1. Family carrying two different genetic forms of hearing loss. (Redrawn from Principles of Human Genetics by Stern © 1988, 1973, 1960, 1949 by W.H. Freeman and Company. Used with permission.)

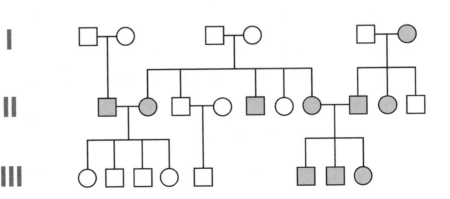

QUANTITATIVE TRAITS

The Mendelian transmission of many human traits is not easily demonstrated because the phenotypes cannot be classified into two or three discrete "types." These are instead quantitative traits for which there seems to be a continuum of expression within the normal population. Classical examples include height and weight. Physiological parameters of great public health significance such as blood pressure or serum cholesterol levels exhibit similar, continuously varying, patterns. In some cases, the variability may be due entirely to environmental conditions, but more typically quantitative traits result from the combined effects of genetics and the environment. In this section, we will explore the genetic mechanisms that give rise to the phenotypic gradations characteristic of quantitative traits.

Multiple Alleles

A potential source of continuously varying traits are single loci with multiple alleles; this allows for various possible allelic combinations that lead to a multiplicity of genotypes. For example, a study of acid phosphatase in the general population of England showed that enzyme levels followed an approximately bell-shaped curve (Figs. 1-13 and 2-2). Analysis of genotypes showed that individuals of genotype *AA* had the lowest enzyme levels, *BB* had higher levels, and the heterozygotes *AB* were intermediate. It also showed that *CB* individuals, although very rare, had the highest enzymatic ac-

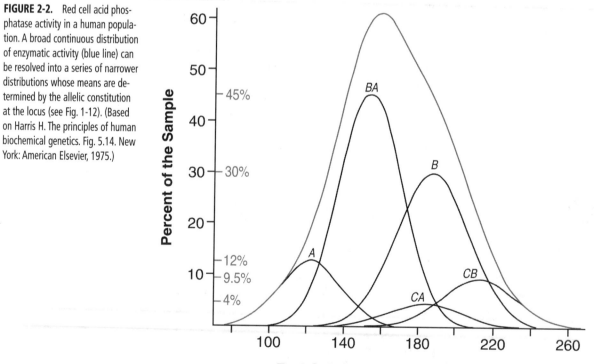

FIGURE 2-2. Red cell acid phosphatase activity in a human population. A broad continuous distribution of enzymatic activity (blue line) can be resolved into a series of narrower distributions whose means are determined by the allelic constitution at the locus (see Fig. 1-12). (Based on Harris H. The principles of human biochemical genetics. Fig. 5.14. New York: American Elsevier, 1975.)

tivity. Thus, different allelic combinations, partial dominance, and some variation in enzyme levels within each genotype (due either to environmental or other genetic causes) combine to produce a population with a broad, continuous distribution of acid phosphatase activity.

Polygenic Inheritance

Another simple model for the inheritance of quantitative traits is polygenic inheritance. In this model, each allele of two or more genes, sometimes called *polygenes*, contributes a certain amount to the variability in question. The model is an extension of the case of one locus with partial dominance (as the previously discussed acid phosphatase alleles *A* and *B*), in combination with additive effects among genes. For example, let us assume the existence of genes that affect body weight in laboratory rats such that the uppercase allele of each gene adds a certain number of grams to a "base" weight. In the F_2 generation of a cross between high-weight and low-weight pure lines, all possible phenotypes appear. Figure 2-3 shows the weight and frequency of various genotypes. Notice that 1) the more genes that are involved, the more weight classes there are in the F_2 generation and the harder it is to distinguish among classes, 2) environmentally generated variation overlaid on the genetic variation would blur the distinction between classes and produce a smooth or continuous distribution, and 3) the shape of the distribution curve depends on how many genes affect the trait—more genes produce a less disperse distribution because the extremes become rarer. This latter property has been used to infer, from the distribution of values in F_2 generations, the number of hypothetical genes that might affect a trait. The examples presented in Figure 2-3 are highly idealized: in a real population, the actual shape of the curve will depend on the frequency of both alleles at each locus (effectively set at 0.5 in this figure), and the contribution of each gene to the trait (in this figure all genes contribute equally).

The polygenic inheritance model is valuable for plant and animal breeding purposes where one has access to pure lines. Estimating the number of genes that affect a trait based on distribution curves such as those shown in Figure 2-3 is important in planning selection strategies. However, this approach has not been very useful in human genetics due to the difficulty in identifying homozygotes for quantitative traits. Also, information obtained in this type of analysis is of limited use to the human geneticist because it neither identifies the genes in question nor suggests their mechanisms of action.

Major and Minor Genes

Polygenes are also called *minor genes* to emphasize the small, quantitative contribution that each makes to the overall phenotype. The term "minor genes" also distinguishes polygenes from *major genes,* those responsible for more pronounced, qualitative effects. For example, in humans, height is probably determined by a number of minor genes that collectively account for much of the variability among normal individuals in a population; in addition, there are major genes, which, when mutant, lead to individuals whose height is outside the normal range of the population; these include the dominant mutation achondroplasia, as well as recessive mutations that lead to inadequate levels of growth hormone and cause dwarfism.

It is likely that the distinction between major and minor genes is a false one and that there are, instead, major and minor alleles. That is to say, the various genes that

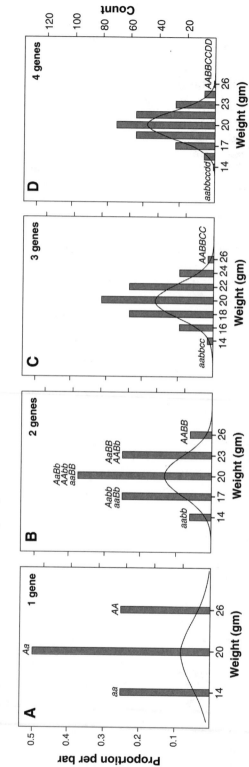

FIGURE 2-3. Weight distribution in the F2 generation of idealized examples in rats, in which one, two, three, or four genes contribute to the trait. The genotype with all dominant alleles weighs 12 grams more than the one with all recessive alleles. All dominant alleles present contribute equally to the weight difference. Thus in A, with only one gene, each A contributes 6 grams while in B with two genes, A and B each contributes 3 grams. (Based on Bodmer and Cavalli-Sforza, 1976.)

affect height may be represented by alleles that differ from one another by small effects on the phenotype or they may be represented by mutant alleles that cause major phenotypic changes.

COMPLEX TRAITS

We deal here with medical, psychological, and developmental syndromes that most likely have strong genetic components but that are not simple Mendelian traits. These include conditions such as ischemic heart disease, insulin-dependent diabetes mellitus (IDDM), schizophrenia, dyslexia, and many others.

Understanding the inheritance of these conditions has eluded human geneticists for many years, in part because many of the syndromes may be caused by various combinations of genetic and environmental factors. The question of incomplete penetrance arises again, but in this case it is complicated by the fact that there may be not just one gene involved, but a combination of alleles in several genes. In this situation even sibs are unlikely to receive identical allelic combinations. Identical twins provide the only opportunities for the observation of incomplete penetrance in these cases because they are genetically identical for all genes, and when they show *discordance* for a complex traits (one sibling showing the syndrome and the other one not), it is an indication of environmental effects. We will see in later chapters that two embryos derived from the same zygote are not necessarily identical, because of mutations or gene inactivation that may affect the two differently.

Other traits may fail to show simple Mendelian inheritance because a coincidence of several genetic factors, polygenes, or combinations of major and minor genes is necessary to produce the syndrome, as will be discussed next under "Liability."

Epidemiological Analysis of Complex Traits

Liability. The severity of complex-trait syndromes varies from case to case, but these traits can be distinguished from traditional quantitative traits because affected individuals present qualitative, diagnosable differences from the normal population. A model has been proposed that assumes the existence of an underlying continuous variable, called a *liability*, associated with each complex-trait syndrome. The concept of liability allows the application of quantitative genetics methods to all-or-none complex traits.

The liability for a given syndrome manifests itself as the normal phenotype in all individuals for whom its value is below a certain threshold, and manifests as the syndrome in question in those individuals with a value above that threshold. A particular liability value is determined, as for other quantitative traits, by a combination of environmental and genetic (multiple alleles, polygenes, major genes, etc.) causes. The hypothetical example in Figure 2-4A shows that *aa* has incomplete penetrance because some individuals of that genotype actually have a liability below the threshold. The low liability could be due to the environment or to genetic constitution at other loci. Carriers of the *A* allele are almost always below the threshold and unaffected. Figure 2-4B presents a case in which a dominant allele increases the liability for a syndrome, although a significant fraction of homozygous recessive individuals have a liability high enough to show the syndrome. The case in Figure 2-4B might represent one of those situations in which 80% of *Bb* individuals show the trait but only 10% of *bb* do.

FIGURE 2-4. Idealized distribution of liability in individuals of various genotypes. The vertical lines mark the mean liability for each genotype; deviations from those means are due to environmental causes or contributions from other genes to this liability. The threshold is the value in the liability spectrum that determines whether the syndrome will be expressed or not; all individuals with a liability greater than the threshold show the trait. The dashed line represents the overall frequency distribution of liability for all genotypes. **(A)** Case in which a recessive allele increases the liability: most *aa* individuals have a liability above the threshold and are affected, but a fraction are below the threshold, which makes this trait display incomplete penetrance. **(B)** A dominant allele increases the liability. Note that if the frequency of the mutant allele (*B*) is very low (as is usually the case), and the threshold for a syndrome is relatively low, the majority of cases in a population may be due to individuals (*bb*) who are not carriers of the high risk allele; an example of this situation is provided by mutations in two breast cancer genes *BRCA1* and *BRCA2*.

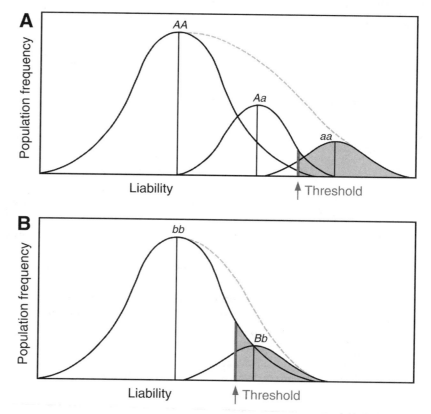

Liability methods are useful to help conceptualize the action of genes, and also to predict the likelihood of certain diseases depending on the presence of risk factors. Thus, with respect to heart disease, smoking, lack of physical activity, obesity, hypertension, the occurrence of the disease in family members, and so on, are all risk factors that contribute to the liability for heart disease. Each factor can be quantified (how much overweight? How many relatives and how closely related?) to assess an overall liability for each individual. This approach is, however, of limited value in determining whether a given trait has a genetic basis or not. Epidemiological techniques such as studies of family clustering and identical twins, on the other hand, can be used to evaluate whether a trait is, at least in part, of genetic origin.

Family clustering.   When the trait under study has a genetic determinant, close relatives of an affected individual have a greater probability of being themselves affected than subjects in the population at large. For example, one study showed that 13% of children with one schizophrenic parent also suffered from schizophrenia. This is a clear example of family clustering because in the general population schizophrenia occurs at a frequency of only 1%. The *relative risk index*, λ, expresses the degree to which a relative of a patient is more likely than a nonrelative to suffer a given syndrome; it is a simple ratio of the two frequencies. Thus, for schizophrenia, $\lambda_o = 13$ (the subscript indicates the relationship in question; in this case, "o" stands for offspring). For IDDM, $\lambda_s = 15$ (sibs of diabetic patients have a probability of 6% of developing diabetes themselves, whereas the prevalence in the general population is 0.4%).

Studies of family clusters can show that some traits tend to run in families, but they stop well short of establishing that the traits are hereditary; environmental factors could just as well be responsible. Adopted children are often used to attempt to clarify this *nature versus nurture* issue. For example, the study of children whose biological parents exhibit a syndrome, but who were raised by normal adoptive parents, helps to distinguish between environmental and genetic causes. Thus, in a study involving mothers whose children were adopted at birth, 5 out of 47 children of schizophrenic biological mothers were diagnosed with schizophrenia, but none of 50 adopted children of normal biological mothers were affected. These data indicate a strong genetic component to the causation of schizophrenia, although an environmental contribution from the biological mothers during intrauterine development cannot be excluded.

Identical twins.   Some of the best evidence for the role of genetic factors in the causation of certain complex traits comes from comparisons of *concordance* in monozygotic (identical) and dizygotic (fraternal) twins; concordance is an expression of the relative frequency with which a trait that occurs in one sibling is also present in the other. Figure 2-5 shows a few such comparisons of concordance. While monozygotic twins are genetically identical with respect to all their genes (except for somatic mutations), dizygotic twins have the same level of genetic similarity as ordinary sibs (but they are more likely to be exposed to similar environmental influences than ordinary sibs). Because we know that dizygotic twins share approximately 50% of their genetic information, we can formulate the following guidelines:

■ If monozygotic twins show close to 100% concordance, but dizygotic twins show significantly less, the trait is determined primarily by genetic mechanisms.

■ If monozygotic twins show a moderate level of concordance (40% to 60%), but one that is significantly higher than for dizygotic twins, both environmental and

FIGURE 2-5.   Comparison of concordance (fraction of sibs of affected probands who were also affected) in monozygotic and dizygotic twins. In most cases the number of pairs of sibs studied was in the hundreds. **(A)** Behavioral traits. **(B)** Physical conditions. (Redrawn with permission from Figs. 1 and 2 in Plomin, R., Owen, M.J. and McGiffin, P. The genetic basis of complex human behaviors. Science 264:1733–1739 [1994]. Copyright 1994 American Association for the Advancement of Science.)

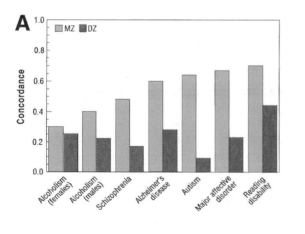

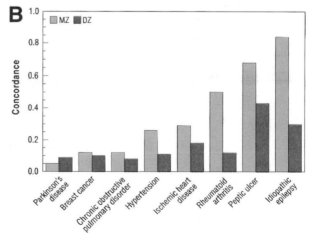

genetic components have a significant role in the genesis of the trait. The increased discordance exhibited by dizygotic twins results because they are more often dissimilar for the pertinent genetic factors.

- If monozygotic twins have a level of concordance that is four or more times greater than that for dizygotic twins, a recessive trait may be involved (a dizygotic twin of an affected individual has only a 25% chance of being homozygous). Alternatively, multiple genes may be involved and certain alleles of each of them are necessary for the trait to occur.
- If monozygotic and dizygotic twins show no significant difference in concordance, there is no detectable genetic contribution and the trait is environmental.

Note that epidemiological studies may help establish that there are genetic influences controlling certain traits, but they do not help to determine which genes are involved or where they are located.

LOCALIZATION OF GENES RESPONSIBLE FOR COMPLEX AND QUANTITATIVE TRAITS

Given that direct genetic analysis of complex traits is usually impossible due to the difficulties discussed above, alternative indirect methods have been developed to achieve the same goals. The aim of these methods is to demonstrate that a gene partly or wholly responsible for a complex trait is located near another gene that is inherited in a simple Mendelian fashion. This demonstration accomplishes two objectives: 1) it provides unequivocal evidence that the trait in question has a genetic determinant and 2) it identifies the approximate location of that genetic determinant. Here we will briefly review three methods: *linkage analysis, allele-sharing methods,* and *allelic-association* or *linkage-disequilibrium.* Each of these procedures requires complex analytical justification of the statistical techniques used, and although computer programs are available for the application of those techniques, a clear understanding of their limitations is necessary to obtain meaningful results. A full discussion is well beyond the scope of this text, so we will limit ourselves to a simplified presentation of the broad principles involved.

Some of the most provocative studies in this area have been attempts to analyze behavioral traits such as schizophrenia, alcoholism, neuroticism, dyslexia, and homosexuality. Studies of twins and family clusters often point to the existence of genetic determinants for some of these traits, but identifying specific genes has proved very difficult. In a number of cases, initial studies appeared statistically convincing only to prove later to be nonreproducible. For example, many studies have obtained statistically significant LOD scores between putative manic depression genes and chromosomal locations; unfortunately, almost two dozen sites have been identified this way with little convincing reproducibility for any one site. The reasons for these inconsistencies have usually been unfathomable, and the finite probability that any statistical test may give an erroneous result cannot be ignored.

Genetic heterogeneity, in which different gene mutations are responsible for similar syndromes, may also be the source of some discrepancies; in other cases the heterogeneity may be in that some cases are genetically determined while others are not. An example of the latter situation is found in breast cancer, where more than 90% of cases are sporadic, and less than 10% of them are due to a hereditary predisposition; mutations in one of two unlinked genes can result in that predisposition.

In any case, sufficient studies of behavioral traits have proved difficult to reproduce, so it is advisable to withhold judgment on the possible role that genes may play on any particular trait until several independent studies yield congruent results, or until specific genes products are identified such that biochemical confirmatory tests are possible.

Linkage Analysis of Complex Traits

Linkage analysis makes use of LOD score estimates for complex traits (see Chapter 1). There are two elements in this approach to genetic analysis:

1. We assume the existence of a *trait locus* for which there are at least two alleles: a high-liability allele that tends to cause the trait under study and a low liability allele found in the majority of the population.
2. We make use of *marker loci* having multiple co-dominant alleles and for which a large fraction of the population is heterozygous. Marker loci may represent RFLPs or other polymorphisms in the DNA sequence that are scattered throughout the genome and that are easily detectable. We introduced this concept in Chapter 1 and we will see more details in Chapter 4.

Affected and unaffected relatives are genotyped with respect to a number of marker loci, and LOD scores between the trait and the various marker loci are estimated. To make those estimates, it is necessary to know the dominance and penetrance of the trait. Although pedigree data can be used to make educated guesses in those regards, dominance and penetrance of the putative trait mutation are unknown quantities. For the sake of the calculations, they are usually treated the same as linkage distance; thus, for any given marker locus a series of penetrance and linkage values are used for the calculation of LOD scores assuming a dominant mutation. Then they are recalculated assuming partial dominance, recessivity, and so on. In principle, if the trait has a genetic basis, and sufficient markers and subjects are tested, a significant LOD score indicating linkage between the trait locus and a marker locus eventually should be obtained. In practice, however, results are often inconclusive, usually due to an insufficient number of subjects. If the number of marker loci is large enough and their distribution is throughout all the chromosomes, the search for linkage is said to be *genome wide*.

As we saw in Chapter 1, it is possible to add data from different families to reach significant LOD values. It is important to bear in mind, however, that complex traits may be caused by different genes in different families (i.e., the trait under study may be of heterogeneous origin). In these cases, combining data can be counterproductive and less informative than when data from each family are analyzed separately. The problem can sometimes be obviated by limiting studies to very large families, yielding significant results based on a single pedigree. In some cases, source heterogeneity may be reflected in phenotypic heterogeneity, and avoiding the latter by more stringent definition of the trait can be helpful in reaching significant linkage values.

In one study of familial breast cancer, for example, linkage of the trait locus—a cancer susceptibility gene designated *BRCA1*—to a particular marker could only be detected when the patients were classified by the age of onset of the disease; mutations in *BRCA1* are responsible for cancers that occur at a younger age, whereas familial cancers that occur at an older age seem to be due to mutations in other genes

EXAMPLE 2.1

Familial psoriasis

A gene responsible for familial psoriasis, a chronic skin inflammation that affects 2% of the population, was localized at the distal end of the long arm of chromosome 17 based on a study of eight white American families with a total of 151 participating relatives, and 65 cases of psoriasis. The marker loci used were a set of polymorphic microsatellites (see Chapter 3) that occur throughout the genome with a density of one every approximately 10 centimorgans. When the genotype for 69 of these markers was determined for all eight families, evidence of linkage between one of them and psoriasis was obtained. Models assuming a dominant mutation with penetrance anywhere between 60% and 99% returned significant indication of linkage to a marker in the long arm of chromosome 17. When the data for all eight families were combined, a maximum LOD score of 5.7 with marker locus *D17S784* at a distance of 15 centimorgans resulted.

Further analysis, however, revealed genetic heterogeneity in these families: four families showed evidence of linkage of psoriasis to *D17S784* (LOD score of 8.44 at 6 centimorgans) and four gave tentative evidence of no linkage to this marker (LOD score of –0.70 at 30 centi-

morgans, probability of linkage <0.20). Linkage to the 17q region was confirmed by analysis of other neighboring markers. This suggests that there are at least two genes for susceptibility to psoriasis. There were no clinical differences in the disease of those families who showed linkage to 17q and those who did not.

The pedigree shown in Figure 2-6 is for PS1, the largest of the eight families studied. The number under the symbol for each individual represents the allelic constitution for the marker locus *D17S784*; numbers in parentheses are deduced genotypes of individuals that were not examined directly. Notice that the trait is usually carried by the chromosome with allele 2. In individual III-5 and her progeny, however, the gene for the trait is on the chromosome with allele 3, probably because of a crossover that occurred in the germ cells of individual II-1.

The authors also used the allele-sharing method, which confirmed the existence of linkage between psoriasis and markers on chromosome 17q. For example, in family PS1 (see Fig. 2-6) the probability was estimated to be <0.0001 that affected pairs of family members would share alleles in marker locus *D17S784* as often as they do, if *D17S784* were not linked to a trait locus.

(see Example 2.2). Sometimes heterogeneity in the sample of families can be revealed by statistical means; when this is possible, the sample may be subdivided into homogeneous groups before further analysis is performed (see Example 2.1). Note that in these cases the data need to be subjected to rigorous statistical tests—it is illegitimate to simply select the families that give high LOD scores and consider them a natural group.

Affected Pedigree Member or Allele-Sharing Methods

Allele-sharing methods are based on identifying alleles of marker loci that tend to be shared by relatives affected by the same syndrome. The most commonly used of these methods are variations of the *sib pair method*. Sibs may share alleles that are *identical by descent* (ibd); for example, the two sibs represented in Figure 2-8A share one ibd allele, both having inherited the *A1* allele from their mother. This means that, unless crossing over occurred, the two sibs also have in common many other alleles, those that are carried on the same chromosome and in the vicinity of *A1*. In Figure 2-8B the

FIGURE 2-6. Pedigree of PS1, one of the families with psoriasis used to map a susceptibility gene. Shading indicates the presence of psoriasis, which behaves as a trait determined by a dominant allele in this pedigree. Female II-6 is half-shaded to indicate that she is an obligate carrier, given her affected offspring. She does not herself show the trait, thus representing an almost certain case of incomplete penetrance. The numbers under each symbol identify the allelic constitution of *S17S784*, a DNA marker closely linked to the psoriasis-causing gene (numbers in parentheses indicate inferred genotypes). (Based on Tomfohrde et al., 1994.)

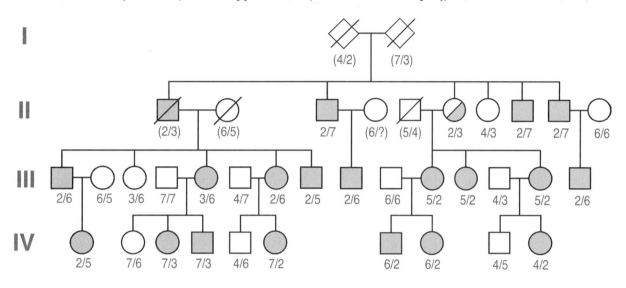

<table>
<tr><td>EXAMPLE 2.2</td></tr>
</table>

Familial breast cancer

Familial breast cancer represents only a small fraction of the total incidence of breast cancer. The 1990 study by Hall and colleagues established the existence of close linkage between a gene that confers susceptibility to breast cancer and marker locus *D17S74* on chromosome 17 at 17q21. The successful model was for a dominant allele with high penetrance—risk of breast cancer of genetically susceptible women is 66% by age 55 and 82% over the entire life span. The study involved 23 families with 329 participating relatives and 146 cases of breast cancer. All relatives were typed with respect to 183 polymorphic marker loci defined by variable number tandem repeats (see Chapter 3). Statistical analysis showed that the families were heterogeneous with respect to the linkage of breast cancer to 17q21. This heterogeneity is most evident when the families are sorted by average age of onset of the disease. Among the seven families in which the age of breast cancer diagnosis was 45 or younger, the LOD score (+5.98) indicated linkage to *D17S74* at a distance of 0.001 centimorgans, whereas the combined LOD scores for the families with late-onset disease were negative (four families illustrating this point are shown in Table 2-1). Analysis of three other markers in this region gave consistent results. By narrowing the definition of the trait from "familial breast cancer" to "early-onset familial breast cancer," it was possible to correlate the trait to a unique genetic entity. A gene designated *BRCA1*, which in carriers increases the likelihood for ovarian, breast, and other cancers, was later identified and cloned.

Figure 2-7 shows a pedigree of a family affected by *BRCA1* mutations—half-shaded individuals are obligate heterozygotes, but disease-free. Molecular analysis showed that individual III-4 is homozygous for the *BRCA1* mutation, but she does not appear more severely affected than her heterozygous relatives; thus, *BRCA1* mutations, like Huntington's

TABLE 2-1. LOD scores for linkage of breast cancer to *D17S74* in four of the 23 families. *M* is the mean age of diagnosis of breast cancer within each family.

Family	*M*	Recombination Fraction				
		0.001	0.10	0.20	0.30	0.40
1	32.7	+2.36	+1.89	+1.38	+0.82	+0.28
4	39.8	+1.14	+0.91	+0.64	+0.35	+0.11
19	55.8	−2.56	−0.93	−0.45	−0.20	−0.05
22	59.4	−0.85	−0.13	+0.04	+0.05	+0.02

disease mutations (see Chapter 1), appear to be fully dominant.

Note that the narrowing down of the trait definition, which helped in this case, may also lead to a loss of discrimination when dealing with pleiotropic mutations. In the case of *BRCA1*, patients with ovarian and other cancers that may also be the consequence of *BRCA1* mutations were not counted as "affected," making the detection of cancer heritability harder to accomplish.

sibs shown share no ibd alleles, and in Figure 2-8C they have two ibd alleles. Thus, at any one autosomal locus, sibs can have zero, one, or two ibd alleles. Absent any selection process, the three classes occur in the Mendelian ratio of 1:2:1.

The significance of ibd alleles can be made clearer by contrasting alleles identical by descent to alleles *identical by state*. In Figure 2-8D, the two sibs have in common the allele *A1*. Note, however, that in this case the two alleles are identical by state (they are both *A1*), but not identical by descent. (We can tell by the identity of the other allele in each child that one sib received *A1* from one parent and the other received it from the other parent.) Finally, in Figure 2-8E the two sibs have two alleles identical by state (they are both *A1/A2*), but we have no way of knowing whether they are also identical by descent. For this type of analysis, then, such a family would be totally uninformative.

Allele-sharing analysis begins with selection of families in which two sibs are concordant for the trait under study. Then the sibs' allelic constitutions for a number of marker loci are determined, a process called *genotyping* or *typing*, and the frequencies of ibd alleles are calculated. Finally, statistical methods are used to judge which, if any, of the marker loci show an excess of ibd alleles shared by affected sibs, relative to what would be expected at random. As the sibs included in such studies are selected because they are affected by the same trait (such as diabetes or schizophrenia), the discovery that they also happen to have ibd alleles for a particular marker locus suggests that the marker is linked to a gene controlling the trait under study.

If the trait gene, *T*, and the marker locus, *L*, are very tightly linked and if *T* is the only gene that can cause the trait in question, sibs who are concordant for the trait will almost always share the same *L* allele; that is, there will be no cases of affected sibs with zero ibd alleles of *L*. If there are two or more genes (*T*, *U*, etc.) that are capable of causing the trait, among 100 concordant sib pairs, only some will exhibit the

FIGURE 2-7. Pedigree of a family affected by *BRCA1* mutations. Individuals with cancer are indicated by filled symbols, with the organ and age given below: *B*, breast; *L*, lung; *O*, ovarian; *T*, throat. Note that many affected individuals did not have breast cancer. II-2 and II-3 were shown to carry the mutation but had not developed cancer at the time of the study. (Redrawn with permission from Fig. 1 in Boyd, M., Harris, F., McFarlane, R., et al. A human *BRCA1* gene knockout. Nature 375:541–542 (1995). Copyright 1995 Macmillan Magazines Limited.)

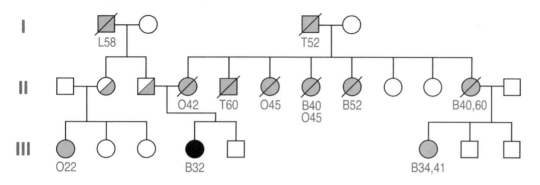

FIGURE 2-8. Some possible ways of allele-sharing between sibs. **(A)** One allele identical by descent. **(B)** Zero alleles ibd. **(C)** Two alleles ibd. **(D)** Zero alleles ibd, but one allele identical by state. **(E)** Two alleles identical by state, but unable to tell if any is ibd.

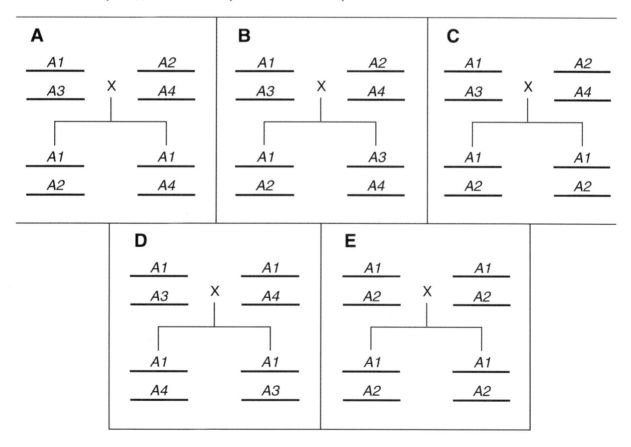

trait because they carry the defective gene *T* allele; in others the responsible gene will be *U*, and so on. Each one of the different trait genes would be linked to a different marker locus, so for any one marker locus we would find not the total absence of zero ibd alleles, but a statistically significant reduction in their number relative to what would be expected from sibs chosen at random (i.e., without regard for whether they carry the trait or not). The occurrence of a certain amount of recombination between the trait gene and the marker locus also tends to increase the frequency of zero ibd alleles among concordant sibs.

A number of statistical methods for sorting out the various confounding processes have been devised. Often results of this type of study are presented in the form of a statistic called the *maximum-likelihood LOD score* (MLS). The MLS may be interpreted, as is a LOD score, as the ratio of the probability that a given observational outcome would occur if the trait and marker loci were linked to the probability that it would occur if they were unlinked. In other words, the MLS represents the odds of linkage expressed as a logarithm.

There are also techniques that allow the use of pairs of relatives other than sibs. They are applicable to any relationship in which there is the potential for ibd alleles: parent–child, grandparent–grandchild, cousins, uncle/aunt–nephew/niece, and so on. One advantage of allele-sharing methods over linkage analysis methods is that it is not necessary to specify the form of inheritance in a model in order to ascertain linkage between the trait and marker loci (see Examples 2.1, 2.3, and 2.4).

Analysis of quantitative trait loci by allele-sharing methods.   The term *quantitative trait locus* (QTL) refers to polygenes, most often those polygenes that have been identified by their linkage to marker loci. As in the case of complex traits, the identification of QTL by linkage to markers became practical with the proliferation of molecular probes that allow the detection of neutral polymorphisms (marker loci) at many sites throughout the genome. In experimental animals, the method involves selection and inbreeding, over many generations, of strains with very high or very low values for a quantitative trait. These selected strains are characterized by a high level of homozygosis because of the inbreeding. When representative individuals of the two strains are crossed to each other and their offspring (F1) are interbred to obtain a second generation (F2), analysis of F2 individuals will indicate if the segregation of the quantitative trait is linked to the segregation of any one of the available molecular markers.

In human genetics, the term QTL is sometimes also used to designate complex trait loci even when the trait in question is not continuously variable, because of the assumption of an underlying liability, which is continuous. QTL analysis is in its infancy and the most common methods used are based on the sib-pair approach or one of its variants. A particularly powerful variant uses extremely discordant sib pairs—that is, sibs expressing values for the trait at opposite ends of the spectrum. For example, one might select sib pairs in which one sib has systolic blood pressure in the top 10% of values found in the population and the other has a value in the lowest 10% and analyze their allelic constitution with respect to a number of marker loci. If one of the marker loci is closely linked to a QTL with a significant effect on blood pressure, sibs at opposite ends of the spectrum would be expected not to have inherited the same allele for the marker in question (i.e., there would be an excess of sib pairs with zero ibd alleles); but they would share freely alleles for marker loci unlinked to the QTL.

EXAMPLE 2.3

Insulin-dependent diabetes mellitus

Insulin-dependent diabetes mellitus (IDDM), or diabetes type 1, is caused by the autoimmune destruction of the islets of Langerhans in the pancreas, and the consequent loss of capacity for insulin synthesis. In a genome-wide search for linkage using the sib pair method, Davies and co-workers identified 20 chromosomal regions that might be involved in IDDM; they then obtained conclusive evidence that four of those regions contribute to the disease. The first step in the study was the analysis of 289 marker loci in 96 United Kingdom sib pairs. The strongest evidence of linkage was for two markers in regions that were suspected, from previous studies, to be important determinants of diabetes. A marker on chromosome 6p21 in the HLA region gave an MLS of 7.3; this HLA-linked gene is now known as *IDDM1*. A marker on chromosome 11p15 produced an MLS of 2.1; this is *IDDM2*, and it is thought to correspond to the structural gene for insulin, which is known to be in that region.

Eighteen other regions gave MLSs greater than 1.0, an MLS of 1.0 indicating a probability of 95% of linkage. Although this is not conclusive evidence of linkage, it provides a useful screen of marker loci for further testing in other families. Two of the 18 regions were further tested by analysis of 102 sib pairs from the United Kingdom and 84 sib pairs from the United States. One of these, a region in 11q, gave an MLS of 3.4 (probability of no linkage <0.0001) and was designated *IDDM4*. For another region, in 6q, the authors judged evidence of linkage to be strong but not overwhelming (MLS = 2.0, probability <0.005); they designated this site *IDDM5*.

It should be noted that MLS values for a given locus are an indirect reflection of the frequency with which a particular site is responsible for the trait among all the sib pairs in the sample. Davies and co-workers were able to estimate that *IDDM1* is responsible for approximately 40% of IDDM incidence; *IDDM2* is responsible for 10%; and *IDDM4* and *IDDM5* probably make similarly small contributions. They also concluded that there are probably no other major genes with as high an incidence as *IDDM1*; Davies et al. predict that many of the remaining 16 possible sites will prove to be false positives.

EXAMPLE 2.4

Male homosexuality

Epidemiological studies suggest that male homosexuality may have a genetic component, because, according to some of these studies, brothers of a homosexual proband are four times more likely to be homosexual than brothers of a heterosexual proband—relative risk index, $\lambda_s = 4$. Also, in one study, concordance among twin brothers of homosexual probands was 52% (29/56) for monozygotic twins, and 22% (12/54) for dizygotic twins. These observations led to a study of homosexuality in nuclear families by Hamer et al. (1993), who noticed that in certain families, brothers, maternal uncles, and nephews (sons of a sister) of gay men were often homosexual, whereas fathers or sons were not (Fig. 2-9). This is a pattern that suggests a possible contribution by the X-chromosome, and the authors set out to analyze allele sharing frequencies of X-chromosome marker loci in 40 pairs of concordant sibs. Markers near the tip of the long arm of the X (Xq28) were indeed shared by homosexual sibs more frequently—statistically significantly so—than would have been expected if they were not linked to a trait locus for homosexuality.

In a second study from the same lab (Hu et al., 1995), a new set of brothers was identified and genotyped for X-chro-

FIGURE 2-9. Partial pedigree of one of the families that suggested that in some cases male homosexuality may be determined by an X-linked factor. Note the presence of the trait in a group of males that carry an X-chromosome potentially identical by descent. (Based on Hamer et al., 1993.)

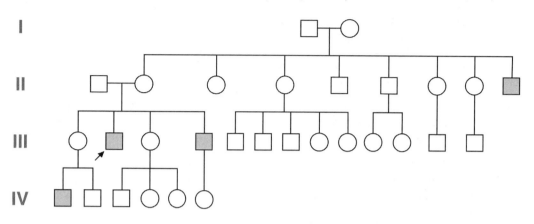

mosome markers. Twenty-two out of 32 concordant gay sib pairs had Xq28 markers that were identical by descent (68%). On the other hand, genotyping of 10 heterosexual brothers of the concordant sib-pairs showed that 9 of them had zero ibd Xq28 markers (one had one allele ibd). In both cases one would expect an average of 50% of sibs having one allele ibd in the absence of linkage, and the overall probability of obtaining these results in the absence of linkage was estimated to be 1% or less. (Note that in the case of X-linked traits two brothers can only have either zero or one ibd allele.)

These studies do not represent samples of the entire male homosexual population. Even if correct, they do not prove that all male homosexuality is influenced by genes on the X, nor do they prove that most cases in the population have a genetic component. What they attempt to show is that in those cases of male homosexuality that appear to be transmitted as an X-linked trait, there is a small section of the X-chromosome that seems to be responsible for the condition, because gay brothers tend to share alleles in that region more often than expected by chance alone. As to the putative gene(s) on Xq28, final proof will require the molecular isolation of the gene(s) in question, analysis of their action, and effects of possible mutations (see "Candidate Genes," and Chapter 4). Also, another study, in which gay males were not preselected by possible maternal inheritance, failed to find linkage to Xq28 markers.

The imaginary pedigree in Figure 2-10 demonstrates how the allele-sharing method could be used for the study of inheritance of blood pressure levels. In this example, p^H is an allele for a QTL that adds a significant amount to the blood pressure of carriers and p^L reduces a carrier's blood pressure. A^1, A^2, A^3, and A^4 and B^1, B^2, B^3, and B^4 are alleles at marker loci that are linked and unlinked, respectively, to the QTL. We would expect that the two brothers represented in Figure 2-10 will inherit zero, one, or two ibd alleles at the B locus according to the rules of independent assortment (i.e., there would be no correlation between blood pressure and the allele inherited). On the other hand, we would expect the brothers to inherit zero ibd alleles at the A

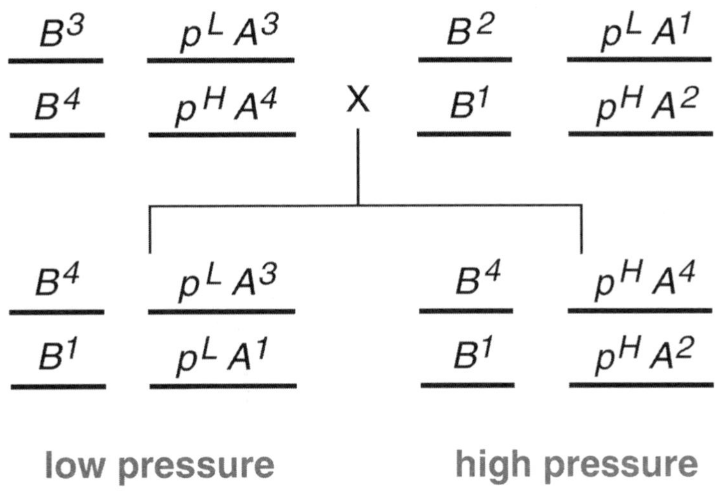

FIGURE 2-10. Hypothetical example of the identification of a quantitative trait locus. Sibs who are discordant for a trait, such as high versus low blood pressure, determined in part by a QTL (*p*), tend to have fewer alleles ibd of loci linked to the QTL.

$$\frac{B^3}{B^4} \quad \frac{p^L A^3}{p^H A^4} \quad X \quad \frac{B^2}{B^1} \quad \frac{p^L A^1}{p^H A^2}$$

$$\frac{B^4}{B^1} \quad \frac{p^L A^3}{p^L A^1} \qquad \frac{B^4}{B^1} \quad \frac{p^H A^4}{p^H A^2}$$

low pressure   **high pressure**

locus much more frequently than two ibd alleles (i.e., there would be a negative correlation between the difference in blood pressure and the number of ibd alleles inherited).

By analyzing many discordant sib pairs with respect to their allelic constitution at a large number of marker loci, we can hope to discover a marker locus that has a statistically significant negative correlation with blood pressure differences; we would thus identify a locus linked to a QTL affecting blood pressure. As in the case of complex traits analyzed by allele-sharing methods, LOD scores can be calculated from this type of data, and an estimate of mapping distance between the QTL and the marker can be obtained.

Identifying a QTL in this fashion may be more or less difficult depending in part on how large a fraction of the total variance in the trait is controlled by the QTL. The number of sib pairs that need to be analyzed varies from a few score—if the QTL is responsible for a large fraction of the total variance (40% to 50%)—to a few hundred—if the QTL accounts for a moderate fraction of the variance (20% to 30%). QTLs with relatively minor effects (less than 10% of total variance) would not be detectable by this method using realistic sample sizes. Naturally, one does not know a priori whether a QTL yet to be detected by linkage to a marker locus has a large or small effect on the trait. The experimental design (number of sib pairs tested) governs whether it will be possible to detect, for example, all QTL with effects of 30% or greater or only those with effects of 60% or greater. In the latter instance we are more likely to detect no QTLs at all, if the trait is controlled by several genes, all with small or moderate effects (Examples 2.5 and 2.7).

Allelic Association

The analytical methods described above—linkage analysis and allele sharing studies—are generally known as nuclear family methods because they rely on studies of closely related individuals. In contrast, allelic associations methods do not involve the analysis of family members. When a mutation at a QTL or complex trait locus occurs

EXAMPLE 2.5

Dyslexia

Dyslexia, or *specific reading disability*, is a mental disorder that results in patients who have an IQ in the normal range but who are significantly deficient in their ability to read. A study by Cardon et al. (1994) localized a QTL for reading disability in chromosome 6p21.3, within the major histocompatibility complex human leukocyte antigen (HLA) region. The Cardon study carried out two independent analyses involving allele-sharing methods. The first examined 114 sib pairs, and the second 50 dizygotic-twin pairs. Each analysis provided evidence of linkage of dyslexia to the HLA region. The probability that the observed frequency of shared alleles in the two groups combined would occur in the absence of such linkage was determined to be 0.0002.

Rather than genotype all participating subjects for a large collection of marker loci that encompass the whole genome (as we saw in other examples), the Cardon study targeted the HLA region based on evidence of a possible association between dyslexia and autoimmune diseases: The incidence of autoimmune diseases is elevated among relatives of dyslexic probands, and reciprocally, the incidence of dyslexia is increased among relatives of patients with autoimmune diseases.

and then spreads through the population, the alleles at tightly linked marker loci that were present on the same chromosome will remain associated with the mutation for many generations. The closer the marker locus and the trait locus, the longer it will take for crossing over to separate them, a phenomenon known as *linkage disequilibrium*. Linkage disequilibrium is the basis for allelic association methods.

Let us assume that there is a mutation in a trait locus that changes allele *T1* to *T2*; assume also that this locus is near marker *B*, a marker that has two equally frequent alleles, *B1* and *B2*; assume further that the mutation occurred in a chromosome having the *B2* allele. For a period of many generations, all *T2* will be associated almost exclusively with *B2*; at the same time, *T1*, the original allele, will occur with *B1* and *B2* with approximately equally frequency. If mutation *T2* causes its carriers to exhibit a particular trait—for example, *T2* carriers are more likely to be found in the top 10% of the distribution for a particular quantitative trait, or they are eight times more likely to suffer a certain syndrome than non-*T2* carriers—individuals at the top of the distribution for that quantitative trait or those affected by that syndrome, will be found to carry an excess of *B2* relative to what we find in the general population.

To discover allelic associations, one determines allele frequencies for a number of markers in individuals at the extremes of a continuously varying distribution (when studying a quantitative trait), or affected by a syndrome (when studying a complex trait), searching for a marker that occurs with greater frequency in the target group than in the general population.

This method can be very powerful, but it is less generally applicable than sib pair analysis. It obviously works best when, to return to the example given above, the mutation *T1* → *T2* occurred only once in the population; otherwise, the data might be complicated by another *T2* mutation having occurred on the chromosome carrying the *B1* allele. Such complications are most likely to be avoided in small and somewhat isolated populations where the genes of a single ancestral individual may become dispersed to a significant fraction of that population. This is known as the *founder effect*, a phenomenon that results when a population at some point in its history is reduced

| EXAMPLE 2.6 | *Nonsyndromal autosomal recessive deafness* |

Nonsyndromal autosomal recessive deafness is the most common form of hereditary deafness; it is associated with no abnormalities beyond the impaired hearing. Pedigree analysis shows that the responsible mutation is recessive, and interethnic marriages in which two deaf parents have only hearing children (genetic complementation, see Fig. 2-1) indicate that there are many loci that can cause this type of deafness. Friedman and colleagues (1995) applied the linkage disequilibrium method to localize *DFNB3*, one of the autosomal recessive hearing-loss genes, to the region 17p–17q12. The study was conducted in Bengkala, a small, isolated village in Bali, where 47 out of 2185 inhabitants suffer from neurosensory deafness. Because deafness is so common, the people of Bengkala developed a sign language that is used by everyone, hearing and non-hearing alike.

Given that cases of normal-hearing progeny born to two deaf parents are unknown (lack of complementation), the authors postulated that all cases of deafness in Bengkala result from a single mutation that spread through the population via a founder effect. To establish the chromosomal location of this gene, the investigators studied 13 unrelated, affected individuals who were the offspring of normal parents. Blood samples were collected and DNA was typed for 148 marker loci that were distributed at approximately 40 centimorgan intervals along the 22 autosomes. Forty-eight unrelated, normal individuals, representative of the population at large, served as a control group. The data were analyzed so as to reveal marker loci that were homozygous more often in the 13 affected individuals than would be expected if those markers were in linkage equilibrium with the deafness locus. The only markers satisfying this criterion were on chromosome 17. Genotyping of still other markers, so as to pinpoint the region, identified three that occur near the centromere; these defined a segment in which the trait locus designated *DFNB3* was located. Note that this method defines a segment within which the trait locus is located; it does not give quantitative information on genetic distances.

Figure 2-11 shows the extended pedigree from a Bengkala family. Despite the existence of excellent records and several large pedigrees, linkage analysis is not practical in this situation because many of the affected individuals are children of homozygous parents and are, therefore, uninformative.

to very small numbers (a *bottleneck*), so that random fluctuations in allele frequencies can have major effects (Example 2.6).

Candidate Gene

In this chapter we have emphasized the localization of putative trait genes by linkage to anonymous marker loci. The ultimate goal of this type of studies is, however, the identification of a specific trait gene so that its biochemical and metabolic function may be understood. As we will see in Chapter 4, the step that goes from an approximate map position to a specific sequence can be extraordinarily difficult, especially when there are few biochemical clues. A strategy that obviates this problem and has been applied with great success in a few cases is the search for the so-called *candidate gene*. This strategy consists of making an educated guess of the gene(s) that may be responsible for a certain trait and then test for linkage (see Examples 2.5 and 2.7).

FIGURE 2-11.   Pedigree of a family presenting an autosomal recessive form of deafness caused by a mutation in *DFNB3*. (Based on Friedman et al., 1995.)

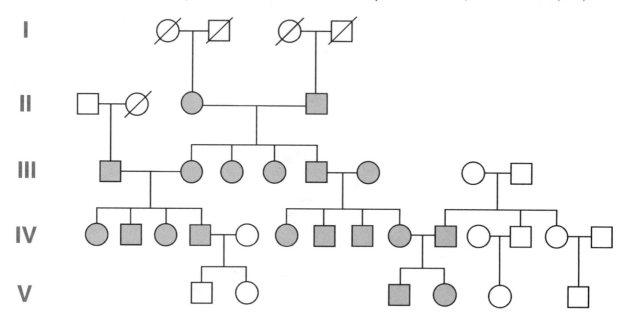

| EXAMPLE 2.7 | *Neuroticism and genes for serotonin metabolism* |

Neuroticism, or anxiety-related traits, is a continuously distributed variable characteristic of the normal human personality that can be measured by standardized personality tests. Studies of twins suggest that 40% to 60% of individual variation in this trait is heritable. Changes in serotonin (5-hydroxytryptamine, 5-HT, a neurotransmitter in the central nervous system) play a role in anxiety and depressive disorders, and the effectiveness 5-HT uptake inhibitors in the treatment of such disorders led Lesch and co-workers to propose the hypothesis that allelic differences in the 5-HT transporter gene (*5-HTT*) may be associated with neuroticism. To test this hypothesis they applied allele association methods and standard psychometric questionnaires. They observed the existence of two allelic forms of the *5-HTT* promoter: the short (S) form of the promoter had a 44 base pair deletion (relative to the long, L, allele) in a segment of repeated sequence-elements located 1 kb upstream of the transcription initiation site. The long allele was present with a frequency of 57% and the short allele with a frequency of 43% in a sample of 505 subjects. The *5-HTT*^{L} allele acted as a stronger promoter, leading to higher steady-state levels of mRNA and 5-HTT. Surprisingly, however, the two alleles are not partially dominant, but rather the *5-HTT*^{S} allele seems to be completely dominant. Thus, cultured cells with the genotypes *5-HTT*^{S}/*5-HTT*^{S} or 5-5-*HTT*^{S}/*HTT*^{L} had rates of uptake of serotonin that were approximately half of the level in *5-HTT*^{L}/*5-HTT*^{L} cells. To assess neuroticism the authors used three different psychometric tests that measured anxiety or neuroticism as well as other personality traits and they found a statistically significant association between the *HTT*^{S} allele and higher neuroticism scores but no association with other personality traits such as extroversion, openness, agreeableness, and conscientiousness. The study also verified that sibs who were discordant for *5-HTT* genotype also showed significant differences in the neuroticism scale. Thus, the differences found in the sample as a whole could not be ascribed to the existence of ethnic subgroups with numerous genetic differences other than the one being studied here, a phenomenon called *population stratification*.

The degree of overlap of this quantitative trait between the two genotypes is very large (Fig. 2-12), and less than 10%

of the genetic variation in neuroticism in the samples is explained by this genetic difference. These facts demonstrate both the power of these methods in detecting small but significant differences, and the level of complexity of the problem when any given factor may have such a small effect.

CONCLUSION: NATURE VERSUS NURTURE

The age old dichotomy concerning the origin of complex traits, especially behavioral ones, is beginning to resolve itself in many cases into nature *and* nurture. A priori, the biomolecular substratum of all human behavior makes it difficult to conceive that polymorphisms and differences in expression of brain proteins—of the type that have been amply documented for the vast majority of proteins studied—could be totally without impact on the psychology of an individual. This preconception seems to be borne out by genetic studies of a few specific traits, as we have seen in this chapter. Most studies so far, however, demonstrate that it is not a simple picture: when it comes to behavior, the *one trait, one gene model* (if there ever was one) is definitely wrong. A number of studies with positive results have shown that discrete genetic entities have very small effects, raising liability scores by 10% to 15%. We can surmise the existence of alleles with effects in the 2% to 5% range, which would, in many cases, be undetectable with current methods.

On the other hand, common sense observation, as well as many rigorous studies have shown that the environment matters, that the mammalian brain and the rest of

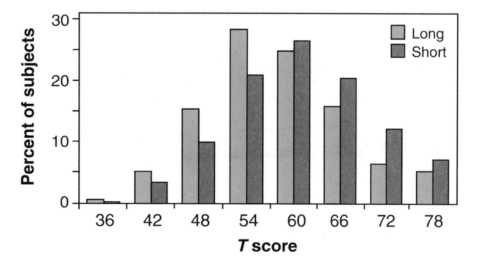

FIGURE 2-12. Distribution of neuroticism (NEO-PI-R Neuroticism T scores) in a sample of 505 individuals, 163 homozygous for the long *5-HTT* allele (color bars) and 342 with at least one short allele (empty bars). The difference between the means is 3.4 units in this scale, which has a very small probability of not being significant ($p = 0.002$). (Redrawn with permission from Fig. 3 in Lesch, K.-P., Bengel, D., Heils, A., et al. Association of anxiety-related traits with a polymorphism in the serotonin transporter gene regulatory region. Science 274:1527–1531 [1996]. Copyright 1996 American Association for the Advancement of Science.)

the nervous system have enormous plasticity, and that stimuli in early stages of development can have lasting effects into adulthood.

Thus, in humans as in other genetic systems, the action of genes molds a phenotype in response to a multitude of environmental cues, and experience and upbringing forms the individual within parameters more or less—for some traits more, for others less—defined by the genes.

EXERCISES

2-1. In the pedigree in Figure 2-6, two traits are indicated: the presence or absence of psoriasis and the allele of marker locus *D17S784* (abbreviated *L*). (A) Assuming that these two traits are on the same chromosome and that psoriasis is caused by a dominant allele (*T*), redraw that pedigree, "flattening" it so that all progeny appears offspring of I-1 and I-2. Assume that individual I-1 was affected. Indicate, below each symbol, the genotype for trait and marker locus (e.g., individual II-1 could be $T\,L^2/t\,L^3$). Note that individual II-6 must be a carrier of *T* because she has affected daughters. (B) Calculate the LOD score for a recombination frequency of 5%. (C) Next, assume that III-9 and III-15 display incomplete penetrance, that they have a normal phenotype and would be assumed to be *t/t*. Recalculate the LOD score. How does incomplete penetrance affect estimates of linkage?

2-2. In Figure 2-6, calculate the frequency of zero, one, and two ibd marker alleles for all possible concordant sib pairs, in which both sibs have psoriasis. Make the same calculations for all discordant and unaffected sib pairs.

2-3. What percent concordance would you expect for monozygotic and dizygotic twins with respect to:

(A) A trait determined by the recessive allele *tt* with 80% penetrance.

(B) A trait determined by the dominant allele *U__* (in another gene) with 50% penetrance.

(C) A trait determined by a combination of *tt* and *U__*, with an overall penetrance of 30%.

Assume that all the matings are between heterozygotes for the recessive trait, and between a heterozygote and a homozygous normal for the dominant trait.

2-4. Given that the familial breast cancer genes (*BRCA1* and *BRCA2*) have such high penetrance (80% to 90% of carriers develop cancer in their lifetimes), how can you explain the low concordance (about 10%) and the small difference in concordance for breast cancer between monozygotic and dizygotic twins (see Fig. 2-5).

2-5. Verify that all gay males in Figure 2-9 *might* carry the same X-chromosome.

REFERENCES

Bailey JM, and Pillard RC. (1991) A genetic study of male sexual orientation. Arch Gen Psychiatry 48:1089–1096.

Bennett ST, and Todd JA. (1996) Human type 1 diabetes and the insulin gene: principles of mapping polygenes. Ann Rev Genet 30: 343–370.

Bodmer WF, and Cavalli-Sforza LL. (1976) Genetics, evolution and man. San Francisco: W.H. Freeman.

Boyd M, Harris F, McFarlane R, Davidson HR, and Black DM. (1995) A human *BRCA1* gene knockout. Nature 375:541–542.

Cardon LR, Smith SD, Fulker DW, Kimberling WJ, Pennington BF, and DeFries JC. (1994) Quantitative trait locus for reading disability on chromosome 6. Science 266:276–279.

Cloninger CR, Adolfsson R, and Svrakic NM. (1996) Mapping genes for human personality. Nat Genet 12:3–4.

Crabbe JC, Wahlsten D, and Dudek BC. (1999) Genetics of mouse behavior: interactions with laboratory environment. Science 284:1670–1672.

Davies JL, Kawaguchi Y, Bennett ST, et al. (1994) A genome-wide search for human type 1 diabetes susceptibility genes. Nature 371:130.

Fausto-Sterling A, and Balaban E. (1993) Genetics and male sexual orientation [comment on Hamer et al., 1993]. Science 261:1257–1259.

Friedman TB, Liang Y, Weber JL, Hinnant JT, Barber TD, Winata S, Arhya IN, and Asher JH Jr. (1995) A gene for congenital, recessive deafness *DFNB3* maps to the pericentromeric region of chromosome 17. Nat Genet 9:86–91 (DFNB3).

Goldman D. (1996) High anxiety [comment on Müller et al., 1996]. Science 274:1483 Erratum in Science 275:741 (1997).

Hall JM, Lee MK, Newman B, Morrow JE, Anderson LA, Huey B, and King M-C. (1990) Linkage of early-onset familial breast cancer to chromosome 17q21. Science 250:1684–1689.

Hamer D. (1993) Sexual orientation [comment on Maddox, 1993]. Nature 365:702.

Hamer DH, Hu S, Magnuson VL, Hu N, and Pattatucci AML. (1993) A linkage between DNA markers on the X chromosome and male sexual orientation. Science 261:321–327. See also Fausto-Sterling and Balaban, 1993; Hamer, 1993; King, 1993; Kruglyak, 1993; Maddox, 1993; Risch et al., 1993.

Harris H. (1975) The principles of human biochemical genetics. New York: American Elsevier.

Hu S, Pattatucci AML, Patterson C, Li L, Fulker DW, Cherny SS, Kruglyak L, and Hamer DH. (1995) Linkage between sexual orientation and chromosome Xq28 in males but not in females. Nat Genet 11:248–256.

King MC. (1993) Human genetics. Sexual orientation and the X [comment on Hamer et al., 1993]. Nature 364:288–289.

Kruglyak L. (1993) Sexual orientation [comment on Maddox, 1993]. Nature 365: 702.

Lander ES, and Botstein D. (1989) Mapping Mendelian factors underlying quantitative traits using RFLP linkage maps. Genetics 121:185–199.

Lander ES, and Schork NJ. (1994) Genetic dissection of complex traits. Science 265:2037–2048.

Lesch K-P, Bengel D, Heils A, Sabol SZ, Greenberg BD, Petri S, Benjamin J, Müller CR, Hamer DH, and Murphy DL. (1996) Association of anxiety-related traits with a polymorphism in the serotonin transporter gene regulatory region. Science 274:1527–1531. See also Goldman, 1996.

Maddox J. (1993) Wilful public misunderstanding of genetics [comment on Hamer et al., 1993]. Nature 364:281.

Plomin R, Owen MJ, and McGiffin P. (1994) The genetic basis of complex human behaviors. Science 264:1733–1739.

Rice G, Anderson C, Risch N, and Ebers G. (1999) Male homosexuality: absence of linkage to microsatellite markers at Xq28. Science 284:665–667.

Risch N, and Botstein D. (1996) A manic depressive history. Nat Genet 12:351–353.

Risch N, and Merikangas K. (1996) The future of genetic studies of complex human diseases. Science 273:1516–1517.

Risch N, Squires-Wheeler E, and Keats BJ. (1993) Male sexual orientation and genetic evidence [comment on Hamer et al., 1993]. Science 262:2063–2065.

Risch N, and Zhang H. (1995) Extreme discordant sib pairs for mapping quantitative trait loci in humans. Science 268:1584–1589.

Rose S. (1995) The rise of neurogenetic determinism [comment]. Nature 373:380–382.

Smith SD, Kelley PM, and Brower AM. (1998) Molecular approaches to the genetic analysis of specific reading disability. Hum Biol 70:239–256.

Stern C. (1949) Principles of human genetics. San Francisco: W.H. Freeman.

Tomfohrde J, Silverman A, Barnes R, Fernandez-Vina MA, Young M, Lory D, Morris L, Wuepper KD, Stastry P, Menter A, and Bowcock A. (1994) Gene for familial psoriasis susceptibility mapped to the distal end of human chromosome 17q. Science 264:1141–1145.

Wickelgren I. (1999) Nurture helps mold able minds. Science 283:1832–1834.

Genome Organization I

For supplemental information on this chapter, see Appendix Figures 4, 5, 6, 7, and 10

In Chapters 1 and 2 we reviewed the methods by which human geneticists establish that certain traits are inherited as Mendelian factors and how the location in a genetic map is determined. This approach gives us the somewhat abstract concept of genes that is provided by genetic analysis. In Chapters 3 and 4 we will review the physical reality of genes (and other DNA elements) as derived from molecular biology, as well as the overall organization of DNA into the genome. In Chapter 5 we will study the chromosomes, visible (at the microscope) subunits of the genome.

The human genome consists of 24 DNA molecules: one for each autosome, one for the X, and one for the Y-chromosome. The size of these molecules ranges approximately from 39×10^6 nucleotide pairs (np) or 39 megabases (Mb) for chromosome 21, to 263 Mb, for chromosome 1. The sequence of these DNA molecules is an irregular juxtaposition of a variety of elements. In this chapter we present the different types or classes of sequences that have been described in the human genome, but it should be noted that the boundaries between classes of sequences are quite ambiguous.

In part, the difficulty in classifying DNA sequences derives from the multitude of methodologies that were applied to define the various categories. Thus, the concepts of repeated and unique sequences were originally derived from the kinetics of denatured DNA reannealing, whereas the concepts of mini- and microsatellite DNA derive from Southern analysis of genomic DNA, polymerase chain reaction (PCR) assays, and sequencing. Finally, the notions of expressed and unexpressed DNA are based on a variety of methods used to identify genes.

DNA in the human genome can be classified in two categories: *genic* or *expressed sequences* and *non-genic* or *unexpressed sequences*. Although conceptually this distinction seems unexceptional, when applied to real cases the edges blur. We would include in the first category all those sequences that form part of, or are involved in, the production of the various RNAs (such as rRNA, mRNA, tRNA, snRNA); it would comprise transcription regulatory elements (promoters and enhancers) and exons, and perhaps also intronic splicing signals. The second category would include the sequences between genes that do not play a regulatory function. But how would we classify introns? One might consider the bulk of long introns as unexpressed sequences (although they are transcribed), but consider small introns, near the lower functional limit of intron sizes, as part of the expressed sequences. Including introns as expressed sequences is justified by experiments that have shown that the expression of many genes is indeed affected if introns are removed.

In any case, sequences may not be directly involved in the process of gene expression but that does not necessarily mean that they lack function. Some of them are known to be important to the structural organization of chromosomes, forming part of centromeres, telomeres, and scaffolding attachment sites. Others may be important in less direct ways; for example, it may be critical for the processing of a particular transcript that it should have a very long intron, even though the specific sequence in that intron may be irrelevant. Or there may be a developmental requirement that mammalian cells should have a DNA content no smaller than approximately 6×10^9 base pairs (bp). Some of these unexpressed sequences have at times been called *junk DNA,* and, in studying the details of their organization, the designation may seem highly appropriate. The fact is, however, that we do not know enough about the human genome to call any part of it junk.

We lack enough information to give definitive answers to many global questions about the human genome, questions such as how many genes there are, or what fraction of the genome is occupied by expressed versus unexpressed sequences. Based on comparisons with other eukaryotes, the best estimate of the number of genes could be taken as 70,000, with the real number probably being anywhere between 50,000 and 100,000. A small sample of 94 fully characterized human genes gave an average size of 10,000 bp (10 kilobases) for these transcription units; considering that this sample represents as little as 0.1% of the total number of genes, we might assume the average of all human genes to fall between 7 kb and 15 kb. Multiplying the estimates of average size by the estimates of number of genes gives the total number of base pairs of human DNA that is transcribed into RNA; thus, we can estimate that the fraction of DNA that is transcribed is between 12% and 50% of the whole genome. Note that *transcribed sequences* include considerably different elements from expressed sequences (as defined above)—transcribed sequences include all introns but exclude transcription regulatory sequences.

UNEXPRESSED DNA, REPETITIVE AND UNIQUE SEQUENCES

DNA that does not participate directly in gene expression can be classified into four categories depending on the repetition and organization of its sequences: 1) clustered, highly repeated DNA, 2) small-cluster, intermediately repeated DNA, 3) dispersed, intermediately repeated, and 4) unique sequence DNA.

Clustered, Highly Repetitive Sequences

Satellite or *highly repeated DNA* represents approximately 10% of the total human genome. It consists of a few families of sequences repeated hundreds of thousands or millions of times, arranged in blocks within which the units have a tandem, or head-to-tail, disposition. They are called satellite DNA because in CsCl gradients they may form minor peaks next to the main band or bulk of the DNA. Most of these highly repetitive sequences are found in centromeric regions.

The main families of highly repetitive sequences. There are five sequence families that constitute most of the highly repetitive DNA; the nomenclature is confusing, and a system based on the size of the unit repeat is widely used. The simplest satellite DNA is made up of a 5-bp repeat and corresponds to satellites 2 and 3; it comprises derivatives of the sequence GGAAT and is present in the centromeric region of most, and probably all, chromosomes. The 42-bp satellite (satellite 1) is AT-rich and found in most chromosomes. The 48-bp satellite and the 68-bp satellite (β DNA) are limited to a few chromosomes. The best studied of these families is the 171-bp satellite (also called α or alphoid DNA), which is present at the primary constriction of all chromosomes.

Alpha DNA is thought to constitute a functional element of human centromeres—involved in kinetochore organization and microtubule attachment—because it is located in the primary constriction and also because many, although not all, of the 171-bp primary units include a 17-bp segment that binds a centromere-specific protein called CENP-B (Fig. 3-1). The arrangement of alpha DNA and the other satellites is known for only a few of the human centromeres (Fig. 3-2).

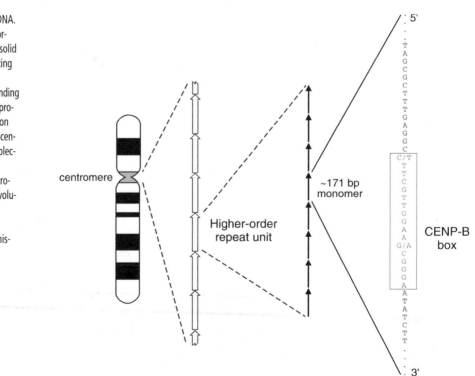

FIGURE 3-1. Alpha satellite DNA. Chromosomal localization and organization of the 171-bp units (solid arrows) into higher order repeating units (hollow arrows). Some monomers contain the 17-bp binding site for the centromere-specific protein CENP-B. (Based on Warburton PE, and Willard HF. Evolution of centromeric alpha satellite DNA: molecular organization within and between human and primate chromosomes. In: Human genome evolution. Jackson M, Strachan T, and Dover G, eds. Oxford: BIOS, 1996:121–145, Fig. 5-1, by permission.)

FIGURE 3-2. Localization of the main satellite DNAs in the centromeric regions of chromosomes 9, 21, and Y. The Y-chromosome is presented in the form of a molecular map to scale. Note that the actual size of each array is polymorphic. (Based on Tyler-Smith and Willard, 1993.)

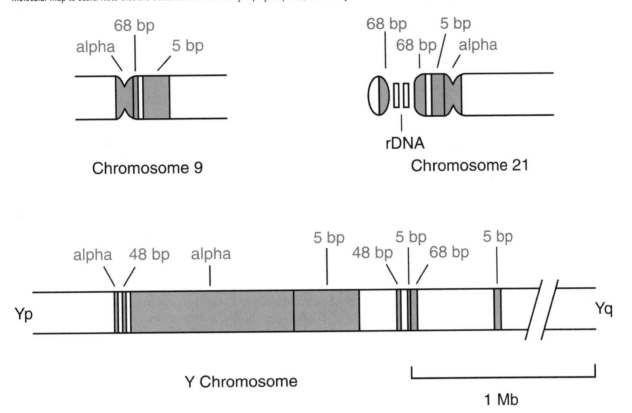

EXAMPLE 3.1

The α family of highly repetitive DNA

Several million bases of DNA in the centromere of each human chromosome are made up of α DNA tandem repeats. The repeats are not identical and their differences are arranged in hierarchical patterns (see Fig. 3-1). For example, on the X-chromosome the units are arranged in sets of twelve 171-bp repeats (Fig. 3-3), such that any two members in one group of 12 could differ by as much as 30% or 40%, but the first member of a group of 12 would be very similar in sequence (within 2% or 3%) to the first member of the next set. Thus, in addition to the basic monomer of approximately 171 nucleotides, there is a higher order unit of 171 × 12 or approximately 2052 nucleotides, and this higher order unit is repeated roughly 5000 times in each

X-chromosome.

The higher order pattern can be visualized by restriction-enzyme digestion; because of the significant differences among monomers within the higher order unit, it is usually possible to find restriction enzymes that cut only once per higher order unit. Thus, X-chromosome α satellite would produce a 2-kb band when digested with *Bam*H1 (Figs. 3-3 and 3-4). Because there are several thousand copies of the higher order unit, a somewhat prominent band would appear in the 2-kb region when *Bam*H1-digested DNA is run on a gel.

In addition to the differences found among basic units within one higher order repeat, there are also differences among units from different chromo-

FIGURE 3-3. Alpha satellite DNA. Chromosome-specific organization of 171-bp units into higher order repeating units and the distribution of some restriction enzyme sites. **(A)** Hypothetical higher order unit of *n* monomers with restriction sites for enzymes *E1* and *E2* in each higher order repeat. **(B)** Repeating units in chromosomes X and 7. (Based on Willard and Waye, 1987.)

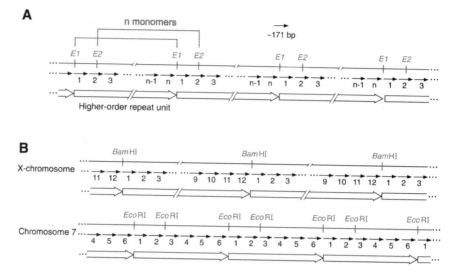

FIGURE 3-4. Analysis of alpha satellite DNA by restriction enzymes. **(A)** Restriction enzyme with a 4-bp recognition site (four-cutter) that happens to cut once within the higher order unit; outside the repeating DNA this type of enzyme cuts every few hundred base pairs. Digestion of DNA cuts the entire array into fragments of a size characteristic for that array (see Table 3-1). **(B)** Same as (A), but for an enzyme that does not have restriction sites in every higher order repeat. Digestion will cut the array into one or a small number of pieces with a small amount of flanking DNA associated. **(C)** Restriction enzyme with an 8-bp recognition site (eight-cutter or rare cutter) without recognition sites in the array. Outside the array, these enzymes typically cut every 15 to 50 kb; they are useful to obtain the array in one fragment, but there is substantial amount of flanking DNA associated. (Based on Warburton and Willard, 1996.)

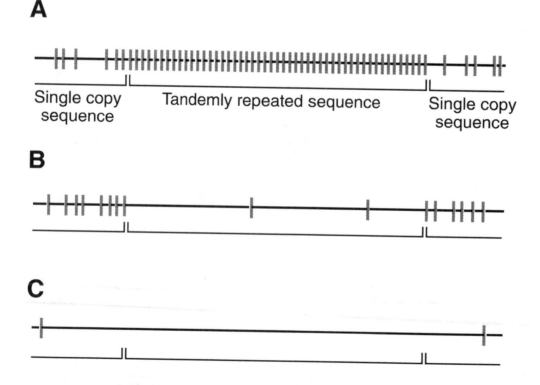

somes. For the most part, each chromosome has a unique arrangement of higher order repeats, with some chromosomes having more than one array (Table 3-1). In some cases the same higher order unit is present in several chromosomes, and this is taken as evidence that there is a certain level of genetic recombination and exchange between some nonhomologous chromosomes. (Especially chromosomes 13, 14, 21, and 22; see also "Ribosomal RNA Genes"; in Chapter 5, see "Robertsonian translocations".) In most cases, the differences found in the α sequences of nonhomologues are extensive enough that it is possible to prepare hybridization probes that are chromosome-specific.

Although homologous chromosomes have the same higher order unit, allelic differences do occur; these are due to the presence/absence of certain restriction sites, which give rise to RFLPs, or due to homologous centromeres having different numbers of higher order repeats, probably as a consequence of unequal crossing over (see Chapter 7). The number of repeats is, thus, constantly changing, which makes the human population very polymorphic and with a high degree of heterozygosity for this locus. This property can be exploited through Southern and PCR analyses to make α DNA an excellent marker locus for the centromeric region of most human chromosomes (see Chapter 1).

Small-Cluster, Intermediately Repeated DNA

Small-cluster, intermediately repeated DNA is a type of sequence divided into two main groups: *variable number tandem repeats* (VNTR), also called *minisatellite* DNA, and *short tandem repeats* (STR) or *microsatellite* DNA. Both of these are short, interspersed, tandem repeats that differ from centromeric highly repetitive DNA in that the repeating unit occurs only from a few to a few hundred times in each block (*small cluster*), and the blocks are distributed throughout the length of the chromosomes rather than being clustered in centromeric regions, although VNTR tend to accumulate in the subterminal region of chromosomes. VNTR or minisatellite DNA is characterized by repeating units between 10 and 200 bases in length, whereas STR or microsatellite DNA has repeating units of 2 to 5 bases.

Minisatellite DNA.   Minisatellite DNA are sequences in which repeating units occur in blocks of up to 50 kb. One of the first minisatellites described was found by Jeffreys and his collaborators in the first intron of the myoglobin gene where it occurs as four tandemly repeating units of 33 bp (Fig. 3-5A). Sequences related to this 33-bp unit were detected at several other sites in the genome; eight of these were cloned and showed that, although they shared a conserved core sequence of 10 to 15 bp, they differed from one another in both the sequence of the repeating unit (outside the core) and its length (between 16 and 64 bp per repeat). At four of those eight sites the repeated sequences proved to be highly polymorphic with respect to the number of repeats; the two alleles of any one individual at any one site were often different, not necessarily in the sequence of the repeats but in the number of repeats present (see Fig. 3-5B). Thus, in DNA from 14 unrelated people they found six different alleles at one site; the various alleles had 20, 24, 26, 27, 31, or 40 copies of a repeating unit of 62 bp. In the same sample they found that at another site there were five different alleles (between 10 and 18 copies of a 64-bp unit) and at a third site there were eight alleles (between 12 and 25 copies of a 37-bp unit). The large number of alleles and high level of polymorphism suggested to Jeffreys and co-workers that minisatellites would be ideal as genetic markers.

TABLE 3-1. Higher-order repeating unit structure of alpha satellite subsets. Shown here are several alphoid subsets, including the length of the higher-order repeating unit in kilobases (kb), the number of monomers (mer) present in each subset, the restriction enzymes that cut once per higher-order repeat, and the chromosome where they are found.

Chromosome	Higher-Order Repeat Unit	Enzyme
1	1.9 kb (11mer)	HindIII, XbaI, SphI
2	0.68 kb (4mer)	HinfI, XbaI, HindIII
3	2.9 kb (17mer)	AccI, HindIII, PvuII
4	3.2 kb (19mer)	MspI
4, 9	1.2 kb (7mer)	PstI, RsaI, TaqI
5, 19	2.25 kb (13mer)	EcoRI, PstI, RsaI
6	2.9 kb (17mer)	BamHI, TaqI
7	2.7 kb (16mer)	HindIII, TaqI
	1.0 kb (6mer)	EcoRI
8	2.5 kb (15mer)	ScaI, HindIII, PstI
9	2.7 kb (16mer)	PstI, RsaI, MspI
10	1.35 kb (8mer)	PstI, RsaI, AccI
11	0.85 kb (5mer)	XbaI
12	1.4 kb (8mer)	HindIII, TaqI, PvuII
13, 21	1.9 kb (11mer)	HindIII, TaqI, Sau3AI
	0.68 kb (4mer)	EcoRI, XbaI
13, 14, 21	3.95 kb (23mer)	RsaI, MspI
14, 22	1.4 kb (8mer)	XbaI, TaqI
15	2.5 kb (14mer)	RsaI, DraI
	4.5 kb (26mer)	AccI, BglII, BstNI
16	1.7 kb (10mer)	Sau3AI, EcoRV
17	2.7 kb (16mer)	EcoRI, PstI, PvuII
18	1.4 kb (8mer)	EcoRI
	1.7 kb (10mer)	HindIII
20	1.0 kb (6mer)	HinfI
22	2.7 kb (16mer)	EcoRI
X	2.0 kb (12mer)	BamHI, PstI, SstI
Y	5.7 kb (33mer)	EcoRI, PstI, HindIII

Only subsets for which the higher-order structure has been determined are included; other cloned subsets not fully described are not included here.
SOURCE: Warburton PE, and Willard HF. Evolution of centromeric alpha satellite DNA: molecular organization within and between human and primate chromosomes. In: Human genome evolution. Jackson M, Strachan T, and Dover G, eds. Oxford: BIOS, 1996:121–145, Table 5-1, by permission.

FIGURE 3-5. Schematic representation of minisatellite DNA. **(A)** The myoglobin gene. Detailed is a segment of the first intron that includes a minisatellite. In the myoglobin gene, the repeated sequence seems to exist always in four repeats, but at other sites in the genome it occurs with variable numbers of repeats. **(B)** Two hypothetical sites containing repeats that include the core of the myoglobin gene minisatellite. The size of the repeating unit outside the core is site-specific; at each locus two alleles are shown. (Based on data from Jeffreys et al., 1985b.)

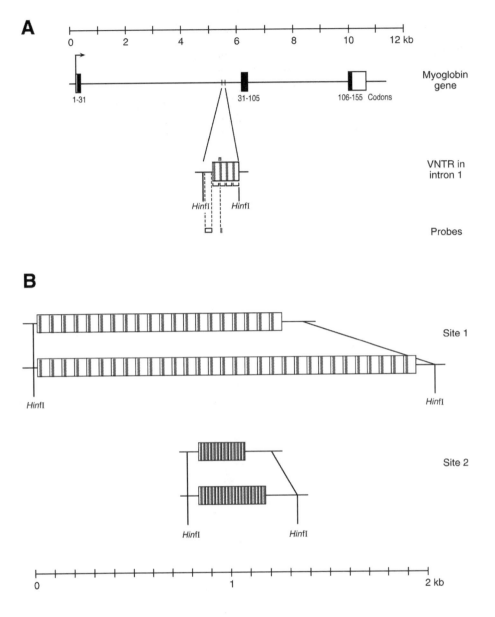

Minisatellites are often studied by Southern analysis, using restriction enzymes that cut outside the repeats. The probe used can be derived either from minisatellite sequences or from flanking, unique sequences (see Fig. 3-5A). Because the same repeating unit may occur at dozens of sites in the genome, the results obtained depend on the probe used. Digestion of genomic DNA with a restriction enzyme that does not cut within the repeating units generates fragments whose size depends largely on the size and number of the repeating units. The size of the flanking, single sequence DNA can be minimized by using enzymes that recognize 4-bp sites such as *Hin*fI in Figure 3-5, rather than enzymes of 6-bp restriction sites (six-cutters).

If, on Southern analysis of this DNA, we use as a probe a fragment that corresponds to the flanking DNA (probe represented by an open box in Fig. 3-5A), there will be only one or two fragments detected, those that correspond to that particular genomic site (the myoglobin gene in this case). If, on the other hand, we use as a probe the core sequence present at all sites (probe represented by the small color box), numerous fragments will be detected, one or two for each site where the repeated sequence occurs in the genome.

Given the large numbers of alleles in the population and large number of sites involved, it is unlikely that a particular combination of alleles, and therefore a particular banding pattern, will occur in two unrelated individuals. The complexity and individuality of these banding patterns gave them the moniker of *DNA fingerprints.* Notice, however, that bands behave as Mendelian factors and banding patterns can be used in ways that are not possible to use fingerprints—to establish paternity, for example. Figure 3-6 is a Southern analysis of DNA from a nuclear family and two unrelated men. DNA was digested with *Hin*fI and the probe was derived from the core repeated sequence of the myoglobin gene minisatellite (see Fig. 3-5A). The twins have

FIGURE 3-6. Southern analysis of minisatellite DNA from a nuclear family (lanes 1–4) and two unrelated men (lanes 5 and 6). Of the bands that can be resolved (arrows), those that are not maternal match bands found in the father but not the other two men. (Redrawn with permission from Fig. 3b in Jeffreys, A.J., Wilson, V. and Thein, S.L. Individual-specific "fingerprints" of human DNA. Nature 316:76–79 [1985]. Copyright 1985 Macmillan Magazines Limited.)

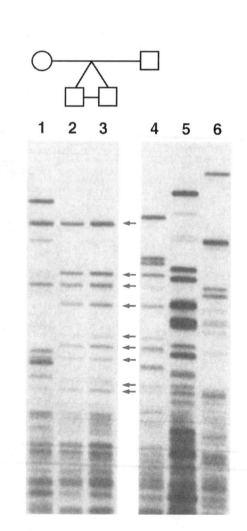

identical patterns, and many of their bands can be traced either to their mother or their father (arrows)—the smaller bands are shared by many individuals and are, therefore, uninformative.

Useful fragment size for this type of analysis is typically between 2 and 20 kb. Mutations that change the number of copies at a locus seem to occur with a frequency of approximately 1 in 1000 gametes. This frequency is high enough to maintain a great deal of polymorphism in the population, but low enough to show stable Mendelian patterns when analyzing single families.

Microsatellite DNA. Simple repetitive elements such as $(CA)_n$, $(AAT)_n$, and $(GATA)_n$, also called *simple sequence DNA,* exist in thousands of loci in the human genome. Using primers that complement the unique sequences outside the tandem repeats, it is possible to amplify by PCR the simple sequence DNA at a specific locus, and measure accurately the number of repeats by gel electrophoresis. The total length of these blocks is usually less than a thousand bases and often between 100 and 300 bases. This type of study demonstrated that microsatellite loci are highly polymorphic in number of repeats; it is not unusual to find loci with 10 to 20 alleles in a given population. We will see in Chapter 7 the molecular basis for this instability in number of repeats.

Because none of those alleles is overwhelmingly predominant, it is also not unusual for a large fraction of the population (>70%) to be heterozygous at many microsatellite loci. These properties make microsatellite sites ideal marker loci. For example, using 2794 $(CA)_n$ sites (a partial sample of the total number of such sites), a genetic map was constructed covering over 90% of the human genome, with $(CA)_n$ sites spaced every few centimorgans (see Chapter 4). It is estimated that there are $(CA)_n$ sites every 100 kb of DNA sequence (more than 300 sites for even the smallest chromosome). Figure 3-7 shows the inheritance, through two generations, of a $(CA)_n$ site located in the cardiac muscle actin gene.

Another application of microsatellites is in forensic identification. This is accomplished by determining the allelic constitution of a DNA sample at multiple microsatellite sites (genotyping). Given the multiplicity of alleles, the use of less than 10 loci is sufficient to provide virtual certainty of whether two DNA samples belong to the same individual or to establish familial relationship (as in paternity cases). Compared to minisatellites, microsatellites can provide greater accuracy and less uncertainty in forensic identification because of the smaller size of microsatellite bands and the fact that Southern blotting is not necessary.

Dispersed, Intermediately Repeated Sequences

Most *dispersed, repeated sequences* correspond to the category of middle repetitive DNA, the number of copies varying between a few and a few thousand. They are characterized for being dispersed as single elements throughout the genome, rather than clustered in tandem repeats, and for being the consequence of transposition. For this reason they are also known as *transposable elements,* and they account for more than 35% of the human genome.

Transposition is a process by which DNA segments become incorporated into new sites in the genome. In humans, most transposition is mediated by an RNA intermediate and the elements involved are known as *retrotransposons.* RNA is copied into a DNA strand (copy DNA or cDNA) by the enzyme reverse transcriptase, and the DNA fragment is then incorporated into the genome by an integrase.

FIGURE 3-7. **(A)** Pedigree of a family in which sister and brother married brother and sister. **(B)** Autoradiograph of a sequencing gel used to fractionate radioactive PCR-amplification products of DNA extracted from individuals immediately above each lane, in the pedigree. The PCR primers were 20-nucleotides long and complementary to unique sequences that flank the $(CA)_n$ segment in the cardiac muscle actin gene. In each lane there are two main bands, which correspond to the amplified repeated segment, and a number of fainter, artifactual "echo bands." (See the appendix for a review of the PCR technique.) The four left lanes correspond to sequencing reactions of a known DNA segment used as size markers. (Redrawn with permission from Fig. 1 in Litt, M. and Luty, J.A. A hypervariable microsatellite revealed by in vitro amplification of a dinucleotide repeat within the cardiac muscle actin gene. Am J Hum Genet 44:397–401 [1989]. Copyright 1989 by the University of Chicago Press.)

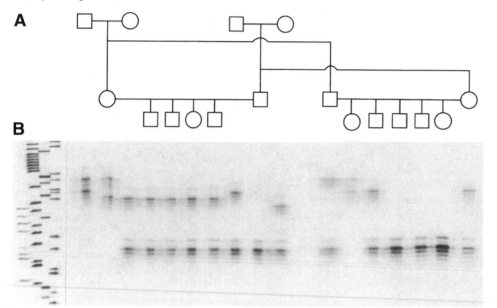

Viral retrotransposons. The prototypical example of retroposition is supplied by RNA animal viruses. These viruses infect the mammalian cell with an RNA molecule that is copied by reverse transcription into cDNA and then incorporated into a chromosome. A segment of DNA sequence known as the *long terminal repeats* (LTRs) is present at both ends of the viral cDNA genome. The LTR that corresponds to the 5' end of the original RNA molecule acts as a promoter, stimulating transcription of the viral DNA and thus creating more RNA molecules. These RNAs can either be processed into mRNAs to produce viral proteins or act as viral genomes, in which case they are wrapped into mature viral particles and shed to infect other cells.

In the course of evolution some viruses become innocuous to the host cell—they lose their extracellular phase and become benign intracellular parasites that retain, nonetheless, the ability to transpose. Several examples of this have been found in animals. In humans there is a retrotransposon called THE1 that seems to belong to this category. In the most general case, viral retrotransposons encode their own reverse transcriptase and integrase, thus controlling their own transposition. Many of these modified elements, however, seem to have lost one or another of these functions and instead make use of enzymes supplied by other retrotransposons.

Nonviral retrotransposons: *Alu*, *L1*, and processed pseudogenes. Retroposition can also occur to nonviral, cellular RNAs of various types. These nonviral retrotransposons are generally characterized by the absence of LTRs and the presence of poly-A tails. Nonviral retrotransposons are classified into *short interspersed nucleotide elements* (SINES) and *long interspersed nucleotide elements* (LINES).

The most common of these sequences—representing 5% to 6% of the genome—is a 500,000-member SINE called the *Alu* family. It is derived from an abundant signal-particle RNA, 7SL RNA, that is involved in the translocation of nascent proteins across the endoplasmic reticulum membrane. Full size *Alu* elements are 300-bp dimers: the right-hand monomer includes approximately 150 bp of the 7SL RNA gene and a poly-A tail, and the left-hand monomer contains approximately 120 bp of the 7SL RNA gene and a poly-A tail (Fig. 3-8). Obviously there was a certain amount of processing (probably at the RNA level) in the development of the *Alu* element from its 7SL precursor gene. The 7SL RNA genes are transcribed by RNA polymerase III and, therefore, contain promoters that are internal to the gene. Once an *Alu* element was formed, a low level of transcription may have provided enough RNA to generate new retrotransposons and thus, the number of *Alu* elements in the genome expanded through time. Many other SINES seem to be derived from transposed tRNA genes. All SINES lack reverse transcriptase and integrase coding sequences, and their transposition depends on enzymes supplied by other transposons.

As mentioned earlier, we do not know what overall function, or effect, these repeated sequences have in the host cell. We do know that in specific cases integration may occur into gene sequences and cause mutations. In the case of a neurofibromatosis I patient, an *Alu* element was identified in an intron, 44 bp upstream of exon 6; this insertion disrupts proper splicing in such a way that the 280-bp exon 6 is skipped and exon 7 is directly attached to exon 5 leading to a shift in the reading frame. Neither parent carries this mutation (paternity was verified by molecular markers), which demonstrates that this is a new insertion. Insertion of an *Alu* element has also been observed in the case of a mutation in the choline esterase gene. *Alu* sequences are also involved in other mutational events such as deletions and duplications (see Chapter 7).

FIGURE 3-8. Possible origin of the *Alu* element. A 7SL gene transcript is spliced so that 155 bp are removed after polyadenylation, giving rise to the right *Alu* monomer. Subsequently, a further 31 bp are deleted, and this gives rise to the left *Alu* monomer. Finally, the two are joined into a dimer and copied into DNA to be inserted into the genome. (Based on Ullu and Tschudi, 1984.)

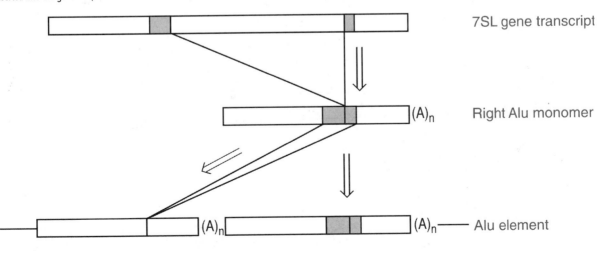

7SL gene transcript

(A)$_n$ Right Alu monomer

(A)$_n$ (A)$_n$ —— Alu element

Another very common repeated sequence is the LINE1 or *L1* family. The full length *L1* consensus is a 6.1-kb sequence with a polyadenylation signal and poly-A tail at the 3' end. There are approximately 100,000 copies of these elements, although most of them are incomplete and many carry point mutations. The full length *L1* consensus sequence codes for at least two proteins, a reverse transcriptase and an endonuclease that plays a role in the integration process. *L1* may be the main source of enzymes for retroposition of "passive" retrotransposons, such as *Alu*, in human cells. Even though some *L1* members do encode for functional proteins, the family as a whole is here classified with unexpressed sequences because over 90% of its members carry 5' truncations; only 3500 *L1* elements are full length and probably a small fraction of them are functional.

In a systematic screen of 240 unrelated patients suffering from hemophilia A, Kazazian and colleagues found two that carried insertions of LINE1 fragments, of 2.3 and 3.8 kb, respectively, in exon 14 of the factor VIII gene. In both cases, the parents of these two patients lacked the insertion; as before, this indicated a new transposition. The Kazazian study analyzed one of the patients in more detail: they synthesized an oligonucleotide that matched 20 bp of the insert in a region where there were three differences between this insert and the consensus *L1*. They then used this probe to clone *L1* elements that matched this particular subfamily and were able to identify one, located in chromosome 22, that matched exactly the sequence of the factor VIII insertion. Thus, they were able to identify the putative precursor of the *L1* element that resulted in the factor VIII insertion.

Many of the same transposable elements are recognizable in other mammalian genomes and this point is well illustrated by the study of the factor VIII insertion. The Kazazian study observed that the *L1* element in chromosome 22 is present, in that same position, in the chimpanzee and gorilla, thus tracing back the location of this *L1* element for at least 6 million years.

Another group of transposable elements is that of the processed pseudogenes. These elements are similar to the *Alu* family in that they are derived from functional RNAs, in this case mRNAs, which are copied into cDNA and then inserted into the genome. As would be expected, they are characterized by the absence of introns, the presence of a poly-A tail, and the absence of any promoter sequences (Fig. 3-9). By genetic drift, pseudogenes tend to accumulate mutations over time. Processed pseudogenes corresponding to many genes have been described. There are numerous families with only one or two pseudogenes, but there are also many cases of 10, 20, or even more pseudogenes. Some examples of the latter type are the genes for glyceraldehyde-3-phosphate dehydrogenase, tubulins, and actin. As a rule, the genomic location of pseudogenes bears no relationship to the location of the original, functional gene; transposition is passive, making use of viral or *L1* enzymes.

Some RNAs accumulate very large numbers of genomic copies by retroposition, while others are limited to just a few copies or none. What determines whether a particular RNA is an active transposon is not clear. One can appreciate how ubiquitous these transposable elements are by considering that, in the α-globin cluster (see below), a 43-kb segment of DNA that includes three active genes contains 43 *Alu* and 11 *L1* elements (see also Fig. 4-19).

Unique Sequence DNA

Unique sequence DNA are simply nonrepetitive sequences that stretch between genes and among the repeated elements we have described. When unique sequence DNA is

FIGURE 3-9. Origin of a processed pseudogene. A mature mRNA is copied into cDNA and this is reinserted into the genome at a new site.

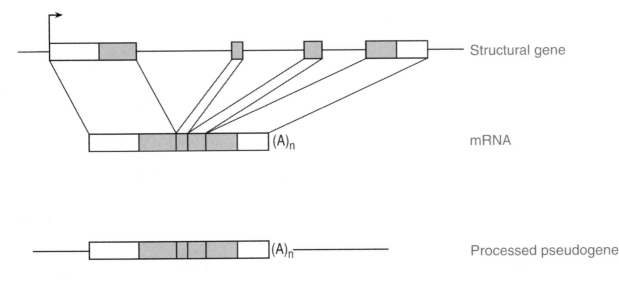

defined experimentally as the fraction of the total genome slowest to renature, it naturally includes unique sequences that are part of genes.

EXPRESSED DNA, GENES AND GENE FAMILIES

The expressed DNA fraction of the genome corresponds to the genes for structural RNAs—such as rRNA, tRNA, and small nuclear RNA (snRNA)—and the exons of RNA polymerase II messenger RNAs. We include in expressed DNA those sequences that are not themselves transcribed but are involved in the control of transcription, such as promoters and enhancers.

A brief aside on nomenclature. Human genetics has not had a long tradition of generally accepted rules of nomenclature. The symbols used in this text follow common usage (see Box 3.1). These are the most generally followed rules:

1. The symbols that represent genes are all uppercase and in italics.
2. The protein product of a gene can be identified by the same symbol as the gene, but in non-italic type.

The Organization of RNA Polymerase II Genes: Introns and Exons

RNA polymerase II genes are typically those that code for proteins. A sample of 94 fully sequenced genes provides a general idea of the structural range spanned by human genes. Of this sample, only 5% have no introns, and the median number of introns is four, meaning that 50% of these genes have fewer than four introns and 50% have more than four. The highest number of introns in this sample is 39, but genes with more than twice as many are known. Human introns tend to be on the large side, especially when compared to those of invertebrate eukaryotes. Their size distribution shows that two-thirds of all introns are between 300 and 3000 bp, with a pretty well-

---

www *Box 3.1   Internet Sites*

The National Center for Biotechnology Information (NCBI) home page provides access to several databases of DNA and protein sequences (Entrez), as well as links to the original literature through Medline (PubMed), information on specific genes (OMIM), and other resources. Entrez allows the search for sequences that have been entered into GenBank, as well as other international databases. Each sequence is identified by an Accession and a name; for example, Accession X57010, name HSCOL2A1G corresponds to the *Homo sapiens collagen 2 alpha 1 gene*. Entrez can be searched by sequence name, Accession, or by keyword (in this example, "collagen"). Sequences found in the database can be displayed in various ways depending on the needs; a convenient format for visual inspection of the main features of a sequence is the GenBank Report. NCBI can be found at:

http://www.ncbi.nlm.nih.gov/

Another site of interest is the home page of the Human Gene Nomenclature Committee, which sets the rules for the allocation of names and symbols to human genes:

http://www.gene.ucl.ac.uk/nomenclature/

---

defined lower end at approximately 90 bp. No upper limit is known; in this sample there are introns as long as 50 kb (Fig. 3-10A).

The presence of introns divide each gene into two or more exons. The position of introns seems to be independent of the coding or noncoding nature of a DNA segment; the first intron is usually near the initiation codon but it can be upstream or downstream of the AUG (see Fig. 3-10B). In terms of size, the exon at the 3' end of the gene is usually longer and more variable in size than the other exons; thus, while 3' exons form a broad distribution peak with two-thirds of them between 200 and 1100 bp (none over 5000), exons located farther upstream have a very narrow distribution, with two-thirds of them between 65 and 230 bp (see Fig. 3-10C and D). In this sample of 559 upstream exons there is only one longer than 1000 bp. The relative length homogeneity of upstream exons may be explained by in vitro experiments that showed that mammalian exons longer than 300 bases are not spliced very efficiently.

From the sample mentioned above, two-thirds of genes produce mRNAs that range in size between 700 and 2500 bases. Although the corresponding size of the transcription units are between 2000 and 15,000 bp, the median gene size of the sample is 5000 bp (i.e., half of the genes are larger than 5000 bp and half are smaller). The distribution is, however, very skewed, with a high concentration of smaller genes and a very long tail of few but very large genes (Fig. 3-10E, gene size distribution). As one would expect, given the constraints in exon size discussed, genes that code for longer polypeptides tend to have more exons than do genes that code for short polypeptides—on average, 25% of the length of primary transcripts is in exons and 75% is in introns.

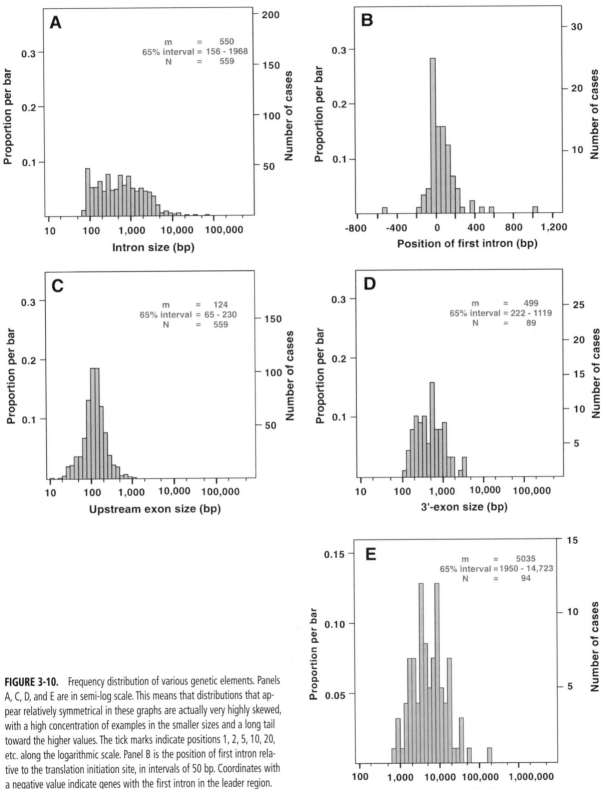

FIGURE 3-10. Frequency distribution of various genetic elements. Panels A, C, D, and E are in semi-log scale. This means that distributions that appear relatively symmetrical in these graphs are actually very highly skewed, with a high concentration of examples in the smaller sizes and a long tail toward the higher values. The tick marks indicate positions 1, 2, 5, 10, 20, etc. along the logarithmic scale. Panel B is the position of first intron relative to the translation initiation site, in intervals of 50 bp. Coordinates with a negative value indicate genes with the first intron in the leader region. (Based on data from Maroni, 1996.)

Examples of the organization of some small human genes are presented in Figure 3-11. The gene for the heat shock protein of 70 kd—*HSP70*, a gene whose transcription is stimulated by high temperatures and whose product acts to protect cells under these conditions—is one of the minority of human genes that lack introns. The remaining three are genes for various peptide hormones. Figure 3-12 includes the entire sequence of the gastrin gene.

Hormone genes illustrate another aspect of gene organization, and that is that the polypeptides encoded by genes are not always identical to the final active form of the corresponding proteins. The coding sequences of all secreted proteins include, at the amino-terminus, a segment of approximately 20 largely hydrophobic amino acids known as the *signal peptide*. While the protein is being synthesized, this signal peptide is threaded through the endoplasmic reticulum membrane so that the newly synthesized protein ends up within a vesicle where the signal peptide is clipped off. The vesicle eventually fuses to the cell membrane and releases its contents in the extracellular space. Direct translation products that include the signal peptide are sometimes designated by addition of the prefix *pre* (for example, preproinsulin).

FIGURE 3-11. Schematic representation of transcription units. The boxes represent those sequences that are present in mature mRNA (exons): shaded boxes represent protein-coding segments, and open boxes represent the 5' and 3' untranslated regions. Thin lines represent untranscribed as well as intron segments; bent arrows indicate the transcription initiation sites. The translation initiation site (ATG) is very near the beginning of exon 2 in the insulin and corticotropin releasing genes, and the translation termination site is very near the 3' end exon 2 in the natriodilatin gene.

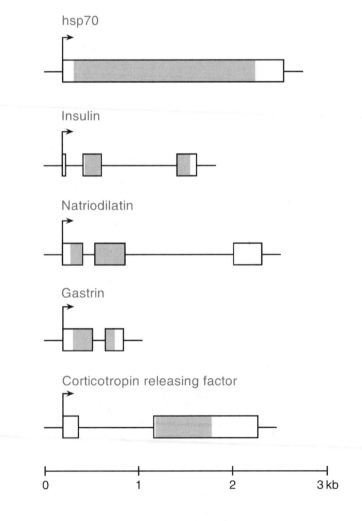

FIGURE 3-12. Gastrin gene sequence. →, transcription initiation site; TATA and polyadenylation boxes are in color; | (A)$_n$, marks the polyadenylation site. The portion of the peptide in italics represents the small active peptide released by proteolysis after synthesis. (Accession X00183).

```
-261   CCTATAGTCCCAGATATTCTGGAGGCTGAGGCAGGAGAATCACTTGAACCCGGGAGGCGG
              .              .              .              .              .              .

-201   AGGTTATAGTGAGCCGAGATCCCACCACTGCACTCCAGCCTAGGCAACAAGAGTGAAACT
                    .              .            ⌐→            .              .              .

-141   CTGTCTAAAAAAAAAAAAAGAAAGAATTGCACACTCATCAGCAGGTAGAGGCCTAGAGCC
              .              .              .              .              .              .

 -81   ACATGGTTCAGTCCCCGCCTCTGGGCCTCTGTGGGGACAGCCTCACCCTTAAGCTAGTCC
              .              .        1     .              .              .              .

 -21   CTTCTCCCCTTTGCAGACGAGATGCAGCGACTATGTGTGTATGTGCTGATCTTTGCACTG
                                        MetGlnArgLeuCysValTyrValLeuIlePheAlaLeu

  39   GCTCTGGCCGCCTTCTCTGAAGCTTCTTGGAAGCCCCGCTCCCAGCAGCCAGATGCACCC
  14   AlaLeuAlaAlaPheSerGluAlaSerTrpLysProArgSerGlnGlnProAspAlaPro

  99   TTAGGTACAGGGGCCAACAGGGACCTGGAGCTACCCTGGCTGGAGCAGCAGGGCCCAGCC
  34   LeuGlyThrGlyAlaAsnArgAspLeuGluLeuProTrpLeuGluGlnGlnGlyProAla

 159   TCTCATCATCGAAGGCAGCTGGGACCCCAGGGTCCCCCACACCTCGTGGCAGGTAGGAGC
  54   SerHisHisArgArgGlnLeuGlyProGlnGlyProProHisLeuValAlaA

 219   TGCTGACTGCCCTGCTTGCCTCACTTGGCCATGTTTGGCCAAGGTCTCCCCAGACTGGCT
              .              .              .              .              .              .

 279   CTGACTTCAGTTCCTAGAAGGTAGGCATCCTTCCCCCATTCTCGCCTCTCTCCCCTCCTC
              .              .              .              .              .              .

 339   AGACCCGTCCAAGAAGCAGGGACCATGGCTGGAGGAAGAAGAAGAAGCCTATGGATGGAT
  71   spProSerLysLysGlnGlyProTrpLeuGluGluGluGluGluAlaTyrGlyTrpMe

 399   GGACTTCGGCCGCCGCAGTGCTGAGGATGAGAACTAACAATCCTAGAACCAAGCTTCAGA
  90   tAspPheGlyArgArgSerAlaGluAspGluAsnEnd

 459   GCCTAGCCACCTCCCACCCCACCTCCAGCCCTGTCCCCTGAAAAACTGATCAAAAATAAA
              .              .              .              .              .              .

 519   CTAGTTTCCAGTGGATCAATGGACTGTGTCAGTGTTGTAGGGCAGAGGA
                        └
                         (A)$_n$
```

Another example of post-translational processing of the polypeptide sequence is provided by genes involved in the synthesis of multiple functional peptides. Thus, preproinsulin is a 109 aa polypeptide; after removal of the 25 aa signal peptide, the 84 aa proinsulin folds and two disulfide bonds are formed between Cys-7 and Cys-70 and between Cys-19 and Cys-83. Once these disulfide bonds covalently link the two ends of the protein, digestion of the peptide bonds next to amino acids 30 and 63 release a 33 aa middle fragment and produce insulin. So insulin is made up of two chains (A and B) but both are encoded by the same gene.

EXAMPLE 3.2

Collagen

The preprocollagen genes are examples of a certain correspondence between the modular nature of proteins (see Example 3.3) and modular gene organization. This correspondence is especially evident in the triple helical central domain where the protein's 3-amino-acid-repeats (Gly-X-Y, see Chapter 6) correspond to exons that encode whole number of such repeats (usually five or six). Figure 3-13A represents the first 15 exons of the gene for the α1 chain of collagen type II, in which exons 8 to 15 show the characteristic exon sizes of 45 and 54 bp. The entire gene encompasses over 50 exons. Scale representations of such very large genes are not possible because of the small size of exons compared to the overall size of the gene. Figure 3-13B shows the sequence of one of those exons (Accession X57010).

EXAMPLE 3.3

Cystic fibrosis

Figure 3-14A shows the organization of *CF*, the cystic fibrosis gene, which codes for the cystic fibrosis transmembrane conductance regulator (CFTR), an ion transporter (see Example 6.8). The 27 exons of *CF* are spread over nearly 250,000 bp; the introns are so extensive that attempting to visualize the entire transcription unit reduces all exons to narrow lines—in this case the mature mRNA is only approximately 2.6% of the total size of the transcript. The size of individual exons can be seen in Figure 3-14B; note that they are mostly quite small and relatively uniform in size. The two large exons are number 14, which seems to have been "imported" as a unit to code for the R-domain, and the 3′ terminal exon that, as we saw earlier, is usually the longest.

CFTR illustrates another general principle of gene organization. Large proteins are often made up of a combination of regions or domains, each of which plays a specific role in the overall function of the protein. Compare, for example the organization of CFTR (Fig. 6-14) with that of a hemoglobin monomer (see Fig. 6-10). CFTR includes six membrane-spanning domains, two nucleotide-binding domains, and a regulatory region; hemoglobin is a single-domain protein in which the polypeptide chain forms a cage within which the heme group is located. In the CFTR gene, as in many other cases, a functional protein domain is composed of neighboring amino acids, rather than being composed of amino acids that lie far apart from each other along the polypeptide chain but that are then brought together by the folding of the chain. This type of modular organization is sometimes reflected in the organization of the corresponding gene. In this fashion, a protein may gain a new function by its gene acquiring a DNA segment, perhaps from another gene, which codes for the proper polypeptide segment. This seems to have occurred in the evolution of CFTR, where we find the R, or regulatory region, absent from other members of this family of transport ATPases. Most transport ATPases only have the transmembrane and nucleotide-binding domains.

Figure 3-14B shows the correspondence between exons and various regions of the CFTR protein. Unlike the collagen gene, the boundaries between protein features fail to match the boundaries between exons. The one exception is the correspondence between exon 14 and the R-domain (see also Fig. 6-14) (Accession NM_000492).

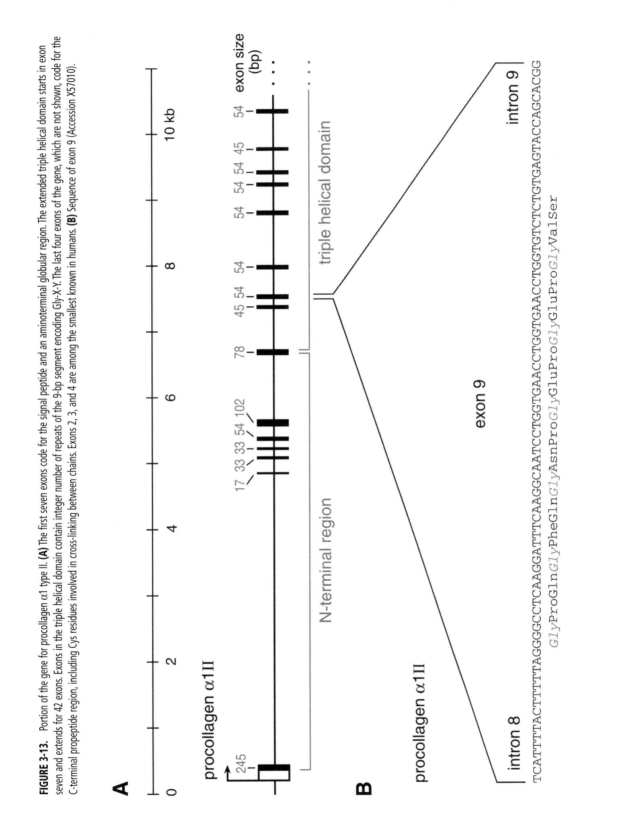

FIGURE 3-13. Portion of the gene for procollagen α1 type II. **(A)** The first seven exons code for the signal peptide and an aminoterminal globular region. The extended triple helical domain starts in exon seven and extends for 42 exons. Exons in the triple helical domain contain integer number of repeats of the 9-bp segment encoding Gly-X-Y. The last four exons of the gene, which are not shown, code for the C-terminal propeptide region, including Cys residues involved in cross-linking between chains. Exons 2, 3, and 4 are among the smallest known in humans. **(B)** Sequence of exon 9 (Accession X57010).

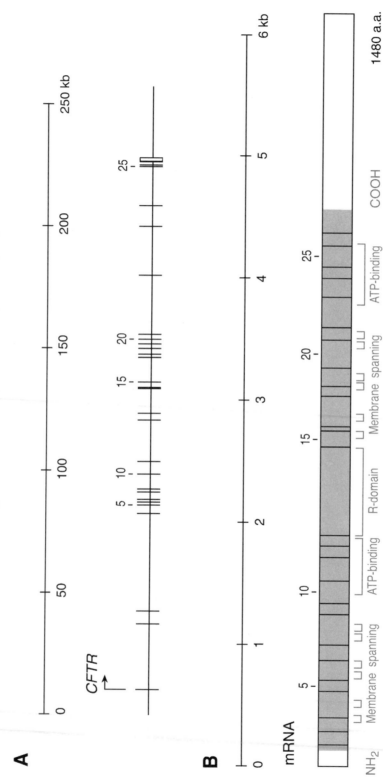

FIGURE 3-14. **(A)** Genomic organization of the cystic fibrosis gene. The transcription initiation site is marked by the bent arrow, and the exons are numbered. **(B)** Messenger RNA, with vertical lines delimiting the 27 exons. Below the mRNA the extent of the various functional regions of the 1480-amino-acids protein are indicated. (Based on Tsui, 1995.)

EXAMPLE 3.4

Duchenne muscular dystrophy

Almost 10 times larger than CF is the gene responsible for Duchenne muscular dystrophy (see Example 6.6). Dystrophin has 79 exons and spans 2,300,000 bp. It has been estimated that it takes 16 hours for RNA polymerase to complete transcription of this gigantic gene. Although the protein is itself one of the largest known, the size of the mRNA is less than 1% of the size of the dystrophin transcript.

The size of this gene appears to be matched by the complexity of its expression. There are three different promoters that control expression of full size transcripts, as well as two internal promoters near the C-terminal end that drive expression of a shorter protein. There are also multiple alternative-splicing variants, perhaps providing tissue specificity (Accession NM_000109).

Along similar lines, in the pronatrodilatin gene, codons 1 to 25 code for the signal peptide, codons 26 to 98 code for the hormone cardiodilatin, which causes smooth muscle relaxation, and codons 99 to 126 code for the atrial natriuretic factor (ANF), which has natriuretic, diuretic, and smooth-muscle relaxant activities. As the name implies, this gene is active in cells of the heart atrium, and, from a single encoded polypeptide, two active factors are released by proteolysis.

As mentioned, these are examples of the smallest known human genes. Most genes are considerably larger than those in Figure 3-11. The genes for collagen, the cystic fibrosis transmembrane conductance regulator (CFTR), and dystrophin are at the other end of the distribution, among the largest genes that have been studied. Table 3-2 summarizes the organization of several human genes.

Gene Families

Many human genes occur not as single copy elements but in families of more or less closely related sequences. The various members of a gene family may be found clus-

TABLE 3-2. Size and organization of some human genes. Columns two, three, and four give the size in base pairs of the coding region, the mature mRNA, and the entire gene. Column five is a ratio and column six is the number of exons.

Genes	Codons	mRNA	Transcript	mRNA/Transcript	Exons
gastrin	306	516	646	0.80	2
β-globin	444	626	1556	0.37	3
GAPDH	1011	1284	3856	0.333	9
visual pigment	1095	1236	13,372	0.092	6
retinoblastoma	2790	4979	177,000	0.028	27
CFTR	4443	6500	250,000	0.026	27
dystrophin	11,058	14,000	2,300,000	0.006	79

Abbreviations: CFTR, cystic fibrosis transmembrane conductance regulator; GAPDH, glyceraldehyde-3-phosphate dehydrogenase; visual pigment, cone photoreceptor pigment sensitive to green light.

EXAMPLE **3.5**

The globin gene family

There are a total of eight globin genes in two subfamilies (Chapter 6). The α globin subfamily consists of three genes clustered in chromosome 16 (Fig. 3-15), and the β globin subfamily consists of five genes clustered in chromosome 11 (Fig. 3-16). These multiple copies appeared over a period of hundreds of millions of years and illustrate the process of evolution by gene duplication and divergence. Thus, a progenitor of the vertebrates may have had a single globin gene that underwent duplication; this was probably followed by a translocation that separated the two copies. These two globin genes subsequently acquired slightly different properties and became the predecessors of the modern α and β globin genes subfamilies. In each subfamily, other duplications gave rise to copies that specialized, by mutation, into the adult and fetal forms that are present in mammals. Finally, more recent duplications in primates produced the adult genes α1 and α2 and the fetal genes γ^ε and γ^A.

An indirect indication of the different age of the various duplications can be found in the fact that the older the duplication, the greater the difference in amino acid sequence. The amino acid sequences of α1 and α2 globins are identical, but these polypeptides differ from the embryonic ς globin at 60 out of 145 positions (Fig. 3-17 and Table 3-3). Note that in addition to the coding sequence, the general organization of the genes (number and position of the introns) was also conserved; this is usually the case in multigene families.

Sequence comparisons have shown that the pseudogene ψα1 (see Fig. 3-15) of primates seems to correspond to one of the two active α genes in other mammals. In primates, presumably to compensate for the loss of this second α gene, natural selection preserved a new duplication of α that gave rise to the α1, α2 pair that we see today (OMIM 141800).

FIGURE 3-15.   Alpha globin gene cluster in chromosome 16. In the expressed genes, transcription is from left to right. Here, ψα1, ψα2, and ψς are three pseudogenes. The status of the θ globin gene is somewhat uncertain; it is expressed at very low levels, if at all.

α globin cluster

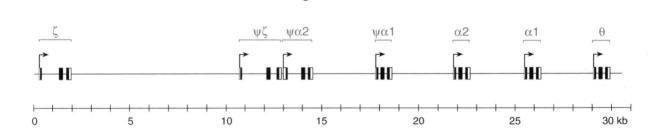

FIGURE 3-16.   Beta globin gene cluster in chromosome 11. In all cases, transcription is from left to right except for ψβ, which is an unexpressed pseudogene.

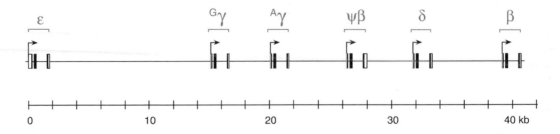

FIGURE 3-17. Globin amino acid sequences in various pairwise comparisons. Symbols in color represent positions in which the two sequences differ. Note that in the comparison of the α and β sequences allowance was made for five gaps, positions where deletions occurred in one chain or additions in the other. The number and extent of these gaps is so as to optimize the overall match between the sequences. Arrowheads point to the positions where introns are found.

A

```
          10        20        30        40        50        60        70        80
MSLTKTERTIIVSMWAKISTQADTIGTETLERLFLSHPQTKTYFPHFDLHPGSAQLRAHGSKVVAAVGDAVKSIDDIGGA
MVLSPADKTNVKAAWGKVGAHAGEYGAEALERMFLSFPTTKTYFPHFDLSHGSAQVKGHGKKVADALTNAVAHVDDMPNA
                                  ∧

          90       100       110       120       130       140
LSKLSELHAYILRVDPVNFKLLSHCLLVTLAARFPADFTAEAHAAWDKFLSVVSSVLTEKYR    ζ globin
LSALSDLHAHKLRVDPVNFKLLSHCLLVTLAAHLPAEFTPAVHASLDKFLASVSTVLTSKYR    α globin
                 ∧
```

B

```
          10        20        30        40        50        60        70        80
MVHFTAEEKAAVTSLWSKMNVEEAGGEALGRLLVVYPWTQRFFDSFGNLSSPSAILGNPKVKAHGKKVLTSFGDAIKNMD
MVHLTPEEKSAVTALWGKVNVDEVGGEALGRLLVVYPWTQRFFESFGDLSTPDAVMGNPKVKAHGKKVLGAFSDGLAHLD
                                 ∧

          90       100       110       120       130       140
NLKPAFAKLSELHCDKLHVDPENFKLLGNVMVIILATHFGKEFTPEVQAAWQKLVSAVAIALAHKYH    ε globin
NLKGTFATLSELHCDKLHVDPENFRLLGNVLVCVLAHHFGKEFTPPVQAAYQKVVAGVANALAHKYH    β globin
                       ∧
```

C

```
          10        20        30        40        50        60        70        80
MVLSPADKTNVKAA WGKVGAHAGEYGAEALERMFLSFPTTKTYFPHF DLSHGSAQVKGHGKKVADALTNAVAHVDDMP
MVHLTPEEKSAVTALWGKVNVD   EVGGEALGRLLVVYPWTQRFFESFGDLSTPDAVM GNPKVKAHGKKVLGAFSDGLA
                                ∧

          90       100       110       120       130       140
NALSALSDL    HAHKLRVDPVNFKLLSHCLLVTLAAHLPAEFTPAVHASLDKFLASVSTVLTSKYR    α globin
HLDNLKGTFATLSELHCDKLHVDPENFRLLGNVLVCVLAHHFGKEFTPPVQAAYQKVVAGVANALAHKYH    β globin
                       ∧
```

TABLE 3-3. Pairwise comparisons of the amino acid sequences of several members of the globin family. Percent sequence identity represents the fraction of positions in the chain occupied by the same amino acid.

Globin Sequences Compared	Percent Sequence Identity
α and β chains	34
Embryonic and adult chains:	
ζ and α	59
ε and β	76
Adult chains:	
α1 and α2	100
δ and β	76

EXAMPLE 3.6

The retinal visual pigments

An example of finely specialized genes within a family is provided by the four photoreceptor pigments of the retina. Closely related genes on the X-chromosome code for cone pigments that detect green and red light. Two more distantly related autosomal genes code for a blue-sensitive cone pigment (on chromosome 7) and for opsin, the protein in rhodopsin, the rod pigment that is sensitive to dim light (on chromosome 3).

Unfortunately, there is no widely used good nomenclature system. All four proteins can be considered opsins and they all bind to the chromophore retinal through a Lys residue that occupies homologous positions. A commonly used designation is rhodopsin (or opsin) for the rod pigment, and green, red, and blue pigments for the cone pigments. The general organization of these genes is shown in Figure 3-18.

The X-chromosome cluster includes the red pigment gene and one or more copies of the green pigment gene (Fig. 3-19). There is 98% identity in the nucleotide sequence of these repeating units—note that the repeating unit includes both genic and intergenic sequences. At the amino acid level, there is 94% identity between the red and green pigments. These differences are mostly concentrated in exon 5, and three amino acid substitutions seem responsible for the different spectral absorption: Ala180Ser, Tyr277Phe, and Thr285Ala.

The green and red pigments, in turn, show approximately 43% identity with blue pigment and 41% with opsin. The identity between opsin and blue pigment is 42%. Figure 5-20 is a direct comparison of the sequences of exon 3 of the blue pigment and opsin genes.

One may speculate that the gene family for visual pigments may be going through a period of transition in humans. As in the case of the globin gene family, we can imagine ancestral duplications that gave rise to rhodopsin, blue pigment, and red/green pigment; the last two would be responsible for dichromatic vision. A more recent duplication of the red/green pigment gene with subsequent specialization into red pigment and green pigment genes led to our present trichromatic vision; the three color vision genes are present in Old World monkeys, but New World monkeys, which are more distantly related to humans, have only two pigment genes.

Finally, some of us have multiple copies of the green pigment gene. This duplication is so recent that it has not yet been "fixed"—the human population is polymorphic for this duplication, some people having multiple copies and some only one. We can imagine that appropriate amino acid substitutions in one of these extra copies of the green pigment gene might enable it to absorb light in a region of the spectrum not well used by our present pigments. Individuals carrying such a mutation would enjoy a more discriminating form of chromic vision, and, if this proves a selective advantage, a few million years from now "normal" human vision might be tetrachromatic.

tered in a chromosomal region or at multiple loci scattered throughout the genome. In some cases, the existence of multiple copies of a gene can be explained by the need for large amounts of product synthesized over short periods of time, as with the histone and ribosomal genes. In other cases, different members of a family acquire specialized functions by subtle amino acid substitutions in the course of the evolutionary process, as with the globins and visual pigments genes.

Mutations may also inactivate some of the multiple gene copies in a family. The mutations might originally affect protein structure, translation, or transcription, but once the mutant gene stops contributing to cell function, other mutations accumulate, leading eventually to loss of transcription. Such "dead" genes are called *pseudogenes*, but

FIGURE 3-18. Exonic organization of the visual pigment genes. The green and red genes are nearly identical in terms of exon and intron sizes and are represented by a single diagram. The lines linking the various genes highlight exon homologies. The numbers at the beginning and end of each coding segment represent codon numbers. The first 20 amino acids are part of the signal sequence. Downstream of codon 21 all exons have been conserved, at least in size, but the aminoterminal end seems to have been replaced in the green/red genes in which there is an extra exon. Numbers in italics indicate the first 37 residues of the green/red genes; residue 38 of this gene corresponds to residue 21 of opsin, and from this point forward the numbering system of opsin is used. Similarly, the blue pigment coding region is three codons shorter and we are assuming here that these are the first three codons, in order to maximize the match or alignment among genes. The 5' untranslated region of the blue pigment gene seems to be only 6 bp long and cannot be resolved in the figure.

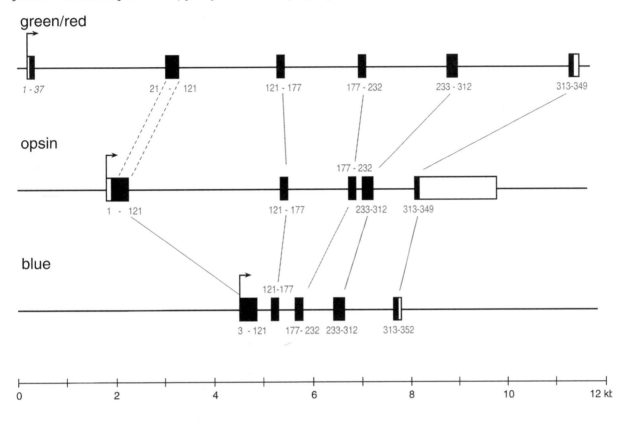

FIGURE 3-19. X-chromosome cluster of visual pigment genes. The straight line segment represents the duplicated region, the wavy lines at both ends are unique flanking sequences. The narrow brackets upstream of the red gene represent a few hundred nucleotide pairs responsible for transcriptional control of the entire cluster. The long brackets define the repeating 39-kb units that include a green pigment gene and 27 kb of flanking DNA. Individual chromosomes may have one or more green pigment genes; in the latter case, not all green genes are active, apparently because the regulatory region's effect fails to reach the most distant genes. (Based on Nathans, 1994.)

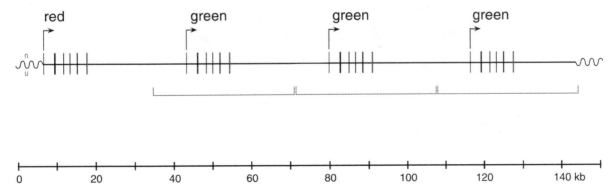

FIGURE 3-20.   Comparison of the nucleotide sequence of the opsin gene (Accession R84449) to that of the blue pigment gene (Accession M13295). Shown here is exon 3 of these genes. If differences in the two sequences lead to amino acid differences, the amino acid in opsin is indicated below its nucleotide sequence. Note how the regions of identity tend to occur in clusters, possibly representing regions of the protein more sensitive to change, and therefore more conserved. A similar comparison of the red and green pigment sequences would show almost no differences.

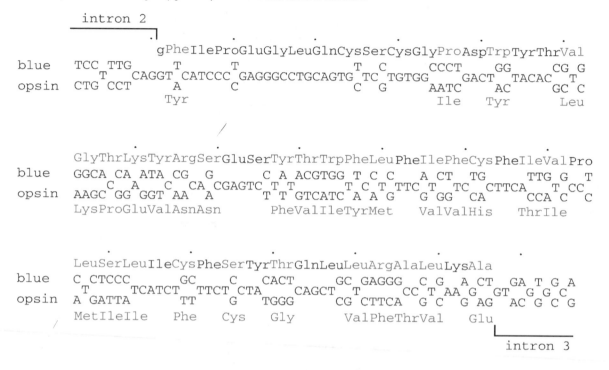

note that their structure is very different from that of the processed pseudogenes discussed in an earlier section. In this case, pseudogenes maintain promoters and introns and lack poly-A tails.

To name a few other examples of gene families, there seem to be eight different actin genes; at least four of which code for variously specialized muscle actins and two code for cytoskeletal actins. There are also 10 to 20 pseudogenes and processed pseudogenes (OMIM 102540 and others). Similarly, there are at least five active tubulin genes (OMIM 301850 and others)—and approximately 10 pseudogenes—and over 30 collagen genes in 10 different chromosomes. The peptide hormones are grouped into several families; thus, the growth hormone (GH) family includes placental lactogen hormone (85% sequence similarity with GH) and prolactin (30% sequence similarity with GH).

Ribosomal RNA Genes

The ribosomal RNAs, 5.8S, 18S, and 28S, are encoded by a gene family with unique characteristics. It comprises approximately 200 gene copies, per haploid genome, distributed among five loci in the middle of the short arm of the acrocentric chromosomes 13, 14, 15, 21, and 22. The genes are in very regular head-to-tail arrays in regions of the chromosomes called the *nucleolus organizers;* the transcribed region is approximately 13 kb and the intergenic spacers 30 kb (Fig. 3-21).

Compared to other gene families, rRNA genes are very little differentiated from each other, and the mechanism that maintains this uniformity is not completely understood. It is thought that it is genetic recombination and exchange involving genes in different clusters that constantly homogenize the family. This is possible because

FIGURE 3-21.   Organization of rRNA genes. **(A)** Elements in each repeating unit. 5' ETS and 3' ETS, external transcribed spacers; ITS1 and ITS2, internal transcribed spacers; 18S, 5.8S, and 28S, coding regions for the corresponding rRNAs. **(B)** Organization of the tandem repeats. UT, transcribed unit; IGS, intergenic or nontranscribed spacer. (Accession U13369.)

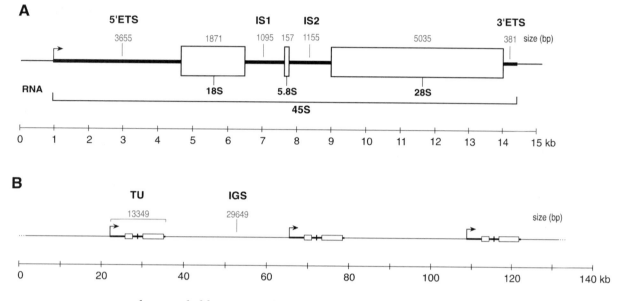

there probably are no other genes between the nucleolus organizer and the telomere of these acrocentric arms. The genes are positioned so that the transcription initiation sites are nearer the telomere and transcription is toward the centromere.

RNA polymerase I produces a 13,349-base transcript, called the 45S precursor that is then processed by RNases in the nucleolus to release the three rRNAs, which are then assembled into the ribosomal subunits.

CONCLUSION

This chapter presents a very summary report of the various kinds of sequences found in the human genome. Such a global view of the genome is somewhat artificial, because it is not the result of any experimental method of analysis, but rather a collage of results obtained with a diversity of approaches. It is, to some extent, an enumeration of individual cases. We leave for the next chapter a review of some of the methods used by molecular biologists and geneticists to isolate and characterize the sequences we have discussed here. It is the relative generality of some of those methods that gives much of this field, now coalescing under the rubric of *genomics*, a measure of unity.

EXERCISES

3-1.   Use Figure 3-6 to test paternity. The arrows point to bands that can clearly be considered as derived from either the mother or the father; label the maternal bands, M, and the remaining bands, P. Check which of the men in lanes 4, 5, and 6 is capable of contributing the P bands—which man is more likely to be the father. Why is this method more powerful than the traditional blood-type test to establish paternity? Can you tell which M band is allelic of any given P band?

3-2.   Analyze Figure 3-7. Mark in pencil the two bands that represent the (CA)$_n$ alleles in each lane. Use the sequencing ladder to determine the size differences between the various alleles; designate the shortest allele *n*, and each of the others *n + x*, where *x* is the number of nucleotides by which each alleles differs from the shortest. Write down the allelic constitution of each member of the pedigree (for example, *n + 1/n + 3*) and verify that the alleles behave as co-dominant markers.

3-3.   An individual is genotyped at four independently segregating microsatellite loci and is found to have the genotype A1/A5; B3/B4; C2/C3; D1/D5. Assume that this individual is a member of a "panmictic" population (people select mates without reference to their genetic background, as if gametes found each other at random). Also assume that at the four loci the allele frequencies were as follows: A1 = 0.12, A5 = 0.20, B3 = 0.5, B4 = 0.15, C2 = 0.10, C3 = 0.25, D1 = 0.18, D5 = 0.3. What is the likelihood that another individual would be found who has the same constitution at these four sites? How reasonable is the assumption of panmixis in a human population? How can it be approximated?

3-4.   In the section on nonviral retrotransposons there is an example of a mutation in the neurofibromatosis I gene. Use a diagram to represent this mutation and its effect. Could you have predicted that skipping of exon 6 would lead to a frameshift?

3-5.   Diagram the relationships among preproinsulin, proinsulin, and insulin. What are the sizes of the two insulin chains? What are the similarities and differences between a gene such as prepronatrodilatin and a polycistronic operon in bacteria?

3-6.   Based on the gastrin sequence in Figure 3-12, what do you expect to be the possible consequences of a G→A base substitution at position 340 of this gene?

3-7.   Using Entrez on the NCBI home page, find an entry that includes the sequence of one of the genes given as examples in this chapter.

(Note: you should obtain the gene sequence and not the mRNA sequence.) Print out the corresponding GenBank report (at least the front portion before the sequence proper starts) and identify as many as possible of the gene features, such as the transcription initiation site, polyadenylation site, exons, and introns. Find each of these elements in the sequence.

3-8.   Using Entrez on the NCBI home page, find an entry that includes the complete sequence of a gene. Print out the corresponding GenBank report (at least the front portion before the sequence proper starts) and identify the transcription initiation site, polyadenylation site, exons, introns, and so on. Make a diagram like the one in Figure 3-11, indicating the various elements and their sizes. Find each of these elements in the sequence.

3-9.   Make a photocopy of a table of codons (Appendix Fig. 9) and use Figures 3-12, 3-13, and 3-19 as a random sampling of codon usage in humans. Write down in the table the number of times each codon is used. Are all synonymous codons for a given amino acid used with equal frequency? Are codons with a certain base in the third position used more often than others?

REFERENCES

Berg DE, Howe MM, eds. (1989) Mobile DNA. Washington, DC: American Society for Microbiology.

Britten RJ. (1996) Evolution of *Alu* retroposons. In: Human genome evolution, Jackson M, Strachan T, and Dover G, eds. Oxford: BIOS, 211–228.

Bruford MW, and Wayne R. (1993) Microsatellites and their application to population studies. Curr Opin Gene Devel 3:939–943.

Dombroski BA, Mathias SL, Nanthakumar E, Scott AF, and Kazazian HH Jr. (1991) Isolation of an active human transposable element. Science 254:1805–1808.

Feng Q, Moran JV, Kazazian HH Jr, and Boeke JD. (1996) Human *L1* retrotransposon encodes a conserved endonuclease required for retrotransposition. Cell 87:905–916.

Gonzalez IL, and Sylvester JE. (1995) Complete sequence of the 43 kb human ribosomal DNA repeat:

analysis of the intergenic spacer. Genomics 27:320–328.

Jeffreys AJ, Wilson V, and Thein SL. (1985a) Hypervariable "minisatellite" regions in human DNA. Nature 314:67–73.

Jeffreys AJ, Wilson V, and Thein SL. (1985b). Individual-specific "fingerprints" of human DNA. Nature 316:76–79.

Litt M, and Luty JA. (1989) A hypervariable microsatellite revealed by in vitro amplification of a dinucleotide repeat within the cardiac muscle actin gene. Am J Hum Genet 44:397–401.

Marks J, Shaw J-P, Perez-Stable C, Hu W-S, Ayres TM, Shen C, and Shen C-KJ. (1986) The primate α-globin gene family. Cold Spring Harbor Symp Quant Biol 51:499–508.

Maroni G. (1996) The organization of eukaryotic genes. Evol Biol 29:1–19.

Nathans J. (1994) In the eye of the beholder: visual pigments and inherited variation in human vision. Cell 78:357–360.

Tsui L-C. (1995) The cystic fibrosis transmembrane conductance regulator gene. Am J Respir Crit Care Med 151:547–553.

Tyler-Smith C, and Willard HF. (1993) Mammalian chromosome structure. Curr Opin Gene Devel 3:390–397.

Ullu E, and Tschudi C. (1984) *Alu* sequences are processed 7SL RNA genes. Nature 312:171–172.

Warburton PE, and Willard HF. (1996) Evolution of centromeric alpha satellite DNA: molecular organization within and between human and primate chromosomes. In: Human genome evolution. Jackson M, Strachan T, and Dover G, eds. Oxford: BIOS, 121–145.

Weiner AM, Deininger PL, and Efstratiadis A. (1986) Nonviral retroposons: genes, pseudogenes and transposable elements generated by the reverse flow of genetic information. Ann Rev Biochem 55:631–661.

Weissenbach J. (1993) A second generation linkage map of the human genome based on highly informative microsatellite loci. Gene 135:275–278.

Wellauer PK, and Dawid IB. (1979) Isolation and sequence organization of human ribosomal DNA. J Mol Biol 128:289–303.

Willard HF, and Waye JS. (1987) Hierarchical order in chromosome-specific human alpha satellite DNA. Trends Genet 3:192–198.

Genome Organization II

ISOLATION AND CHARACTERIZATION OF HUMAN GENES

For supplemental information on this chapter, see Appendix Figures 7, 8, 9, and 10

In Chapter 3 we saw a picture, albeit incomplete, of the entire genome and the organization of human genes. The path we followed skirted one major question: how are those genes isolated so that their organization can be analyzed? There are many volumes available that discuss in detail the molecular techniques and procedures brought to bear on this problem. In this chapter, the strategies and methodological approaches used, rather than the technical details of laboratory procedures, will be covered.

Expression Cloning

Expression cloning methods are based on an understanding of gene function and start with the following two steps: 1) family studies and genetic analysis of a trait demonstrate that it is probably caused by a mutation in a single gene; 2) detailed study of the phenotype identifies a probable biochemical defect and a specific protein that might be altered. This progression was followed, for instance, in the case of sickle cell anemia. Family studies demonstrated its hereditary nature and studies of patients led to the detection of misshaped red blood cells and a deficiency in oxygen transport. Because hemoglobin is the main protein component of red blood cells, the hemoglobins of patients and normal controls were compared and differences in their physico-chemical properties were observed. Eventually, the substitution of valine for glutamic acid in position six of the β globin chain (Glu6Val) was discovered, thus definitely establishing that sickle cell anemia is caused by a mutation in the β globin gene.

Once the protein altered by the mutation is identified, usually two additional steps are involved: 3) the production of DNA copies complementary to mRNA (cDNA) and cloning of the cDNA (see Appendix Fig. A-8). Finally, 4) cDNA is used as a probe to identify clones carrying portions of the gene of interest in a genomic library (see Appendix Fig. A-9). In the case of the β globin gene, isolation of mRNA was relatively easy because reticulocytes—immature red blood cells—make hemoglobin almost exclusively, and a cDNA library of reticulocytes would be largely made up of β globin and α globin clones. By sequencing the insert of a few clones it is possible to identify quickly one that contains the desired cDNA, β globin in this case. Probes derived from the cDNA were then used to isolate clones from a genomic library and also to hybridize to Southern transfers of total genomic DNA digested with various restriction enzymes to prepare a restriction map of the chromosomal neighborhood of the β globin gene (see Appendix Fig. A-7C).

Similar methods were used to clone many genes involved in hereditary defects, when the corresponding gene products were known from biochemical studies, such as hemophilias/clotting factors, Lesch-Nyhan disease/hypoxanthine-guanine phosphoribosyl transferase (HGPRT), and phenylketonuria/phenylalanine hydroxylase. It should be noted that these methods can also be used to clone any gene whose product is known, even if no mutations are available. They have been used to isolate hundreds of human genes.

Expression cloning depends on the ability to isolate a specific cDNA, and this is often the most difficult step. The mRNA of a few abundant proteins can be obtained by enrichment of the proper cell types and proceeding as in the case of hemoglobin. In most cases, however, an mRNA fraction (poly(A)$^+$ RNA) is a mixture of numerous gene products; in the most challenging cases, the specific message in which one is interested may be very rare, a small minority among the multitude of messages produced by most cells. In these cases, the procedure is usually to prepare a *cDNA library* of the cell type or tissue in question and then *screen* the library in search of the clones of interest. The choice of a probe to screen a cDNA library depends on the characteristics of the gene and the extent of our knowledge of it; a few often-used methods are discussed below. Once a cDNA clone is isolated, the nature of the protein coded must be verified; this is often done by direct nucleotide sequencing or by immunoprecipitation of the protein product synthesized with in vitro systems.

Oligonucleotide probes. In many cases, it is possible to isolate a relatively pure protein product and, from this, to obtain a partial amino acid sequence. A stretch of 5 to 10 amino acids is selected, attempting to maximize the number of amino acids with only one or a few alternate codons. By virtual "reverse translation"—and taking into account the degeneracy of the genetic code—all possible nucleotide sequences are deduced and a mixture of oligonucleotides is synthesized. Usually, oligonucleotides that are 15 to 30 nucleotides long and that are made radioactive by phosphorylation perform adequately as probes. A cDNA library is then plated on agar, lifted onto a filter or membrane, and the filter or membrane is subjected to hybridization to the radioactive probe. The plaques or colonies harboring the corresponding sequence are detected by their radioactivity and one can then go back to the original agar plate to take a sample of the colony or plaque that "lit up" in the x-ray film. See Example 4.1 on the factor IX gene.

EXAMPLE 4.1

Isolation and characterization of a cDNA coding for human factor IX

Kurachi and Davie (1982) reported cloning the gene that is defective in patients with hemophilia B, the less common of the two known sex-linked clotting disorders. Two methods were used in these experiments.

In the first method, a radioactively labeled cDNA probe was prepared from baboon liver mRNA that was enriched for factor IX. Enrichment for factor IX was accomplished by injecting a young baboon with goat antibody to factor IX over a 2-day period. This treatment reduced the circulating factor IX coagulant activity to less than 1% of normal, which stimulated transcription and raised the level of the corresponding mRNA. The baboon was then sacrificed and the liver frozen in liquid nitrogen.

Poly(A)-containing RNA was isolated and assayed for factor IX mRNA with a rabbit reticulocyte lysate. [Reticulocytes contain very high levels of protein synthesizing machinery (ribosomes, translation factors, etc.) and extracts can be prepared in which this machinery is used to translate exogenous mRNAs in vitro—this is called a reticulocyte lysate.] Using a factor IX–specific antibody, the authors were able to tell that the liver mRNA fraction they added to the reticulocyte lysate stimulated translation of factor IX. They could also tell that, when the animal had previously been treated with factor IX antibodies, the level of mRNA increased approximately fivefold when compared to that of control animals.

The mRNA for factor IX was further enriched 20-fold by specific immunoprecipitation of the liver polysomes with goat antibodies. Cells were lysed gently and polysomes were isolated by centrifugation in a sucrose gradient. This polysome preparation was then treated with goat antibodies against factor IX so that those polysomes associated with nascent factor IX polypeptides would precipitate preferentially. The final factor IX mRNA level was approximately 2% of this mRNA preparation. The mRNA was used to synthesize a radioactively labeled cDNA probe.

The second probe was a mixture of synthetic 14-nucleotide-long oligonucleotides with 12 different DNA sequences. These sequences were obtained by reverse translation of the amino acid sequence of Met-Lys-Gly-Lys-Tyr and can be abbreviated: T-A-T/C-T-T-C/T/G-C-C-T/C-T-T-C-A-T. The oligonucleotides were radio labeled with T4 kinase and [γ-^{32}P]ATP.

Overall, approximately 18,000 clones from a human liver cDNA library were screened using as probes the synthetic oligonucleotide mixture and the single-stranded DNA prepared from mRNA enriched for baboon factor IX. Four positive clones were identified.

Technical note. Chemical oligonucleotide synthesis is carried out in steps. The base incorporated at any position is determined by the nucleotide added to the reaction at that step. Thus, to synthesize a mixture of oligonucleotides as was done here is not more laborious than to synthesize a single oligonucleotide of that length: when reaching the third position, where some oligos should have a C and some a T, the nucleotide added to the reaction is an equimolar mixture of dCTP and dTTP.

Libraries in expression vectors. *Expression vectors* are viruses or plasmids capable of directing the transcription and translation of the human DNA insert (Fig. 4-1). When a cDNA library in an expression vector is plated on agar, each plaque or colony produces small amounts of the human protein encoded by the insert in that clone. Because, as a rule, each clone harbors a different insert, each plaque or colony secretes a unique protein. If the investigator has access to a reasonably pure preparation of the protein product, an antibody may be prepared against that protein and used in a binding assay to identify the clone producing the corresponding antigen.

In other cases there may be other protein-specific assays. If the gene sought codes for a protein with an enzymatic activity, for example, it may be possible to use a colorimetric assay to detect the colony or plaque producing the enzyme. See Example 4.2 on the von Willebrand factor gene.

Functional assays. In some cases the function of a human protein can be detected in model systems such as yeast or amphibian eggs, and the correct cDNA can thus be identified. For example, yeast cells with conditional mutations in cell cycle genes were rescued by the corresponding human cDNAs. Yeast strains were prepared such that they would grow in galactose medium but not in glucose medium, due to a defi-

FIGURE 4-1. An expression vector. This could be either a plasmid or a virus. In addition to a selectable marker (here antibiotic resistance) and an origin of replication, an expression vector needs to include the promoter and transcription initiation site of a gene that is functional in the normal host of the vector and a cloning site (vertical line) into which the sequences to be expressed are inserted. Attention must be paid that, in the resulting chimeric gene, the same reading frame is maintained.

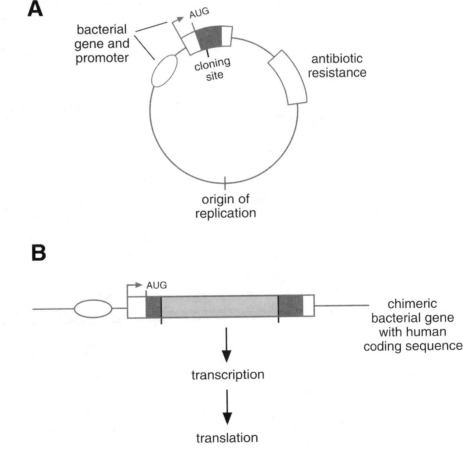

Isolation of cDNA clones for human von Willebrand factor

Von Willebrand factor (vWF) is a multimeric protein-carrier of factor VIII, which is an important participant in blood clotting. Mutations in this gene are often dominant and give rise to von Willebrand disease, an autosomal bleeding disorder. To isolate vWF cDNA, Ginsburg et al. (1985) used the λgt11 bacteriophage expression vector. In this system, the cloned cDNA fragment is spliced into the 3' end of the bacterial β-galactosidase gene (also carried by the phage) such that proteins are produced as fusion products at the COOH-terminus of β-galactosidase. Two large λgt11 libraries were prepared using cDNA derived from cultured human umbilical vein endothelial cells (a known source of vWF). To detect recombinant phage encoding the fusion product of vWF and β-galactosidase, the authors used a vWF-specific antibody to screen viral plaques growing on *Escherichia coli*. Out of approximately 3×10^6 clones from the above λgt11 libraries, nine plaques gave positive results on a primary screen and one was strongly positive upon rescreening.

ciency of one or another cyclin (cyclins are proteins that regulate progression through the cell cycle). A human cDNA library was prepared in a yeast expression vector—a vector capable of replication and gene expression in yeast cells. When a cyclin-deficient strain was transformed with this library and grown in glucose, only those yeast cells that had acquired the corresponding human cyclin gene were capable of cell division and growth. In this fashion the human cDNAs for several cyclins were cloned. See also Example 4.3 on the metal-ion transporter gene, and the cloning of excision repair cross-complementing genes in the xeroderma pigmentosum example in Chapter 7.

Differential colony or plaque hybridization (+/– hybridization). In the differential colony or plaque hybridization procedure, poly(A)⁺ RNA is isolated from cells subjected to a treatment that induces expression of a particular gene; cDNA is prepared and used to make both a cDNA library and a radioactive probe, the "+" probe. A cDNA probe is also prepared with mRNA from uninduced cells, the "–" probe. The cDNA library is plated out on agar medium and duplicate filter lifts prepared from each bacterial plate. One of the filters is hybridized to the + probe and the other to the – probe. Autoradiographs of the two filters are superimposed and the position of each colony is inspected; most will appear labeled by both probes. Inducible RNAs, however, will be in clones labeled by the + but not the – probe (see Example 4.4 on platelet-derived growth factor inducible genes).

Even in the absence of specific induction, this and similar methods can be used to isolate clones of RNAs present in one RNA preparation, but not in another. For example, differentiated muscle cells can be compared to undifferentiated muscle precursor cells.

Cloning by Homology to Other Species

There is a great deal of conservation of cellular functions among all living forms. This was exemplified by the case of human cyclin genes rescuing mutant yeast cells that lacked their own cyclins (see "Functional assays"). Proteins that perform similar functions from distantly related groups, such as mammals and insects or yeast, will, as a rule, have similar three-dimensional structure and will share the same amino acid

EXAMPLE 4.3

Cloning of a mammalian proton-coupled metal-ion transporter

Gunshin et al. (1997) reported the cloning of a protein involved in the transport of metal ions, including iron, across the plasma membrane. Once the clone was obtained, the protein was designated DCT1 (for divalent cation transporter number 1). *DCT1* cDNA was cloned from rats fed on a low-iron diet. The rational was that iron depletion would stimulate the synthesis of transporter proteins by increasing the level of the corresponding mRNA. Total RNA was isolated from the intestinal mucosa of iron-deficient rats and mRNA was purified by separating poly(A) containing RNA. The poly(A)$^+$ RNA was run on an agarose gel which was subsequently sliced transversely, each slice containing RNA of different size ranges. The various RNA fractions were extracted from the gel slices and a sample of each was injected into the amphibian *Xenopus* oocytes. Three days after RNA injection the oocytes were incubated for 1 hour into a buffered medium supplemented with radioactive ionic iron. Those oocytes injected with mRNA that encoded an iron-transporter protein took up more radioactivity than the others; in this case,

that occurred with an RNA fraction of approximately 4500 bases in length.

That 4.5 kilobase (kb) poly(A)$^+$ RNA fraction was used to make cDNA using a system that leaves a *Not*I site at the poly(A) end and a *Sal*I site at the 5' end of the cDNA. This allows directional insertion of the cDNA into the pSPORT1 plasmid at a site next to a prokaryotic promoter, which drives in vitro transcription of the inserted cDNA (as in Fig. 4-1). Once the cDNA was ligated to the vector, competent (capable of taking up DNA) *E. coli* cells were transformed with these recombinant plasmids and grown up into a cDNA library. The library was then plated on agar medium and colonies were combined in groups of 200 to extract plasmid DNA and synthesize RNA in vitro. This RNA was then injected into oocytes as before to see which group of plasmids stimulated iron uptake.

The study described the cloning of a rat gene. Except for the added difficulties of obtaining the original tissue samples (duodenal scrapes from iron-deficient individuals), the procedures would be the same for human genes.

EXAMPLE 4.4

Cloning of gene sequences regulated by platelet-derived growth factor

Platelet-derived growth factor (PDGF) is a polypeptide, released by platelets during blood coagulation, that stimulates the proliferation of connective tissue cells and thus helps in wound healing. Cochran et al. (1983) described the cloning not of a specific, pre-identified gene, but of all possible genes whose transcription is stimulated by PDGF. The method used is differential hybridization or "+/– hybridization."

cDNA was prepared from cells treated with partially purified PDGF (+ cDNA) and a cDNA library was subsequently made. Screening for PDGF-inducible gene sequences by differential colony hybridization consisted of hybridizing replica filter lifts of the library with + cDNA probes and – cDNA probes (probes derived from

cDNA of cells untreated with PDGF). Colonies that gave positive hybridization signals when probed with + cDNA, but not when probed with quiescent cell cDNA, were picked as presumptive PDGF-inducible genes. All such clones were re-screened twice.

Of approximately 8000 clones screened in this manner, 55 were scored as clearly inducible after three rounds of screening. Forty-six of those clones were analyzed further and they turned out to represent five independent gene sequences. Note that the number of times a particular sequence is isolated from a library depends on how abundant the corresponding mRNA was in the RNA preparation from which the library was made.

residues at some key positions. However, overall, protein—and, as a consequence, nucleic acid—sequences will usually be quite divergent. Among more closely related organisms, such as mammals, many proteins show a great degree of sequence conservation (Fig. 4-2); this occurs to such an extent that cDNA from one species can be used to hybridize to a gene of another species. In this fashion, if a clone is available for a mouse or rat gene, a probe can often be prepared to identify a human clone for the corresponding (*cognate*) human gene. (See the use of baboon sequences in Example 4.1 on the factor IX gene.)

Proteins such as rat and human β globins that derive, through evolution, from a common ancestral gene are said to be *homologous*. The level of similarity (122 out of 147 aa, or 83%) is a function of two variables: the length of time since the last common ancestral species, and the degree of conservatism of the protein. A *conservative protein* is one that changes very slowly with time. Usually conservatism is due to functional constraints such that most changes render the protein inactive. To use as heterologous probes, then, it is necessary to select either very conservative genes or genes from closely related species. Sequence conservation is usually not distributed uniformly along the length of the protein. If more than one non-human sequences are known, sequence comparisons can be used to prepare a probe from a region of the protein where sequence conservation is greatest.

Reverse Genetics

We started this chapter with a simplified paradigm of the flow of the research process (Fig. 4-3, from A to B to C), and the cloning of the genes for factor IX and von Willebrand's clotting factor (see Examples 4.1 and 4.2) follow this general pattern. However, cloning by homology to other species, as well as expression cloning can be used to isolate genes even in the absence of mutant alleles. That is to say, genetic analysis is not a prerequisite for molecular analysis; all that is needed is either a well-characterized product or a sequence from a not-too-distantly-related species.

Thus, one could argue that in the example of the cloning of the proton-coupled metal-ion transporter, the path followed was from B to C in Figure 4-3; in the example of the PDGF-inducible genes, the path starts at C. When dealing with genes and products for which no mutations are available (i.e., there is *no genetics*), sequence information obtained by molecular analysis can be used to propose hypothesis as to the biochemical function of the product, or to facilitate its isolation. In some experimental organisms it is also possible to produce mutations in vitro, to replace the normal gene with a mutant allele, and thus to have an opportunity to understand better the

FIGURE 4-2. Comparison of rat and human β globin protein sequences. Notice that in some regions there is greater similarity than in other regions.

```
        10          20          30          40          50          60          70          80
MVHLT DAEK AAV NGLWGKVN PD DVGGEALGRLLVVYPWTQR YF DSFGDLS SASA IMGNPKVKAHGKKV INAF NDGL KHLD
MVHLT PEEK SAV TALWGKVN VD EVGGEALGRLLVVYPWTQR FF ESFGDLS TPDA VMGNPKVKAHGKKV LGAF SDGL AHLD


        90         100         110         120         130         140
NLKGTFA HLSELHCDKLHVDPENFRLLGNMIV IVL GHH LGKEFTP CAQAA FQKVVAGVA SALAHKYH      rat
NLKGTFA TLSELHCDKLHVDPENFRLLGNVLV CVL AHH FGKEFTP PVQAA YQKVVAGVANALAHKYH      human
```

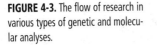

FIGURE 4-3. The flow of research in various types of genetic and molecular analyses.

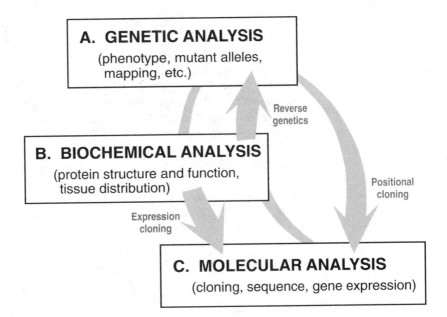

function of the normal product by studying the consequences of its absence. These are the so-called *knockout* mutations in yeast and especially in mice that are so important to understanding the function of human genes. This flow of the research path, from C to B and A has been called *reverse genetics*. Strictly speaking, reverse genetics can be applied only in experimental organisms; in humans, a form of reverse genetics is applied when studying genetic diseases in which the phenotype does not reveal the nature of the biochemical defect and no protein products are known for the gene (such as Huntington's disease). In these cases, a cloned sequence obtained by *positional cloning* (from A to C in Fig. 4-3) provides a *candidate gene* that can be tested to see whether it corresponds to the one mutant in that particular genetic disease. In this case, sequence information then can sometimes be used to make educated guesses as to what the function of the protein might be, and how its mutation might explain the phenotype of the disease (see Examples 4.5 and 4.6 on cystic fibrosis and Huntington's disease later in this chapter).

Positional Cloning

It was apparent in the early 1980s that many human genes, even though they were fairly well characterized from Mendelian analyses, were proving to be quite difficult to analyze biochemically. This inability to characterize the gene products frustrated any attempt to clone the corresponding genes by the traditional methods of expression cloning. One example of this was cystic fibrosis, a disease known to affect glandular epithelia and suspected to involve electrolyte transport. Extensive efforts to identify the protein product affected by this mutation were unsuccessful. Other equally intractable mutations were the ones responsible for Huntington's disease, a dominant trait caused by progressive neural degeneration, and Duchenne muscular dystrophy, a sex-linked recessive disease accompanied by muscular degeneration.

Because these were well-defined Mendelian traits, a strategy was developed that consisted of the genetic mapping of these traits using, as markers, polymorphisms that can be recognized at the DNA level, such as restriction fragment length poly-morphisms (RFLPs). Using the RFLP probe, DNA fragments in that region of the chromosome could be isolated from genomic libraries and analyzed in search of candidate genes in the neighborhood.

Even when a molecular marker is identified at a relatively short genetic distance, distances in nucleotides can be enormous—1% recombination corresponds roughly to 1×10^6 base pair (1 megabase pair). As a reference, consider that cosmids—fairly high capacity bacterial vectors—can only carry pieces of DNA of approximately 4×10^4 base pair (bp). Thus, starting from the molecular marker—the entry point—it is necessary to clone a large amount of DNA before the trait gene is found. *Chromosome walking* is a method used to clone, in an orderly fashion, DNA segments along the chromosome starting at any point for which we have a probe (Fig. 4-4). One limitation to the speed of chromosome walking is the relatively small size of the fragments

FIGURE 4-4. Chromosome walking.

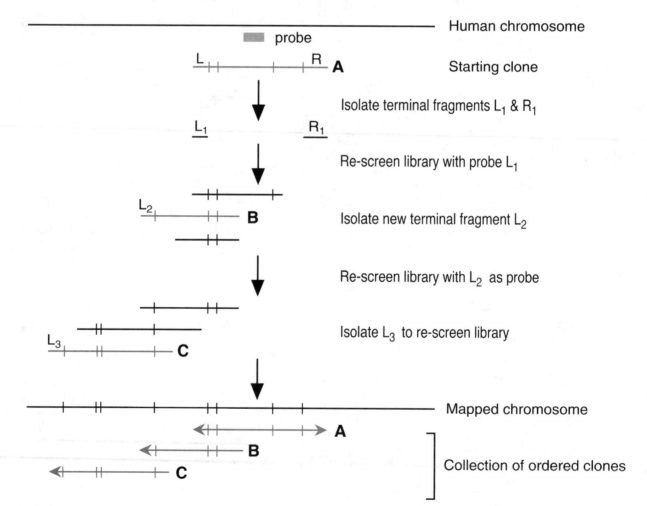

that can be efficiently cloned; another limitation is the difficulty of walking through the repeated sequences that are scattered throughout the genome. These limitations were circumvented to some extent by a method of *chromosome jumping* (Fig. 4-5), which allows advances of several hundred thousand base pairs at a time, and also allows one to skip over repeated sequences and other hard-to-clone segments.

Chromosome walking and jumping. Chromosome walking and jumping are techniques to clone chromosomal regions in the neighborhood of a site defined by an available probe, and make possible the search for genes in those regions.

For *chromosome walking,* human DNA is cut into large fragments (by partial digestion with a restriction enzyme, for example) and cloned into a cosmid vector, which can carry fragments of up to approximately 40 kb. This library is screened with the probe. Among the positive clones, one with a large insert is selected for restriction enzyme mapping (see Fig. 4-4A). This permits the identification of the two ends of the clone, which are in turn subcloned and used as probes to rescreen the library. Some of the clones detected by the left-hand probe (L_1) will extend considerably to the left; as before, one of them is selected for mapping (see Fig. 4-4B). In this case only the left-hand terminal segment is subcloned (L_2), and used once again to screen the library. In this fashion the walk proceeds leftward. Similarly, a clone identified by R_1 is used to initiate a rightward walk. Once the walk has advanced a significant distance, probes can be used to establish the orientation of "left" and "right" relative to chromosomal landmarks (such as centromere and telomere); this can be done by methods such as linkage mapping and in situ hybridization.

In *chromosome jumping,* genomic DNA is digested with a rare-cutting enzyme (such as *Not*I) that produces very large fragments (see Fig. 4-5). These fragments are circularized by ligation to a plasmid that carries a suppressor tRNA mutation and redigested with a second restriction enzyme (a six-cutter such as *Bam*HI). This digestion leaves two relatively short genomic DNA segments attached to the plasmid. A lambda (λ) phage vector carrying amber mutations is then introduced; this phage can only propagate in the presence of the suppressor mutations carried by the plasmid. Ligation to the lambda vector then results in a clone that includes the plasmid and two genomic fragments that were originally hundreds of thousands of base pairs apart in the genome. When the probe, complementary to the left-hand end of the fragment, detects a clone, it makes available to us, in the same recombinant molecule, a DNA fragment that was hundreds of kilobases away in the genome.

By a combination of chromosome walking and jumping, a region of the chromosome is cloned and characterized by restriction mapping. This leads to what is called a *contig map*—a composite map of contiguous segments assembled from partly overlapping clones—such as the collection of ordered clones at the bottom of Figure 4-4. The contig is a purely physical map of DNA sequence landmarks (the restriction sites, for example) with no functional information. Thus, the next question is to determine where in the contig map the gene of interest is located. This is usually an arduous task and there is no single, foolproof method to accomplish it, though several procedures have been developed, and, in combination, they usually prove successful. The objectives are, first, to localize any genes that may exist in the region of the contig, and second, to determine whether any of those genes correspond to the mutant gene causing the syndrome that was mapped to that region in the first place. In almost all cases several of the methods are used, the results of one experiment informing the design of the next. A summary of some of those methods follows.

FIGURE 4-5. Chromosome jumping. (Based on Poutska et al., 1987.)

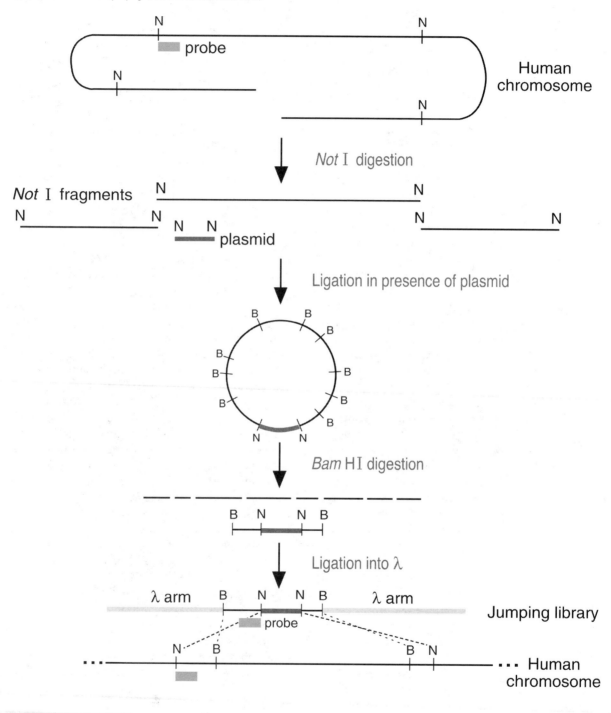

Heterologous hybridization. Heterologous hybridization relies on the evolutionary conservation of coding sequences, mentioned earlier in this chapter, to identify segments of the contig in which there is a gene or parts of a gene. It consists of preparing a Southern blot using, in multiple lanes, DNA from a variety of species digested with some restriction enzyme (the *zoo blot*). As a hybridization probe uses DNA from the contig region, often it is possible to use an entire phage or cosmid insert. Still, because the contig is usually very long, many independent probes need to be prepared to cover as much as possible of its length.

As was discussed in Chapter 4, most of the DNA on a chromosome, or any region of a chromosome, constitutes introns and spaces between genes. These are sequences that are not under selective pressure and therefore diverge very quickly during evolution. Thus, most probes, representing these non-exonic sequences, fail to hybridize to samples from non-human DNA. When a probe does hybridize to the zoo blot, it identifies a section of the contig that includes a conserved sequence, possibly an exon of some conserved gene (see Fig. 4-8).

Chromosomal rearrangements. Some disease-causing mutations are chromosomal rearrangements, deletions, inversions, or translocations. If this is the case, by establishing the position on the contig where the chromosomal break occurred, we can pinpoint the site of the affected gene—although sometimes chromosomal rearrangements can affect the expression of genes located at some distance from the breakpoint. The position of the breakpoint can be determined because on one side of it the restriction maps of normal and mutant chromosomes are the same, but on the other side they are completely different. Translocations were useful in the cloning of the genes responsible for Duchenne muscular dystrophy (see Fig. 5-13) and neurofibromatosis I (see Figs. 4-10B and 5-7).

CpG islands. The dinucleotide combination CG is rare in mammalian genomes because the C in them is often methylated and this leads to C to T transitions with relatively high frequency (see Chapter 7). One exception to this rule is in regions immediately upstream of the transcription initiation site of many genes, where runs rich in C and G seem to play a role in transcription. Thus, stretches with a high proportion of C and G (*CpG island*) often mark the region of the transcription initiation site of a gene.

Open reading frames. When inspecting the nucleotide sequence of an internal exon, in the proper reading frame three codons that are never found are UAG, UGA, and UGG, the three termination codons. These triplets do appear simply by chance in other, noncoding sequences.

Intron sequences that have random G, A, T, C composition present one of the termination codons approximately once every 64 bp (or three times every 64 triplets). Even coding sequences, when read in the wrong frames, have frequent termination signals. We can use these properties to find exons in an anonymous sequence simply by looking for long stretches of DNA sequence without termination codons (*open reading frames*). Unfortunately, short segments of DNA may be devoid of termination signals by pure chance, and as we saw in Chapter 3, mammalian exons are small, with only 17% of them longer than 230 bp (see Fig. 3-10). This method to find genes is more usefully applied to species with longer exons or no introns, such as yeast and bacteria.

Hybridization to mRNA. Another way to identify transcribed sequences is to prepare probes from the genomic region under investigation and use them to hybridize to RNA. This approach can be particularly informative if the RNA is extracted from *target tissues,* tissues and organs affected by the disease. A positive result with any one probe would lend support to the hypothesis that the probe comes from a segment of DNA that contains the gene we are after. This method can be applied in two forms: Northern analysis and cDNA library screen.

In *Northern analysis,* poly(A)$^+$ RNA is isolated from various sources—human cell lines and samples of target as well as other tissues—fractionated by size by gel electrophoresis, transferred to a membrane, and hybridized to one of the labeled probes. A positive result provides information on the approximate size of the mRNA as well as its tissue distribution. The main drawback of this method is that the intensity of the signal (the darkness of a putative band on an x-ray film) is proportional to the prevalence of the mRNA in question. Because often these hard-to-find genes produce low levels of transcripts, the resulting bands are very weak and maybe indistinguishable from background noise.

A more sensitive method is to use poly(A)$^+$ RNA to prepare a *cDNA library* (Appendix Fig. A-8), which is then screened with genomic DNA probes. All RNA sequences have a chance to be represented in the library. The rarer sequences are represented fewer times, but if the clone happens to be present on a membrane, the signal produced will be just as strong as that of a very prevalent sequence. In other words, the difficulty of detection of a rare sequence is converted from a weak signal (in the Northern blot method) to an infrequent event. Because library screening is relatively efficient, it is possible to screen many thousands of colonies or plaques looking for a positive one.

Exon amplification. The amplification method of identifying exons (also known as *exon trapping*) relies on detection of functional splicing signals flanking exon sequences. A plasmid, pSLP1, was engineered so as to include the origin of replication and transcription promoter of the simian virus SV40 (Fig. 4-6). Next to these sequences is a portion of the rabbit β globin gene within which has been inserted a segment of the HIV *tat* gene, including its intron. The intron contains a *Bam*HI cloning site, into which the genomic sequences to be tested are cloned. Upon transfection of this plasmid DNA, with its genomic insert, into cultured monkey cells, RNA is abundantly synthesized, under the control of the SV40 promoter, and processed. If the genomic insert did not contain any splicing signals, it is entirely removed from the transcript by the RNA splicing process. If, on the other hand, the genomic insert included an exon, then the splicing machinery recognizes its accepting and donor splice sites and the exon is retained in the processed RNA (see Fig. 4-6). To detect whether an exon is included or not in the transcript, RNA extracted from the cultured cells a few days after transfection is subjected to polymerase chain reaction (PCR, see Appendix Fig. A-10) using as primers oligonucleotides complementary to the β globin sequences. In the absence of an inserted exon, a 429-bp PCR product is expected; when the genomic DNA includes an exon the PCR product is longer—429 bp + the size of the exon. Once an exon is thus identified, the corresponding PCR fragment can be used as a probe to tag genomic or cDNA clones.

Mutant DNA sequence. Once a segment of DNA is identified as part of a gene, its nucleotide sequence from normal controls as well as from a patient can provide definitive proof that this particular gene is the one responsible for the disease under

FIGURE 4-6. Exon amplification: plasmid pSLP1 showing the processing of transcripts with and without exons. (Adapted from Buckler et al., 1991.)

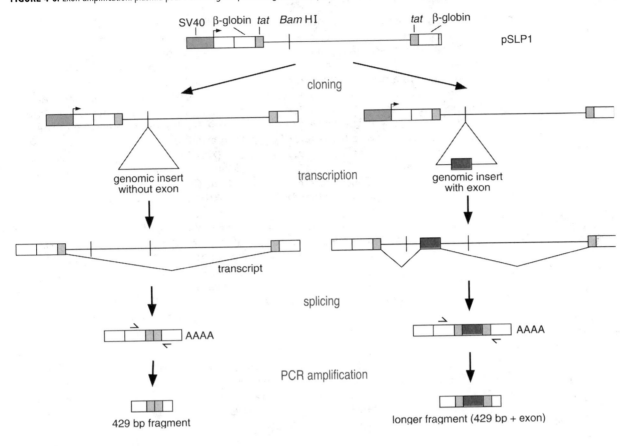

study. If the mutation is an out-of-frame deletion or insertion, a nonsense mutation, or some other mutation with a predictably catastrophic effect on the protein product, this observation would go a long way toward identifying the gene. There are many other cases of more subtle changes—such as conservative amino acid substitutions and base substitutions in promoter regions—in which it is impossible to predict whether the change is responsible for the phenotype of the patient or whether it is just an innocuous polymorphism on a neighboring gene that has no bearing on the disease. In these cases, sequencing of multiple individuals—both patients and controls—usually resolves the issue.

Candidate gene. In some cases there may be information, either from other species or from expression studies, that point to a particular gene already in hand as a candidate gene. We have already seen the use of candidate genes in Chapter 2 to try to identify genes responsible for complex traits; for example, when a marker on chromosome 11p15 was found to be associated with insulin-dependent diabetes, the structural gene for insulin, which was known to be in that region, immediately presented itself as a candidate gene for *IDDM2*. In the context of positional cloning, a probe from such a gene can be used to verify whether it hybridizes to the contig being searched.

EXAMPLE 4.5

The cystic fibrosis gene

Chapter 6 will discuss the biochemical basis of cystic fibrosis (CF) and its inheritance. Studies from several families published in 1985 showed that the CF trait is linked to a polymorphism in the enzyme paraoxonase. These studies were of importance because they provided the first strong evidence that many, and perhaps all, cases of CF were the result of mutations in the same gene. This suggested that the disease was monogenous, a simple Mendelian trait. Independently, linkage of *CF* to an RFLP associated it to chromosome 7, and an extraordinary collaboration of several labs in different countries ensued, culminating with the cloning of the gene 4 years later.

After it was established that the *CF* gene was on chromosome 7, much closer linkage to two markers, the proto-onco-

gene *MET* and RFLP locus *D7S8* (identified by probe J3.11), defined a relatively small region on the long arm of chromosome 7. The gene order *7cen—MET–CF–D7S8–7qter* was determined by linkage analysis (where *7cen* indicates the centromere of chromosome 7, and *7qter* represents the end or telomere of the long arm 7q). Another RFLP marker locus, *D7S340* (identified by the probe TM58), was mapped and it appeared to fall between *MET* and *CF*. Using this information, long range restriction maps, chromosome-7-specific libraries, chromosome walking, and chromosome jumping later generated a contig map that extended from *D7S340* toward *D7S8*. All together, 280 kb of DNA were cloned and physically mapped (Fig. 4-7).

At this point the strategy switched from cloning and physical mapping of

FIGURE 4-7. Restriction map of the region of chromosome 7 near the *CF* gene. Only the sites for *Eco*RI (above the line, unlabeled) and the rare cutters *Not*I and *Xho*I (below the line) are shown. Horizontal arrows above the map indicate the partly overlapping clones of the chromosome walks. Arcs indicate chromosomal jumps. Open horizontal boxes indicate probes used, and vertical color bars indicate *CF* exons. (Redrawn with permission from Fig. 1 in Rommens JM, Iannuzzi MC, Kerem B, et al. Identification of the cystic fibrosis gene: chromosome walking and jumping. Science 245:1059–1065 [1989]. Copyright 1989 American Association for the Advancement of Science.)

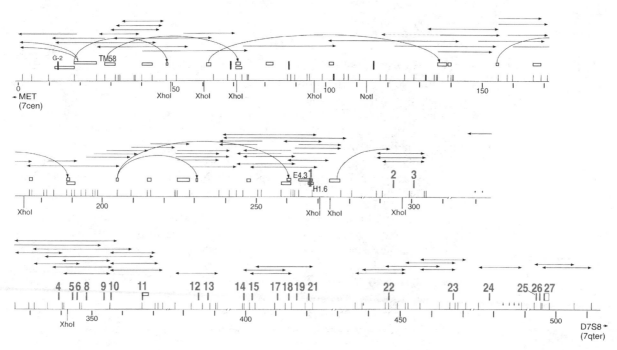

contiguous DNA segments to searching for gene sequences among the segments already mapped. The following criteria were used to establish whether a cloned DNA segment might contain a candidate gene:

1. Detection of cross-hybridizing sequences in other species
2. Presence of CpG islands
3. Presence of mRNA transcripts (or the corresponding cDNAs) in tissues affected by the disease
4. Presence of open reading frames

Four regions in the 280-kb span gave distinct cross-species hybridization signals, but three of them could be ruled out as *CF* by various criteria; the fourth turned out to be part of *CF*. Two probes (E4.3 and H1.6) cross-hybridized to bovine genomic DNA (Fig. 4-8). When these same probes were used on Northern blots of target tissues, however, no hybridization was detected. Because the segment covered by the probes was relatively short, the next step was to sequence the area. A long GC-rich region was found, but no long open reading frames. Screening of seven cDNA libraries with H1.6 eventually yielded a single clone from a library of cultured sweat gland epithelial cells. Sequencing of the 920-bp insert revealed that only 113 bp at the 5' end matched sequences in H1.6. The genomic site originally identified by

H1.6 turned out to be the short first *CF* exon. The difficulty in detecting RNAs and cDNA clones can, in part, be explained by how limited is the segment of overlap between H1.6 and the transcript (see also the technical note below).

This first cDNA cloned, used as a probe, allowed the isolation of other, partly overlapping cDNA clones, and eventually, a complete cDNA contig was generated. With these new cDNA probes the genomic libraries were rescreened and the original genomic contig extended in the direction of *D7S8* to include a total of 500 kb (see Fig. 4-7). Thus, the 27 exons of *CF* spanning 250 kb were eventually identified and sequenced and a 3 bp deletion in exon 11, ΔF_{508}, the most common *CF* mutation in the white population, was described. Note that only 24 exons are portrayed in Figure 4-7; the three exons corresponding to numbers 7, 16, and 20 in Figure 3-14 were missed in this first report and only later identified.

When using cDNAs as a probe, a 6.5-kb transcript was detected in samples from pancreas, trachea, lung, colon, and sweat glands (all organs affected by CF) but not in samples from brain, placenta, or kidney.

Technical note: A difficulty with cDNA libraries is that they are often prepared by priming the poly(A) tail of mRNAs

FIGURE 4-8. Cross-species hybridization of the human probe E4.3 (see Fig. 4-7). DNA of each species was digested with *Eco*RI (R), *Pst*I (P), or *Hind*III (H). Note the strong signals in the human and bovine lanes and weak and uncertain signals in the mouse and hamster lanes. (Redrawn with permission from Fig. 3B in Rommens JM, Iannuzzi MC, Kerem B, et al. Identification of the cystic fibrosis gene: chromosome walking and jumping. Science 245:1059–1065 [1989]. Copyright 1989 American Association for the Advancement of Science.)

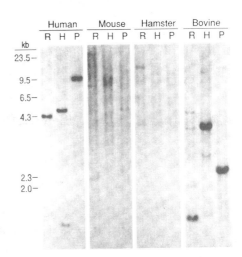

with oligo(dT). That is, reverse transcription is started at the 3' end and, given the tendency of reverse transcriptase in vitro to fall off the template before copying is complete, 5' end sequences tend to be underrepresented in these libraries. An approach taken to solve this problem is to use short, random-sequence oligonucleotides as primers for first strand synthesis so that all regions of the mRNA are more equally represented in the library.

EXAMPLE 4.6

Cloning the Huntington's disease gene (HDH)

As we saw in the Huntington's disease Example 1.3, Gusella and his collaborators obtained significant LOD score values indicating linkage of *HDH* to *D4S10*, a locus defined by the probe G8.

Hybridization of the G8 probe to a panel of mouse–human cell hybrids demonstrated that *D4S10*, and therefore *HDH*, is on chromosome 4 (see Chapter 5 for details). Further analysis showed *D4S10* to be at 4p16.3, in the terminal region of the short arm of chromosome 4. New chromosome 4 probes were found and more families were studied; a few recombinants were found and these led researchers to link *HDH* with the very tip of chromosome 4. It took very exhaustive study of this terminal region for the researchers to discover that they had been misled, and the gene was probably several megabases (Mb) into the chromosome, closer to their original marker *D4S10*. The correct site of *HDH* was finally bracketed by markers *D4S10* and *D4S98* (2.2 Mb), and later by *D4S180* and *D4S182* (500 Kb) (Fig. 4-9). Ten years after the original observation of linkage between *HDH* and a *D4S10*, the gene itself was cloned and sequenced.

Identification of *HDH* within the *D4S180–D4S182* interval was done by preparing a 16-cosmid contig of the segment, which was then subjected to exon amplification. Four genes were identified and one of these, IT15, turned out to be *HDH* (see Fig. 4-9). This was demonstrated by cloning cDNAs for a 10.4-kb mRNA and sequencing them. Toward the 5' end of the coding region there is a CAG repeat with 11 to 34 copies in the normal population but over 42 copies in affected individuals (see "Trinucleotide repeat diseases" in Chapter 7).

THE HUMAN GENOME PROJECT

The first molecular descriptions of a few human hereditary diseases, in the late 1970s and early 1980s, made clear that this would be a very powerful way to tackle all such diseases, and the only hopeful avenue to many of them. By the mid-1980s scientists could contemplate as a distinct possibility the cloning of genes for which no product was known. However, at the same time, it was becoming clear that, absent a convenient molecular probe, identifying and cloning the many human genes involved in diseases would be a very difficult and onerous task: positional cloning and reverse genetics—as demonstrated by the two examples presented here—worked, but at a very high price.

From 1984 to 1985 several scientists and the administrators at the U.S. Department of Energy and National Institutes of Health called for the sequencing of the entire human genome. Obtaining the sequence of individual genes was proving to be revolutionary in advancing our understanding of cellular functions and developmental

FIGURE 4-9. Long-range restriction map of 2.2 Mb of chromosome 4 to which the Huntington's disease gene was localized. The names above the map are four marker loci used in the genetic mapping of *HDH*. Vertical lines mark *Not*I sites, one of three rare-cutting restriction enzymes used. (Sites for the other two are not shown.) The group of colored horizontal lines below the map represent a set of 16 cosmid clones that constitute a "minimum" contig; this set was selected from an original collection of 64 overlapping cosmids that spanned the 500 kb between *D4S180* and *D4S182*. The arrows at the bottom represent the four transcription units identified by exon amplification; IT15 corresponds to *HDH* (the 5'-3' orientation of IT10C3 had not been established). (Adapted from The Huntington's Disease Collaborative Research Group, 1993.)

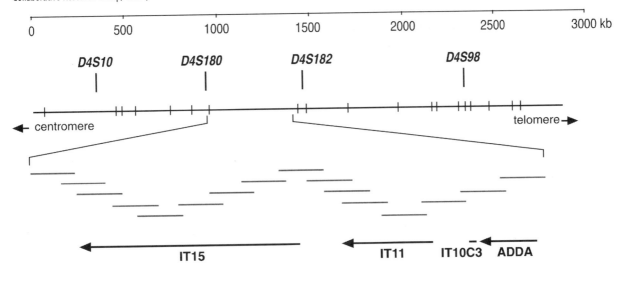

processes. Taking a global approach and sequencing all human genes appeared to provide economies of scale and move us faster toward the goal to which science seemed to be inching along. So, for example, if a disease gene were localized to a particular region of a chromosome, the sequence of candidate genes would already be available to help identify which one was responsible.

Heated debate followed this proposal. Most arguments were not about the desirability of knowing the sequence of all human genes, but rather about funding priorities—that there would be a change in the way biological science had been traditionally carried out (big science projects versus individual scientist labs). Some feared that funds would be syphoned off from all areas of biology and concentrated in one area of human biology. Others questioned the validity of sequencing the entire genome when, as we saw in Chapter 3, coding sequences represent but a small fraction. In a remarkably short period of time, however, many of these fears were allayed and a consensus developed.

A number of projects were initiated in the next few years and the Human Genome Project was officially launched in 1990, when the Department of Energy and the National Institutes of Health submitted a coordinated plan to the U.S. Congress. The Human Genome Project requested $3 billion to complete the human genome sequence in a period of 15 years. This was not conceived as a monolithic enterprise but rather as a broad set of goals. Many scientists and institutions worldwide would be able to contribute to it, and each group of scientists would define and approach specific, manageable parts of the overall goals in their own way.

The Human Genome Project has been likened to the first moon landing; there is a similarity in that, in both cases, the objective was easy to see, but the technologies to achieve it were not at hand. We will review the broad methodological approaches, as well as some of the main techniques, developed between the mid-1980s and the mid-1990s, that made sequencing of the entire human genome an attainable goal.

It was immediately understood by those planning the Human Genome Project that high-resolution genetic and physical maps of the entire genome would be essential intermediate stages from which to launch the final step of sequencing the overwhelmingly large human genome (3×10^9 np).

The Genetic Map

At about the time that talk of sequencing the human genome was starting, an ambitious mapping program, with international participation, was taking shape under the

---

Box 4.1   *Ethical concerns*

In addition to the impact that sequencing the entire human genome might have on biological science research, a number of concerns were raised as to the effect of acquisition of this vast knowledge on society at large. Some of the questions were not of major significance, such as whose DNA would be sequenced. The answer is that the bulk of the sequence will be from a composite of 10 to 20 DNA samples taken randomly from a larger group of male and female donor samples.

Other questions were weightier, however, and many are still under debate. A special grant program was created to study the ethical, legal, and social issues (ELSI) of the Human Genome Project. Some of the issues raised by the increased availability of genetic information include:

- *Use of genetic information by insurers, employers, and others.* Who should have access to this information and how will it be used?
- *Privacy of genetic information.* Who should control it?
- *Psychological impact and social stigmatization.* What issues arise from an individual's genetic differences? How does detailed genetic information affect individuals' perceptions of themselves and their place in society?

- *Genetic testing and population screening.* Should tests be available for specific conditions related to family history (prenatal, carrier, and presymptomatic testing)? Should there be screening for newborns, premarital couples, and occupational groups? Should testing be performed when no treatment is available? Should parents have the right to have their minor children tested for adult-onset diseases?
- *Reproductive issues.* How will informed consent for procedures, use of genetic information in decision making, and reproductive rights be approached?
- *Clinical issues.* How will health service professionals and patients be educated, and how will standards and quality control measures in testing procedures be implemented?
- *Commercialization.* How does the law apply, including property rights (patents, copyrights, and trade secrets) and accessibility of data and materials?
- *Conceptual and philosophical implications.* What is the impact on human responsibility, free will versus genetic determinism, and concepts of disease and health?

A more complete discussion, together with references to some ELSI studies can be found at http://www.ornl.gov/hgmis/resource/elsi.html

direction of Jean Dausset at the Centre d'Étude du Polymorphisme Humaine (CEPH) in France. We have discussed how RFLPs can be used as marker loci to establish linkage to trait loci (see Chapters 1 and 2); the idea advanced by Dausset was to map numerous marker loci *to each other,* thus producing a linkage map covering the entire human genome. Such a map would serve as a frame of reference that would greatly simplify the subsequent localization of specific trait loci.

To develop a linkage map, a group of families was identified. These families, originally 40 but later 61, constitute the CEPH reference panel. They are large, multigenerational families; most of them have at least seven children in the third generation and all four grandparents. Permanent tissue cultures are maintained for all members of each family. To generate data for mapping, the families had to be genotyped with respect to the polymorphisms revealed by numerous probes. Typing work did not need to be carried out all in one lab, nor did the work with different probes need to be coordinated. Thus, to stimulate participation the CEPH made available DNA samples of their entire reference panel to any scientist with an appropriate probe (i.e., one that detected a highly polymorphic locus). The only condition was that they were to undertake the typing of the entire panel. In this fashion, scientists who may have had their own reasons to map a particular group of loci benefited because DNA from large families was readily available to them. In turn, their results contributed to the overall effort because they could be used to calculate LOD scores relating their probes to others, typed by scientists using the same families for other purposes. The key was, then, that everybody was working on the same families.

Making use of highly polymorphic sites, such as microsatellites and variable number tandem repeats (VNTRs), by 1989 scientists had covered the genome with markers at an average distance of 10 centimorgans. The next goal was to develop a 5-centimorgan map and eventually a 1-centimorgan map. At the time, this latter goal was described as "achievable but fairly horrendous" by Maynard Olson (Roberts 1989). To appreciate the vastness of this enterprise we must remember that until the early 1980s the only portion of the human genome for which there was a genetic map at all was the X-chromosome. The vast majority of autosomal traits was not even localized to a particular chromosome. The 1-centimorgan map would require the ordering of approximately 3000 markers.

A high-resolution linkage map would require a great deal of work from many labs, but the means to achieve this goal had been well tested. How would this genetic map help the sequencing program? One centimorgan corresponds, very roughly, to 1 million base pairs of DNA, and a sequencing experiment can generate the nucleotide sequence of fragments 500 to 1000 bp long. How would these tiny fragments be assembled to bridge the three-orders-of-magnitude gap? The answer is in physical maps.

Physical Maps

Physical maps are representations of the DNA molecule in which the markers are sequence traits of the DNA—such as an RFLP—and the distances are measured in number of base pairs. As we saw in the example of the cystic fibrosis gene, a contig is a physical map, a restriction map is also a physical map, and depending on the number and type of enzymes we use, the map can be more or less detailed. The complete sequence of nucleotides would be the ultimate physical map. A goal of the Human Genome Project was to build a map of the entire human genome with markers every 100 kb, which would take 30,000 markers. As in the case of the genetic map, the overall strategy was to prepare lower resolution maps encompassing the entire genome first, and then proceed to increase the resolution by adding markers.

Restriction maps. *Restriction maps* were the first physical maps available, and are still among the most useful. With a single probe and conventional agarose gel electrophoresis, a restriction map may extend for approximately 10 to 15 kb with resolution between 100 and 1000 bp. Using rare-cutting restriction enzymes and pulsed-field gel electrophoresis (PFGE), much longer fragments of DNA can be generated and fractionated so that the corresponding restriction maps may extend between 500 to 2000 kb with a resolution of 50 to 200 kb. By adding new probes, the range of the maps can be extended "indefinitely" by chromosome walking. Unfortunately, the distribution of restriction sites is not uniform; in some regions they are very rare and the fragments produced are too long to be resolved, while in other regions they are too frequent, thus producing short fragments (less than 100 kb), limiting the extent of the map. Long-range restriction mapping has been used to map several specific genes and regions and is a very useful complement to other techniques.

Pulsed-field gel electrophoresis and restriction enzymes with infrequent restriction sites (rare-cutters). Pulsed-field gel electrophoresis (PFGE) is a procedure that permits fractionation of very large DNA molecules. The major source of electric charge in DNA molecules is phosphate groups. Because these groups are distributed uniformly along the length of the molecule, electrophoretic mobility in free solution is the same for all molecules regardless of size. In conventional electrophoresis of DNA in agarose gels, size fractionation is achieved because the gel pores have a sieving effect on the molecules: smaller molecules can fit through most pores and therefore take a more direct route through the gel, while larger molecules need to find larger pores and therefore follow a more indirect, longer path. The upper limit of fractionation in conventional agarose gel electrophoresis is approximately 20,000 bp; when DNA molecules reach this size they become too large for even the largest pores and fail to move unless they orient themselves lengthwise so that they can thread their way through the gel—a process aptly known as *reptation*. Once DNA fragments initiate reptation, they advance through the gel at a fixed speed, regardless of their length.

PFGE is based on the premises that 1) very long DNA molecules need to orient themselves to move in the gel and 2) the longer the molecule, the longer it takes for orientation to occur. In PFGE then, the direction of the electrical field is changed periodically so that DNA molecules relax and reorient themselves in the new direction; shorter molecules do this quicker, so they have a chance to advance farther than longer molecules, which spend more time reorienting and less time moving (Fig. 4-10). Molecules in various ranges between 20 kb and more than 1000 kb can be separated by adjusting the duration of the pulses (usually between 1 second and 5 minutes).

Most restriction enzymes are not adequate to generate specific DNA fragments of such great sizes because they cut every few hundred nucleotide pairs (four-cutters) or every few thousand (most six-cutters). A few enzymes exist, however, with much rarer restriction sites; these are often six-, seven-, or eight-cutters with GC-rich restriction sites, such as

*Bss*HII: GCGCGC
*Not*I: GCGGCCGC

FIGURE 4-10. Pulse field gel electrophoresis. **(A)** Diagram of a PFGE apparatus. Several electrode geometries have been used; here, the electrical field alternates between the upper-right to lower-left direction and the upper-left to lower-right direction. The overall direction of migration is down (diagonal relative to the position of the electrodes); the small arrows indicate the alternating directions of molecular motions. **(B)** Neurofibromatosis 1 gene fragments in t(17;19) (q11.2;q13.2) PFGE analysis of very large DNA fragments from a family discussed in Figures 5-7 and 5-8. Note the size of fragments separated by this procedure, compared to regular electrophoresis (Fig. 4-8). DNA was digested with *Not*I and probed with fragments derived either from exons 22 and 23 of the gene or from exon 24. In normal individuals both probes detect a 290-kb fragment; in persons with neurofibromatosis who carry the translocation in addition to the normal fragment, E22-23 detects a 160-kb fragment, and E24 detects a 450-kb fragment. The sample from individual I-1 was derived from cancer cells, and there seems to be a secondary deletion that removed the 450-kb fragment. (See the discussion of loss of heterozygosity in Chapter 8.) Some extra bands are probably the result of incomplete digestion of the DNA by *Not*I. (Reproduced by permission from Fahsold R, Habash T, Trautmann U, Haustein A, and Pfeiffer RA. Familial reciprocal translocation t(17;19) (q11.2;q13.2) associated with neurofibromatosis type 1, including one patient with non-Hodgkin lymphoma and an additional t(14;20) in B lymphocytes. Hum Genet 96:65–69 [1995]; Fig. 1.)

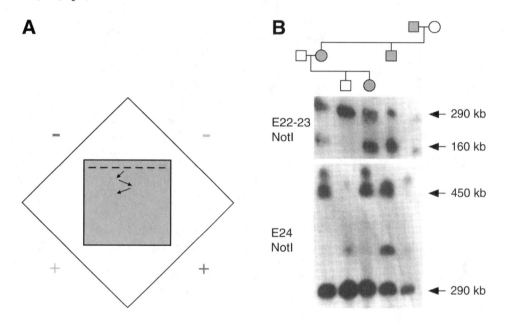

Contigs. Restriction maps are often combined with *contig maps,* in which the structure of a region of the genome is represented by a set of partly overlapping clones. Pure contig maps are difficult to produce because any segment of a region that is unavailable in clones is a discontinuity. The chromosome walk in Figure 4-7 is an example of a hybrid contig/restriction map and so is Figure 4-9. The practical limit of uninterrupted contigs seems to be in the 200- to 500-kb size range, which is still short of what is needed to fill the gaps between markers in the genetic map. Another problem with pure contig maps is that they are defined by, and depend on, the permanent existence of the particular array of clones used to generate it.

Fluorescence in situ hybridization. Two other techniques that had been available for many years were developed and refined so that they could be used for whole genome mapping. One of those techniques is in situ hybridization, the other, mapping by x-ray chromosome breakage.

In situ hybridization is based on the denaturation of DNA in chromosome spreads followed by hybridization with a labeled (either radioactive or fluorescent) probe. After excess probe is washed away, the slides are observed under the microscope to identify the location of the probe. *Fluorescence in situ hybridization* (FISH) is now used to assign probes to specific chromosomes and also to order them along the cytogenetic map. This is accomplished by determining the position of each fluorescent spot, measured from one end of the chromosome, relative to the length of the entire chromosome (fractional length). With proper statistical treatment, the method can have a resolution of approximately 2% of the length of the chromosome—in the order of 1 million to 3 million base pairs (Fig. 4-11).

In the original form of in situ hybridization, the probe is synthesized in the presence of ^{3}H-nucleotides and the radioactivity, bound to the chromosome by hybridization, is detected by covering the slide with a thin layer of photographic emulsion. After proper exposure (several weeks to several months), the slides are developed and the localization of the silver grains can be visualized in the microscope. Two modifications improved the resolution and sensitivity of this technique. One was the use of fluorescent dyes conjugated to the probe, rather than radioactivity, which increased resolution; the other was to increase the length of the probes from a few hundred to a few thousand base pairs, which improved sensitivity. The greater probe length, however, brought with it a new problem: almost any piece of human DNA of that size includes some repeated sequences (Alu, simple sequence DNA, etc.), and these repeated sequences would direct the probe to hundreds of sites in all chromosomes generating a very high level of background. To reduce this background, unlabeled repeated DNA is used, in excess, to compete out the labeled repeated sequences in the probe and thus suppress their hybridization. Another problem that had to be overcome was the necessity to be able to stain the chromosomes so that they could be identified, after the probe was hybridized.

Fiber FISH is a newer technique in which mechanically stretched chromosomes, interphase chromatin, or even naked DNA are used as substrate for in situ hybridization. The stretched fibers have lost all morphologically identifying features, but by being less condensed than metaphase chromosomes, the longitudinal resolution is much better. Once the localization of two or more genes to a particular chromosome

FIGURE 4-11. Schematic representation of fluorescence in situ hybridization of several chromosome 11 probes. The position of each probe along the chromosome is determined relative to the total length of the chromosome in each instance. The arrows point to the position of the centromere. (Based on Lichter, 1990.)

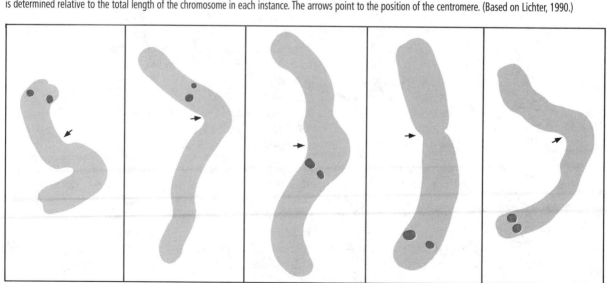

has been established, fiber FISH can be used to determine the relative order of the genes along the chromosomes. The resolution of in situ hybridization on metaphase chromosomes is of the order of more than 1 Mb, but markers as close to less than 100 kb can be discriminated by fiber FISH. Probes labeled with different color dyes are used to identify the position of each one. In Plate 1A, for example, the distances between the *PMP22* gene and the flanking repeats are approximately 0.5 and 1.0 Mb.

Radiation hybrid mapping. *Radiation hybrid mapping* is another technique that plays an important role in obtaining a physical map of the human genome. When chromosome breaks are induced by x-irradiation, the likelihood that a break will occur between two markers depends on the distance between the markers. Thus, using frequency of separation as an estimator of distance (Fig. 4-12), a map can be constructed—distance is expressed in centirays (cR), by analogy to centimorgan. In one set of experiments on chromosome 21, after exposure to 8000 rad of x-rays, 1cR corresponded to a distance of 52 to 56 kb.

Chinese-hamster/human cell hybrids have the tendency to lose human chromosomes (see "Somatic Cell Genetics" in Chapter 5). By careful monitoring, it is possible to select hybrid lines that have lost all human chromosomes except one. Let us say that we have several DNA probes (A, B, C, etc.) that are known to be localized on the same human chromosome and we want to order them on a map. The first step is to subject a culture of cell hybrids with the appropriate human chromosome to very high doses of x-rays. The treatment breaks the chromosomes at multiple sites and is lethal, but the chromosome fragments are recovered by fusing the treated cells with healthy Chinese hamster cells. In this way, multiple cell clones are generated from each treatment. Many of the human fragments are lost in the process, but a fraction of clones (30% to 60% in one experiment) retain one or another fragment of human chromosome (see Fig. 4-12A).

The final step is to determine whether each clone hybridizes to one or more of the probes we are trying to map. If probes A and B are often found in the same clones, this indicates that they tend to be contained by the same fragments, and therefore that they are closely spaced. If B and C, on the other hand, are less frequently together, this indicates that they are farther apart and there is a higher probability that a chromosome break occurs between them (see Fig. 4-12).

Note that the principles applied are similar to those used in genetic mapping, where we use the frequency of separation of genes by meiotic recombination to estimate distance. As with genetic mapping, in radiation mapping, LOD scores are calculated to estimate the likelihood of linkage. One significant difference between the two methods is that the frequency of meiotic recombination per unit length of DNA varies from one chromosomal region to another, while the likelihood of chromosome breakage by x-irradiation seems to be more constant and the map generated is a better representation of the physical distance between markers (see Table 4-1).

Other Tools

In addition to physical and genetic mapping techniques, other developments were instrumental in significant advances of the Human Genome Project:

1. The ability to prepare chromosome-specific DNA libraries—by isolating metaphase chromosomes en masse and fractionating them into single chromosome preparations by flow sorting.

FIGURE 4-12. Radiation hybrid mapping. **(A)** Hypothetical mapping of a human chromosome (in color) with three markers, *A* and *B,* which are close together, and *C,* which is farther away. **(B)** Southern analysis of chromosome 21 DNA from radiation hybrids (human/Chinese hamster). Each radiation hybrid contains a fragment of human chromosome 21. DNA was digested with *Eco*RI prior to electrophoresis and the membranes were hybridized with five probes derived from chromosome 21 (listed on the extreme left). All five probes hybridize to DNA fragments from human cells (first lane) or from the original hybrid line that carried a whole chromosome 21 (CHG3)—from which the radiation hybrids were derived—but not to fragments from the hamster cell line (third lane). Most RH clones contain only a subset of bands. Some pairs of bands have a high degree of concordance (S1 and S11), which indicates their proximity; others have less (S16 and S39). (Redrawn from Fig. 1 in Cox, D.R., Burmeister, M., Price, E.R., et al. Radiation hybrid mapping: A somatic cell genetic method for constructing high-resolution maps of mammalian chromosomes. Science 250:245–250 [1990]. Copyright 1990 American Association for the Advancement of Science.)

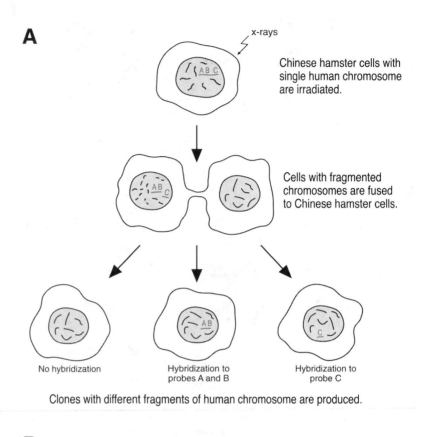

A

Chinese hamster cells with single human chromosome are irradiated.

Cells with fragmented chromosomes are fused to Chinese hamster cells.

No hybridization

Hybridization to probes A and B

Hybridization to probe C

Clones with different fragments of human chromosome are produced.

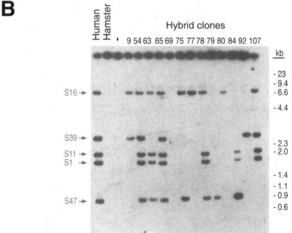

B

2. The use, as markers, of short-tandem-repeat polymorphisms (STRP) which can be easily detected by PCR and whose frequency of heterozygosity often exceeds 70%. (See Box 3.3, "Microsatellite DNA," in Chapter 3.)
3. The development of yeast artificial chromosomes (YACs). These "mini-chromosomes" are cloning vectors that contain yeast centromere and telomere sequences, replication initiation sites, and a selective marker so that, when

introduced into yeast cells, they are replicated and stably transmitted to the progeny as a regular chromosome. Fragments of foreign (human) DNA longer than 1000 kb can be carried by YACs, compared to 40 kb carried by cosmids.

4. The development of statistical methods and computer programs that permit efficient mapping of multiple markers and reduce huge amounts of data on partly overlapping physical distances into comprehensive physical maps. Similarly, the development of computer programs capable of handling large amounts of sequence data has also been key.

The Sequence-Tagged Site (STS) Proposal

The diversity of methods to produce physical maps gives robustness to the mapping projects because it makes possible to confirm the map of a given region by two or more approaches, or when one method fails in one region, another can be brought in. This same diversity, however, creates a weakness—the methods are, in some sense, incompatible with one another. Thus, if one lab creates a map of region A by the radiation hybrid method and another lab produces a long-range restriction enzyme map of region B, even if the two regions are neighbors, there is no easy way to merge both maps into one. Another cloud that loomed over the early stages of the genome project related to the necessity to maintain, catalog, and make available to interested investigators the thousands of clones containing the probes used to identify marker sites. Even worse, some envisioned that it would be necessary to reduce the entire human genome to conventional contigs, one for each chromosome. Figure 4-9 shows that even after most unnecessary redundancy has been eliminated, it takes 16 cosmid clones to cover 500 kb of DNA, or approximately 30 kb per clone. At this rate, it would take 100,000 clones to cover the 3×10^9 bp of the entire human genome. These would be 100,000 samples of DNA to maintain in freezers (certainly replicated several times in repositories around the world) to grow when necessary and make available to the scientific community.

The sequence-tagged site (STS) proposal was a very simple, practical, and elegant way of addressing those problems and it was quickly adopted by many. The proposal was to ask researchers who mapped a region of a human chromosome to sequence 500 bp segments (called STS) in that region. Thus, different teams would be free to use the mapping methods that they thought best, but all results would be reported in the common language of STSs and data from different labs could be merged into a single map.

Another benefit of this type of data collection is that short oligonucleotides matching unique sequences in the STS can be synthesized. Using these oligonucleotides as PCR primers, anybody in the world interested in working in that region would be able to generate their own probes without the need to request clones from the original lab, thus voiding the need to maintain large contig collections (but see below).

STSs in combination with YACs can be used to generate large-scale maps by screening YAC libraries with a collection of STSs. Hybridization of two or more STSs to the same YAC indicates proximity of the STSs in the genome. Hybridization of the same STS to two YACs indicates that the YACs represent segments of the genome that are partly overlapping. In this fashion a map can be assembled. (Precautions need to be taken because YAC clones are often chimaeras—two independent fragments of human DNA become ligated in the same YAC.) Thousands of STS-containing YACs, that have been ordered into regional contigs, are stored worldwide (Table 4-1).

TABLE 4-1 Various mapping methods, the maximum range over which they are usually applied, and their resolution.

Mapping Method	Range	Resolution
Cosmid contig and restriction enzyme maps	up to 500 kb	0.2–2.0 kb
Rare-cutters and PFGE	500–2000 kb	50–200 kb
YAC contig tagged with STS	whole chromosome (>50 Mb)	100–200 kb
Radiation hybrid	whole chromosome (>50 Mb)	50–100 kb
Fluorescence in situ hybridization	whole chromosome (>50 Mb)	1000–3000 kb
Linkage analysis	whole chromosome (>50 Mb)	1000–2000 kb

Advances in the Human Genome Project

Progress in the genetic maps was very fast, thanks in large measure to the CEPH families. Two comprehensive sets of linkage maps of the 22 autosomes and the X-chromosome were published in 1992. One of these included 1676 markers; 1317 consisted of RFLP assayed by Southern blot hybridization, 339 were microsatellite polymorphisms assayed by PCR, and 17 were protein polymorphisms assayed by gel electrophoresis, serological methods, or enzymatic assay. It was estimated that this map covered 92% of the physical length of the chromosomes with a resolution of 3 centimorgans. A subset of the markers was used in FISH assays to correlate the genetic map to the chromosomal banding pattern (Fig. 4-13). The other set of maps was built entirely of microsatellite markers (see "Microsatellite DNA" in Chapter 3 and Example 4.7). In 1994 one of the goals of the Human Genome Project was met when a comprehensive map was published with 5840 loci spaced an average distance of 0.7 centimorgans.

Progress on the physical maps was slower, but in 1992, the entire long arm of chromosome 21 (21q, approximately 1.5% of the whole genome, see Example 4.8) and 40% of the X-chromosome were spanned with overlapping STS-tagged YAC clones. By 1995 most of the genome was mapped with a resolution between 100 kb and 300

EXAMPLE 4.7

Microsatellite DNA

Microsatellite DNA were discussed in Chapter 3. The whole-genome map based on short-tandem repeats polymorphisms (STRP) was built by genotyping a subset of eight CEPH families with respect to nearly 814 such markers. Chromosome assignment was made by establishing linkage between these markers and other markers that had previously been assigned to specific chromosomes. LOD scores were calculated for each pair of markers assigned to the same chromosome. Using these recombination estimates, a subset of 4 to 18 loci per chromosome, spaced approximately every 10 to 20 centimorgans, was chosen to prepare a framework map for each chromosome; odds greater than 1000:1 were used to ascertain their order. The framework maps were then used as a starting point to order a total of 2335 microsatellite markers. The resulting maps had markers spaced every 1.6 centimorgans, on average.

FIGURE 4-13. Genetic map of chromosome 7. The total length in centimorgans (cM) is indicated on top; it was obtained by adding all intervals and is probably an overestimate. Loci positioned with LOD scores of 3 or more are indicated to the right of the vertical bar and distances in centimorgans to the left. Some of the markers were also positioned along the cytogenetic map by FISH. The 500 kb around the CFTR locus shown in Figure 4-7 represent 0.3% of the entire length of this chromosome, approximately the width of the CFTR bar in this figure. (Adapted from NIH/CEPH Collaborative Mapping Group, 1992.)

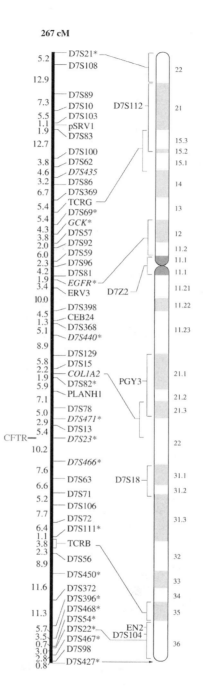

kb. By 1997, the entire X-chromosome was covered, with 2100 landmarks spaced every 75 kb; the rest of the genome was well on its way to a fairly complete, high-resolution physical map, and large-scale (megabase level) sequencing efforts were under way.

Identification and localization of genes. Although identification of human genes was the original motivation for launching the Human Genome Project, it was not

EXAMPLE 4.8

Contig of chromosome 21q

The 810 YAC clones, containing at least one STS each, were ordered into one continuous array (Fig.4-14). There is a great deal of redundancy in the 810 clones; the minimum array necessary to cover the entire 21q is estimated to be 55 clones of approximately 1 Mb each. The chromosome was ultimately subdivided into 198 segments, each characterized by the presence of an STS. The average spacing of STSs was approximately 200 Kb.

The 198 STS sequences were obtained from previously cloned 21q genes and fragments, and from anonymous clones found in chromosome-21–specific libraries. To order the STSs into an array it was necessary to find clones carrying two or more STSs. Three YAC libraries were screened by PCR for the presence of chromosome 21q STSs. The first one contained about 70,000 clones with average size 470 kb, corresponding to 9.4 human genome equivalents. This library was stored as an array of individual clones in 736 96-well plates.

Rather than test all 70,000 clones individually for the presence of STSs, screening was done by PCR in a two-step pooling procedure. First, the entire library was reduced to 92 primary pools, each pool being the combination of eight plates ($92 \times 8 = 736$). Each pool was tested by PCR, with a particular pair of primers, for the presence of an STS. If an STS product was generated, the second step was to identify the candidate clone in the positive primary pool identified by testing 28 smaller pools. These smaller pools contained DNA from clones of individual plates (8 pools), individual rows (8 pools), and individual columns (12 pools) of the corresponding 8-plate set (Fig. 4-15). Candidate positive clones were usually then checked individually. A computer program was used to find partly overlapping clones (which shared at least one STS) and align them into a single contig (Figs. 4-14 and 4-16) (Chumakov et al., 1992).

FIGURE 4-14. STS content map of a small portion of the 21q contiguous array of contiguous YAC clones. Horizontal lines represent clones and filled circles represent STSs. Because in most cases the spacing between STSs is unknown, the distance between them is shown as constant. Above the array is the corresponding section of chromosome 21q showing its banding pattern. Below the chromosome is a numerical designation for each STS (the designations for individual clones were omitted). To visualize the scale of this map, note that *D21Z1, D21S215*, and *D21S120* are the first three named STSs in the whole chromosome map of Figure 4-15. (Redrawn with permission from Fig. 1 in Chumakov I, et al. Continuum of overlapping clones spanning the entire human chromosome 21q. Nature 359:380–387 [1992]. Copyright 1992 Macmillan Magazines Limited.)

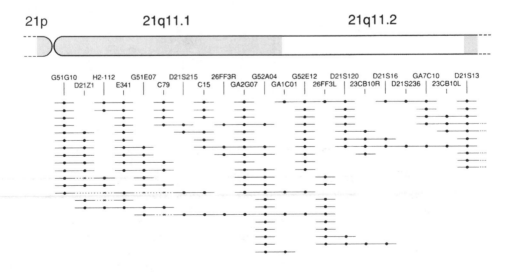

FIGURE 4-15. Screening of an ordered library to identify clones that contain a particular STS. A primary screen of the 92 primary pools, of 8 plates each, identified this group of 8 plates as one that contained a clone with the STS. To determine the identity of that clone, the 768 clones (8 × 8 × 12) were grouped into 28 pools. In this particular hypothetical case we see that, of all the column pools, only the one corresponding to column number 10 gave a positive result. Likewise, the pools for plate 5 and row 7 were positive, and the three pools intersect at the colored well, thus identifying the one clone with the STS in question. In this fashion, the identity of the clone carrying the STS could be established in 28 assays instead of 768, which would have been needed if each clone had been tested individually. If we include the primary screen, the entire library of 70,656 clones was tested with only 120 assays (92 primary pools + 28 secondary pools).

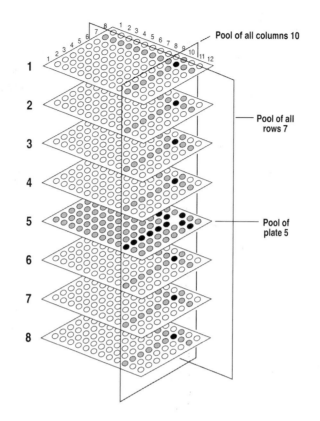

made one of its explicit goals until 1993. In the foregoing discussion of genomic maps, the emphasis was placed almost entirely on the mapping of marker features rather than genes. Even at the time of the early discussions on the Human Genome Project, however, a few scientists put forward the idea of sequencing expressed regions (through their cDNAs) rather than tackling the entire genome from the outset. However, this approach did not win support, and the criticisms centered around the apparent lack of technological means to carry it out. There was also a fear that, once the expressed sequences were determined, interest in completing the rest of the genome would diminish and the bulk of it would never be done, thus leaving an extensive portion of our genetic endowment unexplored.

Working independently of the Human Genome Project, and starting in the late 1980s, Craig Venter and his colleagues, then at the National Institutes of Health (NIH), started to isolate and sequence small fragments of DNA obtained from cDNA libraries. The information on these *expressed sequence tags* (ESTs) was accumulated in a database with appropriate software to compare new ESTs to old ones and determine whether they were new hits or repeats of old ones. By 1992, Venter and co-workers had accumulated over 7000 ESTs when they moved to The Institute for Genomic Research (TIGR), a nonprofit research laboratory associated with private industry. Working with scores of automated sequencing machines and 300 cDNA libraries from a variety of human tissues and disease states, these scientists continued accumulating information on ESTs, and assembling them into cDNA contigs by identifying partly overlapping sequences (*tentative human consensus* or THC). By 1995

FIGURE 4-16. An STS-YAC high-resolution map of chromosome 21. STSs are represented by short horizontal bars arbitrarily shown as equidistant. Each bar corresponds to one of the named STSs in Figure 4-14—three of the names near the top are indicated for guiding purposes. To establish the correspondence between this and other maps, some of the STSs were also mapped by radiation hybridization (RH mapping). RH map distances are in centirays (cR) with an 8000 rad dose. STSs were also positioned relative to breakpoints of chromosomal rearrangements (designations in color). All STSs that were localized between two breakpoints are arbitrarily given a position midway between the two breakpoints; this is indicated by diagonal lines that subtend a number of STSs. Some of these STSs were also connected to *Not*I and genetic linkage maps (not shown). (Redrawn with permission from Fig. 2 in Chumakov I, et al. Continuum of overlapping clones spanning the entire human chromosome 21q. Nature 359:380–387 [1992]. Copyright 1992 Macmillan Magazines Limited.)

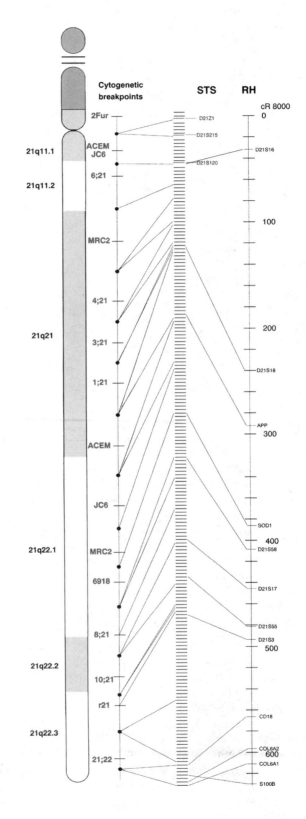

Plate 1. Fiber FISH. Stretched chromatin fibers in which the DNA is uniformly stained with a blue fluorescent dye (DAPI); in situ hybridization, with two probes, labels the *PMP22* gene and flanking repeats with different colors. See Chapter 7, "Errors in Recombination," for details. (Reprinted with permission from Rautenstrauss B, Fuchs C, and Liehr T, et al. [1997] Visualization of the CMT1A duplication and HNPP deletion by FISH on stretched chromosome fibers. J Peripheral Nervous Sys 2:319–322.)

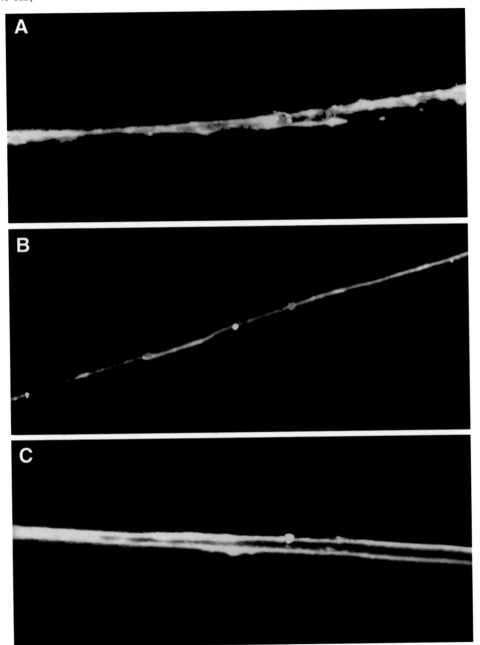

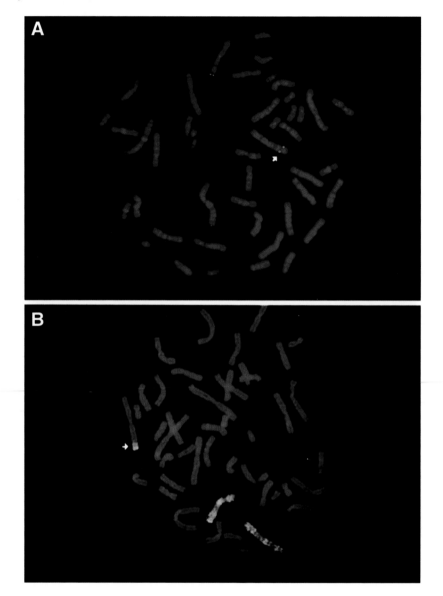

Plate 3. Detection of Di George syndrome, a small interstitial deletion in chromosome 22 using a specific probe. See Chapter 5, "Use of FISH to Identify Chromosomal Rearrangements," for details. (Courtesy of Dr. Kathleen Rao and Dr. Kathleen Kaiser-Rogers, Cytogenetics Laboratory, Department of Pediatrics, University of North Carolina.)

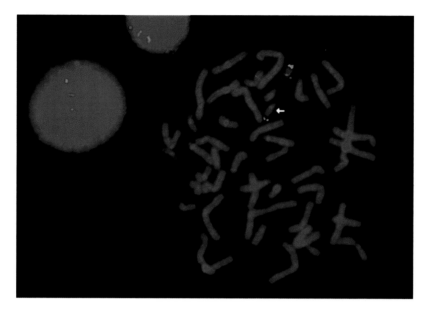

Plate 4. Amniotic fluid interphase nucleus hybridized to probes for chromosomes 13 and 21. This is a screening assay, and results are always confirmed by complete karyotyping. The three signals for each probe, in this case, were confirmed with a diagnosis of triploidy. See Chapter 5, "Use of FISH to Identify Chromosomal Rearrangements," for details. (Courtesy of Dr. Kathleen Rao and Dr. Kathleen Kaiser-Rogers, Cytogenetics Laboratory, Department of Pediatrics, University of North Carolina.)

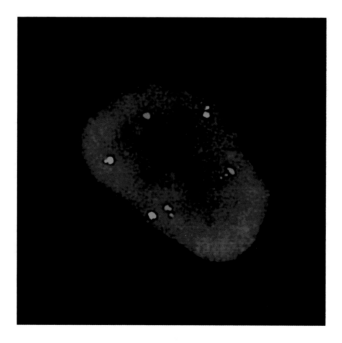

Plate 5. Interphase assay for a 15;17 translocation. A translocation with breakpoints at 15q22 and 17q21.1 is specifically associated with acute promyelocytic leukemia as a consequence of the fusion of the *PML* and *RARA* (retinoic acid receptor alpha) genes. Red fluorescence identifies the *PML* probe and blue-green fluorescence identifies the *RARA* probe; a normal cell on the upper left displays four separate sites of hybridization, while in the other nucleus there is an area of overlap between the two probes (yellow) marking the translocated chromosomes. See Chapter 8, "Chimeric Proteins." (Courtesy of Dr. Kathleen Rao and Dr. Kathleen Kaiser-Rogers, Cytogenetics Laboratory, Department of Pediatrics, University of North Carolina.)

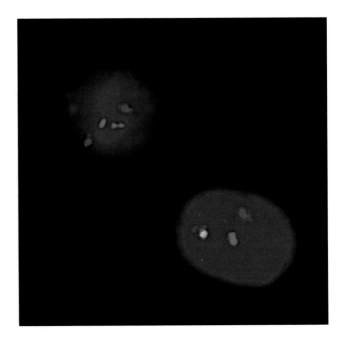

Plate 6. Detection of oncogene *NMYC* gene amplification. The red *NMYC* probe hybridizes to chromosome 2p, at the normal locus (in positions corresponding to eight o'clock and eleven o'clock on the metaphase spread), as well as to the short arm of chromosome 7 into which the enormously amplified gene was inserted, creating a *homogeneously staining region*. This tissue sample is from a neuroblastoma, a cancer of neurons of the sympathetic system; *NMYC* amplification is usually associated with poor response to treatment. See Chapter 8, "Gene Amplification." (Courtesy of Dr. Kathleen Rao and Dr. Kathleen Kaiser-Rogers, Cytogenetics Laboratory, Department of Pediatrics, University of North Carolina.)

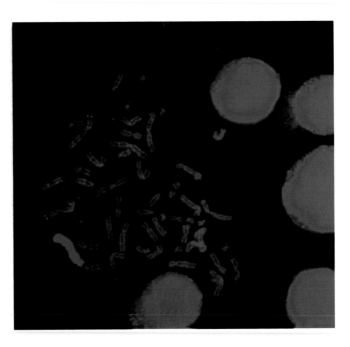

they had 88,000 independent THC and EST units. It is not yet possible to ascertain how many genes they represent because an undetermined fraction of them correspond to nonoverlapping sections of the same gene. The majority of these data were made available to the scientific community and integrated into whole genome maps.

Genome sequencing. In sequencing DNA, a small fragment (1 to 2 kb) is cloned in a special, sequencing vector, and a primer, complementary to the vector, is used to extend the sequencing reactions into the cloned DNA. The products of the sequencing reactions are then run in a high-resolution polyacrylamide gel and the sequence is read off the gel. The amount of sequence that can be read from each clone is usually limited to 500 to 700 bp. (This 500- to 700-bp sequence obtained from a set of reactions in a gel is sometimes called a *read*.) As a rule, scientists are interested in obtaining the sequence of longer stretches of DNA; these stretches range from a few kilobases when studying individual genes, to several megabases in whole chromosome sequencing projects.

The task then is to assemble those short, fragmentary sequences into a complete, single sequence covering the entire region of interest. The region of interest may be contained in one or a few plasmid or lambda clones if it is small. If it is more exten-

---

| **Box 4.2** | *You're going to patent my genes?!* |

With the potential for acquiring the sequence for large portions of the human genome, the question of patenting human genes, always controversial, assumes new urgency. At one extreme, the mere idea of commercializing our genetic endowment will appear repugnant to some; at the other extreme, the notion that we should not derive benefits from our knowledge about the human genome (benefits in the form of diagnostic or therapeutic tools) is abhorrent to others. Intellectual property laws (patent laws) are an attempt to regulate the process and protect the right of inventors to have a return on their work and investments. This, in turn, would stimulate applied research and the development of new products that benefit society at large.

It should be pointed out that it is not possible to patent somebody's actual genes. But courts have ruled that it is possible to patent reagents derived from those genes (such as DNA fragments prepared in vitro to use as a probe), or even the sequence of those genes if a clear practical application can be proposed.

In general terms, to qualify for a patent an invention must be novel, nonobvious, and practical. Wholesale patenting of thousands of genes or cDNAs will probably not be attempted, because they fail the third requirement. Although many of those sequences may eventually yield practical applications, given our present state of ignorance about the function of most genes it is impossible to predict what the application might be. The change in strategy by the Human Genome Project in deciding to put in the public domain as much sequence as possible, as soon as possible, to some extent is seen as an effort to prevent private labs from patenting large numbers of sequences because, by the first requirement, something already in the public domain cannot be patented.

The dichotomy is not between purely altruistic scientists on one side and greedy pharmaceutical companies on the other—most biotechnology companies are started by scientists. There are probably some of each type at the extremes, but the grey area in the middle includes the majority: people driven by intellectual curiosity, by the desire to make a difference in the fight against disease, by the desire for recognition, and by profit motives. All four impulses mix in various proportions in individuals.

sive, special vectors, such as cosmids, BACs (bacterial artificial chromosomes), or PACs (P1 artificial chromosomes), are necessary—genomic sequencing is typically carried out in clones of 40 to 200 kb. We will refer to these clones that contain the entire region of interest, as the *source clones*.

There are two strategies that can be used to assemble a collection of reads into a comprehensive sequence of the source clone: 1) directed sequencing and 2) random (or shotgun) sequencing.

Directed sequencing. In *directed sequencing* the source clone is subcloned into smaller fragments that are ordered into a contig, and these subclones are in turn fragmented into approximately 1-kb pieces which are themselves cloned and ordered (Fig. 4-17). This last cloning step is into the sequencing vector. Because the sequencing clones are already ordered into a contig, the overall sequence comes out directly from the individual sequences. This approach is most often used in projects limited to fragments of a few kilobase pairs such as cDNAs or isolated exons. To sequence in this fashion a 40-kb source clone, it is necessary to order an array of several dozen clones, and this is a very laborious process.

Random (shotgun) sequencing. *Random* (or *shotgun*) *sequencing* obviates the necessity to preassemble a contig of short segments by the redundant sequencing of randomly produced fragments. The final assembly is based on the partial overlapping of individual sequences. The source clone is fragmented *at random* into 1- to 2-kb pieces, which are cloned directly into the sequencing vector, and reads are ob-

FIGURE 4-17. Schematic representation of the subcloning steps in the directed sequencing procedure. The process is illustrated with subclone 4 only; the other four "ordered clones" would have to be similarly subcloned. There would be a total of approximately 20 sequencing clones with an average size of 1 kb.

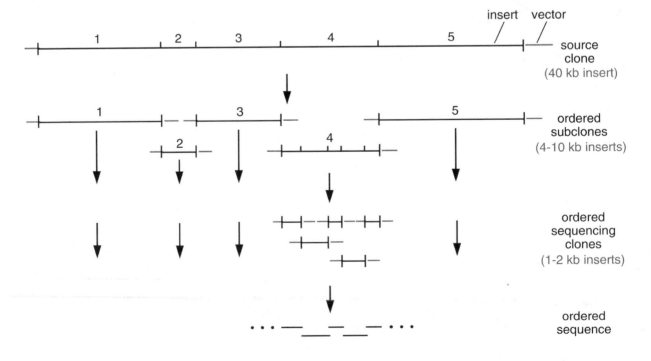

FIGURE 4-18. Schematic representation of random sequencing. From a 40-kb insert it would be necessary to obtain approximately 200 unordered (partly overlapping) sequencing clones, each of which would be on average 1 kb, and there would be a fivefold redundancy.

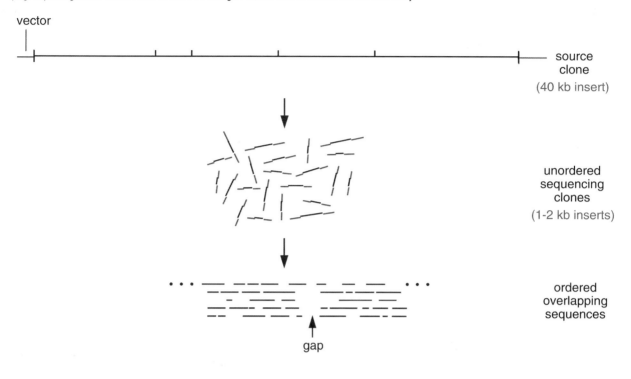

tained (Fig. 4-18). This strategy is based on the principle of sequencing enough random fragments that by chance most sections of the source clone will be covered—it can be calculated that if reads equivalent in length to five times the source clones are sequenced, the probability that a particular base will not be sequenced is less than 1%. As much as possible of the overall sequence is assembled by computer programs that search for regions of overlap among all the partial sequences. Once this level of coverage is accomplished (the resulting sequence is sometimes called the *unfinished sequence*), the gaps are completed by directed sequencing—this finishing process can sometimes be quite time consuming.

Even though this strategy demands several times more sequencing than directed sequencing, it is found to be much more economical for extensive regions because it avoids the need to subclone an ordered array, the sequencing steps being more easily automated than subcloning and mapping. The entire chromosome of the bacterium *Haemophilus influenzae*, almost 2 Mb in size, has been sequenced using this strategy, and large segments of the human genome are being approached this way.

CONCLUSION: THE END IN SIGHT

By 1998, large-scale sequencing was under way in a number of labs and an estimated 10% of the genome was finished. Celera Genomics, a private lab under the direction of Craig Venter, announced that by massive random sequencing they would generate the human genome sequence in a couple of years and patent the more promising stretches. In response, administrators of the Human Genome Project and scientists in the main publicly funded labs undertook a strategic shift. The original goal had been to sequence 99.99% of the human genome with 10-fold redundancy (each stretch would be sequenced 10 times) to reduce errors. The new goals were to complete a "rough draft" of 90% of the genome with fivefold redundancy by the spring of the year 2000 and to refine it into the original goal by 2003. At the same time, all the publicly supported labs involved agreed to release sequence data daily through the databanks (Table 4-2). This approach would maintain most of the information on the human genome in the public domain.

It should be clear that although completing the human genome sequence is an important landmark, it is not really the finish line. To derive benefits from this information it will be necessary to know details about the function and properties of gene products, and about the regulation of gene expression that is known for only a handful of genes now. At present, 99% of the work needed to understand the human genome at this next level is yet to be done.

---

www | **Box 4.3 *Internet sites***

Sites that include a map with over 30,000 human genes already localized as well as general information on the human genome can be found at

> http://www.ncbi.nlm.nih.gov/genemap/
> http://www.ncbi.nlm.nih.gov/genome/guide/

Information on the Human Genome Project, sponsored in the United States by the Department of Energy and the National Institutes of Health can be found at

> http://www.nhgri.nih.gov:80/HGP/
> http://www.ornl.gov/hgmis/

The second site includes a page called "Medicine and the New Genetics" that deals specifically with genetic diseases, support groups, genetic testing and counseling, and many other topics of interest to health care professionals. There are also links to frequently asked questions about the Human Genome Project as well as reports of current progress.

Databases for radiation hybrid mapping can be found at two sites:

> http://www-shgc.stanford.edu/RH/rhserver_form2.html
> http://www.sanger.ac.uk/Users/cari/RHsc/RHserver.shtml

A database of DNA patents is being kept at:

> http://www.geneticmedicine.org/

EXAMPLE 4.9

Sequence of chromosomes 21 and 22

The first chromosome sequence essentially complete was that of chromosome 22, whose 97% complete sequence was published in 1999 by one of the groups in the Human Genome Project. That is to say, the sequence corresponds to 22q; 22p seems to be made up exclusively of rDNA and other repeated sequences. The completed sequence covers 33.4 Mb (10^6 bases) of an estimated 34.5 Mb total in 22q (there are 11 relatively minor gaps that could not be cloned or sequenced). Approximately 42% of the sequence represents tandem and interspersed repeated sequences (the *Alu* family accounts for 17% and *L1* accounts for 10%). Identified transcription units (exons and introns) add up to probably between 40% and 50% of the sequence. Identifying genes from sequence data is a difficult task. In this case they were able to find 247 "known genes," i.e., genes for which human DNA or protein sequences were already available. They also identified 150 "related genes," i.e., genes with sequence similarity to known genes from human or other species. Finally they found 148 "predicted genes," which had sequence similarity only to some ESTs. These 545 genes correspond to sequences that to some extent were already known and account for 39% of the sequence. What about "new genes"? Computer programs are notoriously unreliable in predicting the presence of a gene from raw sequence; after applying several tests and correcting for various errors, the authors predict that there may be another 100 genes, so far unidentified, in chromosome 22q (Fig. 4-19). The second human chromosome sequenced was number 21. It was rather surprising that in the 33.5 Mb of 21q they were able to identify only 220 genes, less than half those observed in a comparable amount of DNA in chromosome 22q.

FIGURE 4-19. Summary obtained from the complete sequence of a segment of chromosome 22q. The centromere is at coordinate 0, and the 22q telomere at 34,500 kb, thus, this 3-million-base segment includes approximately 10% of the long arm. Transcript (+) represents genes transcribed from left to right, and (-) from right to left (on the opposite strand). Note that transcription units are represented by solid boxes, without differentiation of exons and introns. The first letter of the labels of identified genes lines up with the corresponding box. Gray boxes indicate "predicted" genes. *Alu* and *L1* repeats are so frequent that, at this level of resolution, they appear as solid blocks in parts. This portion of the chromosomal sequence was assembled from a contig of 42 partly overlapping segments cloned into bacterial artificial chromosomes (BAC) and other high capacity vectors (Average insert size, approximately 70 kb). (Based on Dunham et al., 1999.)

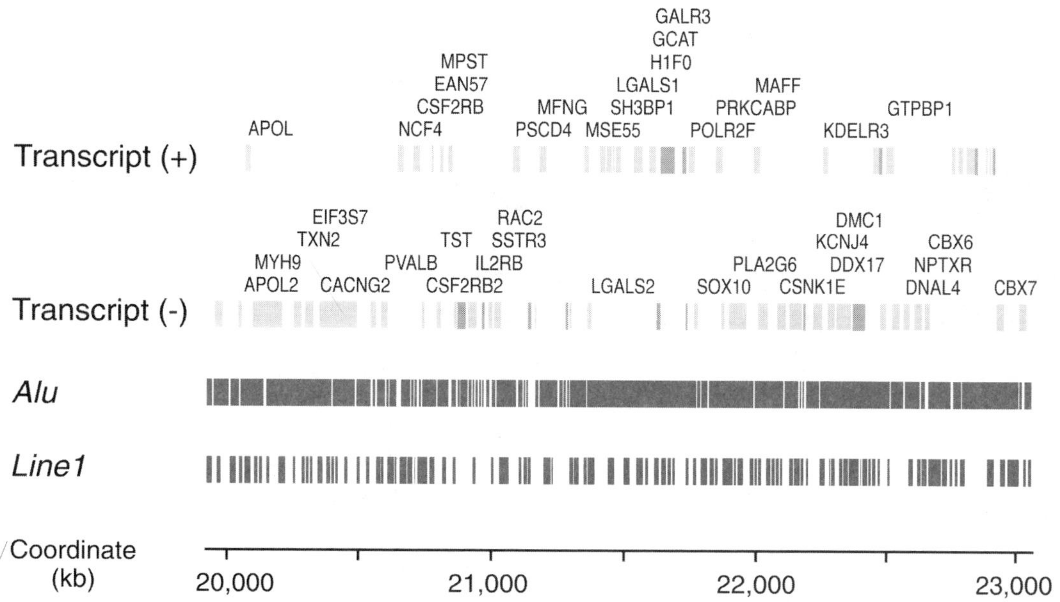

TABLE 4-2. The principal nonprofit centers involved in the sequencing of the human genome (chromosome assignments are not final).

DNA Sequencing Center	Chromosome Assignment
Baylor College of Medicine (USA)	3, 12, X
Massachusetts Institute of Technology/Whitehead Center (USA)	17, others
Department of Energy (USA)	5, 16, 19
Washington University (USA)	2, 3, 7, 11, 15, 18, Y
Sanger Centre (UK)	1, 6, 9, 10, 13, 20, 22, X
Genoscope (France)	14
Institute of Molecular Biotechnology (Germany)	8, 21
Tokyo University (Japan)	8, 18, 21, 22

EXERCISES

4-1. Compare the level of sequence similarity (% amino acid identity) in the pairwise comparisons human α globin versus human β globin, and rat β globin versus human β globin (see Figs. 3-17 and 4-2). Do you find it surprising that the number of differences between the two human genes is greater than the number of differences between the rat and human genes? How can you explain this observation?

4-2. Construct a three-generation pedigree with at least six sibs in the third generation. Imagine it is a member of the CEPH panel and that it has been typed for three probes, A, B, and C. Each probe detects the presence or absence of a restriction site with the allele *A* indicating presence of the site detected by probe A, and *a* the absence of that site, and similarly for *B* and *b*, and *C* and *c*. Distribute these six alleles as they might occur if the sites detected by *A* and *B* are closely linked but *C* is on a separate chromosome.

4-3. What is the minimum number of clones needed to have a complete human library in the following vectors (nominal capacity in parentheses): (**A**) bacterial plasmid (5 kb); (**B**) phage lambda (15 kb); (**C**) cosmid (40 kb); (**D**) YAC (1 mb).

4-4. From Figure 4-10B, diagram a *Not*I restriction map of the *NFB1* gene in the region of exons 22 to 24. Similarly, diagram the translocation t(17;19) (q11.2;q13.2).

REFERENCES

Adams MD, Kerlavage AR, Fleischmann RD, et al. (1995) Initial assessment of human gene diversity and expression patterns based upon 83 million nucleotides of cDNA sequence. Nature 377(suppl):3–174.

Buckler AJ, Chang DD, Graw SL, Brook JD, Haber DA, Sharp PA, and Housman DE. (1991) Exon amplification: A strategy to isolate mammalian genes based on RNA splicing. Proc Natl Acad Sci USA 88:4005–4009.

Cantor CR. (1990). Orchestrating the Human Genome Project. Science 248:49–51.

Chumakov I, Rigault P, Guillou S, et al. (1992) Continuum of overlapping clones spanning the entire human chromosome 21q. Nature 359:380–387.

Cochran BH, Reffel AC, and Stiles CD. (1983) Molecular cloning of gene sequences regulated by platelet-derived growth factor. Cell 33:939–947.

Collins F, and Galas D. (1993) A new five-year plan for the U.S. Human Genome Project. Science 262:43–46.

Collins FS. (1995) Positional cloning moves from perditional to traditional. Nature Genetics 9:347–350.

Cooperative Human Linkage Center and Centre d'Etude du Polymorphisme Humain. (1994) A comprehensive human linkage map with centimorgan density. Science 265:2049–2070. An extensive group of international scientists working at many sites.

Cox DR, Burmeister M, Price ER, Kim S, and Myers RM. (1990) Radiation hybrid mapping: a somatic cell genetic method for constructing high-resolution maps of mammalian chromosomes. Science 250:245–250.

Cox DR, Green ED, Lander ES, Cohen D, and Myers RM. (1994) Assessing mapping progress in the human genome project. Science 265:2031–2032.

Dib C, Fauré S, Fizames C, Samson D, Drouot N, Vignal A, Millasseau P, Marc S, Hazan J, Seboun E, Lathrop M, Gyapay G, Morissette J, and Weissenbach J. (1996) A comprehensive genetic map of the human genome based on 5,264 microsatellites. Nature 380:152–154.

Dunham I, Shimizu N, Roe BA, Chissoe S, et al. (1999) The DNA sequence of human chromosome 22. Nature 402:489–495.

Fahsold R, Habash T, Trautmann U, Haustein A, and Pfeiffer RA. (1995) Familial reciprocal translocation t(17;19) (q11.2;q13.2) associated with neurofibromatosis type 1, including one patient with non-Hodgkin lymphoma and an additional t (14;20) in B lymphocytes. Hum Genet 96: 65–69.

The Genome Directory. Supplement to Nature 377 (6547S) (1995) (Review of EST data on as many as 70% of human genes, low resolution YAC contigs of 75% of the human genome, and high-resolution maps of several individual chromosomes.)

Ginsburg D, Handin RI, Bonthron DT, Donlon TA, Bruns GAP, Latt SA, and Orkin SH. (1985) Human von Willebrand factor (vWF): isolation of complementary DNA (cDNA) clones and chromosomal location. Science 228:1401–1406.

Goodfellow PN. (1992) Variation is now the theme. Nature 359:777–778. Discussion of early whole genome maps.

Gunshin H, Mackenzie B, Berger UV, Gunshin Y, Romero MF, Borob WF, Nussberger S, Gollan JL, and Hediger MA. (1997) Cloning and characterization of a mammalian proton-coupled metal-ion transporter. Nature 388:482–488.

Gusella JF, Wexler NS, Conneally PM, Naylor SL, Anderson MA, Tanzi RE, Watkins PC, Ottina K, Wallace MR, Sakaguchi AY, Young AB, Shoulson I, Bonilla E, and Martin JB. (1983) A polymorphic DNA marker genetically linked to Huntington's disease. Nature 306:234–238.

Hattori M, Fujiyama A, Taylor TD, et al. (2000) The DNA sequence of human chromosome 21. Nature 405:311–319.

Heiskanen M, Peltonen L, and Palote A. (1996) Visual mapping by high resolution FISH. Trends Genet 12:379–382.

Hudson TJ, Hudson TJ, Stein LD, Gerety SS, et al. (1995) An STS-based map of the human genome. Science 270:1945–1954. Whole genome map of 15,000 STSs with an average spacing of 200 kb.

The Huntington's Disease Collaborative Research Group. (1993) A novel gene containing a trinucleotide repeat that is expanded and unstable on Huntington's disease chromosomes. Cell 72:971–983. A group of 58 international scientists working at six different sites.

Kerem B, Rommens JM, Buchanan JA, Markiewicz D, Cox TK, Chakravarti A, Buchwald M, and Tsui L-C. (1989) Identification of the cystic fibrosis gene: genetic analysis. Science 245:1073–1080.

Kurachi K, and Davie EW. (1982) Isolation and characterization of a cDNA coding for human factor IX. Proc Natl Acad Sci USA 79:6461–6464.

Lew DJ, Dulic V, and Reed SI. (1991) Isolation of three novel human cyclins by rescue of G1 cyclin (Cln) function in yeast. Cell 66:1197–1206.

Lichter P, Tang CC, Call K, Hermanson G, Evans GA, Housman D, and Ward DC. (1990) High-resolution mapping of human chromosome 11 by in situ hybridization with cosmid clones. Science 247:64–69.

Little P. (1986) Finding the defective gene. Nature 321:558–559.

NIH/CEPH Collaborative Mapping Group. (1992) A comprehensive genetic linkage map of the human genome. Science 258:67–86. An extensive group of international scientists working at many sites.

Olson M, Hood L, Cantor C, and Botstein D. (1989) A common language for physical mapping of the human genome. Science 245:1434–1435. The STS proposal.

Poste G. (1995) The case for genomic patenting. Nature 378:534–536.

Poustka A, Pohl TM, Barlow DP, Frischauf A-M, and Lehrach H. (1987) Construction and use of human chromosome jumping libraries from NotI-digested DNA. Nature 325:353–355.

Riordan JR, Rommens JM, Kerem B, Alon N, Rozmahel R, Grzelczak Z, Zielenski J, Lok S, Plavsic N, Chou J-L, Drumm ML, Iannuzzi MC, Collins FS, and Tsui

L-C. (1989) Identification of the cystic fibrosis gene: cloning and characterization of complementary DNA. Science 245:1066–1073.

Roberts L. (1989) Genome mapping goal now in reach. Science 244:424–425.

Rommens JM, Iannuzzi MC, Kerem B, Drumm ML, Melmer G, Dean M, Rozmahel R, Cole JL, Kennedy D, Hidaka N, Zsiga M, Buchwald M, Riordan JR, Tsui L-C, and Collins FS. (1989) Identification of the cystic fibrosis gene: chromosome walking and jumping. Science 245:1059–1065.

Smith CL, and Cantor CR. (1986) Pulsed-field gel electrophoresis of large DNA molecules. Nature 319:701–702.

Venter JC, Adams MD, Sutton GG, Kerlavage AR, Smith HO, and Hunkapiller M. (1998) Shotgun sequencing of the human genome. Science 280:1540–1542.

Watson JD. (1990) The human genome project: past, present and future. Science 248:44–51.

Weissenbach J, Gyapay G, Dib C, Vignal A, Morissette J, Millasseau P, Vaysseix G, and Lathrop M. (1992) A second generation linkage map of the human genome. Nature 359:794–801.

White R, Leppert M, Bishop DT, Barker D, Berkowitz J, Brown C, Callahan P, Holm T, and Jerominski L. (1985) Construction of linkage maps with DNA markers for human chromosomes. Nature 313:101–105.

5 | *Chromosomes and Karyotypes*

 ## THE MORPHOLOGY OF HUMAN CHROMOSOMES

In addition to genetic and molecular techniques, the visualization of human chromosomes at the light microscope provides another window into human heredity. As with most traditional light-microscope work, the study of human chromosomes required fixation and staining, which were well-established procedures in the 19th century. It was not until the 1950s, however, that several technical developments made possible the kinds of detailed studies needed to correlate changes in chromosome morphology and medical syndromes. Those developments, in use to this day, are

1. The mitogenic activation of peripheral blood lymphocytes by phytohemaglutinins (a kidney bean extract). Peripheral blood is one of the most accessible sources of human live cells; lymphocytes, however, are not actively dividing cells absent special treatments.
2. The blocking of dividing cells in metaphase by means of colchicine. Colchicine and similar drugs interfere with microtubule assembly, which leads to the tempo-

125

rary arrest of dividing cells in metaphase—a stage when chromosomes are most condensed and some of their morphological features are best in evidence.

3. The swelling of dividing cells by hypotonic treatment right before fixation. Hypotonic treatment of live cells causes them to swell so that the chromosomes separate from one another; the cells are then dried on a glass slide, which leads to the spreading and flattening of the chromosomes.

After fixation with acetic acid and alcohol, staining used to be done by the Feulgen reaction, or with orcein and a few other dyes.

With these advances it was established that humans have 46 chromosomes, but it is nearly impossible to study in detail the chromosome morphology directly from a chromosome spread. The standard technique is to photograph and enlarge prints of the spread, and then cut out the chromosomal silhouettes so that they can be arranged by size into a *karyotype*. A karyotype permits us to verify that the 46 human chromosomes are in fact arranged into 23 pairs: 22 pairs of autosomes, and either two Xs in females, or an X and a Y in males.

Other than size, the main feature that can be identified in these preparations is the position of the *primary constriction* or *centromere,* the region of the chromosome that contains the *kinetochore* or spindle-fiber attachment point. (The term "centromere" is also used to refer to the segment of DNA that defines the kinetochore.) The position of the centromere can be expressed quantitatively as a ratio of chromosomal-arm lengths; this permits averaging values from several cells. A ratio of 1 characterizes metacentric chromosomes, and one close to 0, acrocentric ones. In submetacentric and subacrocentric chromosomes, the ratio is in between. The name *telocentric* refers to chromosomes in which the centromere is terminal (ratio = 0), but there are no such chromosomes in the human karyotype.

Aceto-orcein staining does not permit identification of every chromosome. Chromosomes 1, 2, 3, 16, 17, 18, and Y can be identified, the rest can only be differentiated into groups B, C, D, F, and G. Chromosomes 1 and 3 and group F are metacentric, and 2 is nearly so. C chromosomes are submetacentric, B chromosomes are subacrocentric, and D and G are acrocentric.

Other features evident in human chromosomes include the presence of heterochromatin, secondary constrictions, and satellites.

Chromosomal regions near centromeres (most prominently in chromosomes 1, 9, and 16) and the end of the long arm of the Y-chromosome seem to condense prematurely during the cell cycle, or perhaps remain condensed throughout interphase. Because they stain differently from the rest of the chromosomes, these regions are called *heterochromatic.* Nonheterochromatic regions are said to be *euchromatic.* Centromeric (or centric) and Y heterochromatin is also called *constitutive heterochromatin,* because it is always found in a heterochromatic state. As we will see later, one of the X-chromosomes in females is also highly condensed, but because Xs are not always condensed, this is called *facultative heterochromatin.* Heterochromatin seems largely devoid of genetic activity: the DNA sequence in constitutive heterochromatin is highly repetitive in nature and contains few, if any, genes (see Chapter 3), and the genes in the heterochromatic X are inactivated.

Secondary constrictions are pale-staining regions evident when the chromosomes are somewhat extended, but that eventually disappear when they reach maximum condensation in metaphase. They are the sites of the ribosomal RNA genes—which exist in tandemly repeated copies and are active (therefore uncondensed) until late in prophase—and are, for this reason, known as the *nucleolus organizing region* (NO). NO are present in the short arm of acrocentric chromosomes of groups D and G.

Distal to the secondary constriction there is a very short section of chromosome that is known as a *satellite,* because at times it appears detached from the chromosome.

Outside the NO, the short arms of D and G chromosomes are mostly heterochromatic and variable in length in different people. The position of features along the chromosome is identified with respect to the centromere. Thus, *proximal* and *distal* refer to "near to" or "far from" the centromere.

Banding Techniques

When chromosomes are treated with some fluorescent dyes such as quinacrine derivatives, they do not stain uniformly along their length, but rather show a reproducible alternation of bright and dull regions that are called *Q-bright* and *Q-dark bands.* Several other methods were discovered that result in the banding of human chromosomes. One of the most commonly used involves pretreating the chromosomes with a salt solution at 60°C and staining with Giemsa; the resulting bands are called *G-bands* (Fig. 5-1). G-dark bands correspond to Q-bright bands and G-light bands correspond to Q-dark bands.

FIGURE 5-1. G-banded metaphase spread of peripheral blood cells from a normal male. An interphase nucleus from another cell can be seen. (Courtesy of Dr. Kathleen Rao, Cytogenetics Laboratory, Department of Pediatrics, University of North Carolina.)

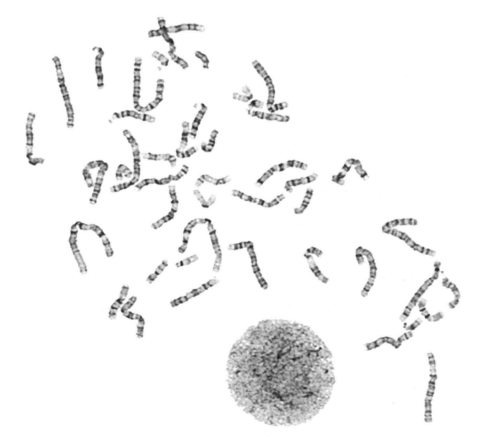

Banding of chromosomes permits the identification of each individual chromosome (Fig. 5-2). In fully condensed metaphase chromosomes, approximately 300 reproducible bands have been described. But, avoiding colchicine and using more extended prometaphase or prophase chromosomes, almost 2000 bands can be seen. High-resolution studies in prophase chromosomes are very laborious, and they are usually carried out only for precise localization of chromosomal rearrangements or molecular probes.

Modifications of these staining techniques allow specific staining of heterochromatic regions (*centric* banding or *C-banding*) or terminal regions (*T-banding*). Most of these techniques were developed in the 1960s. In 1971 the Paris Conference brought together many cytogeneticists who agreed on a standard nomenclature and numbering system for the human karyotype bands (Figs. 5-3 and 5-4).

According to convention, chromosomes are presented vertically with the short arm (*p* for *petit*, "small") up and the long arm (*q*) down. The chromosomal arms extend from the centromere (*cen*) to the *telomeres* or chromosomal ends (the telomeres of

FIGURE 5-2. G-banded metaphase chromosomes from a male arranged in a karyotype. In clinical studies, a technician will typically count chromosomes from 20 nuclei, analyze in detail five of those nuclei for evidence of structural abnormalities, and prepare two karyotypes from each sample. (Courtesy Dr. Kathleen Rao, Cytogenetics Laboratory, Department of Pediatrics, University of North Carolina.)

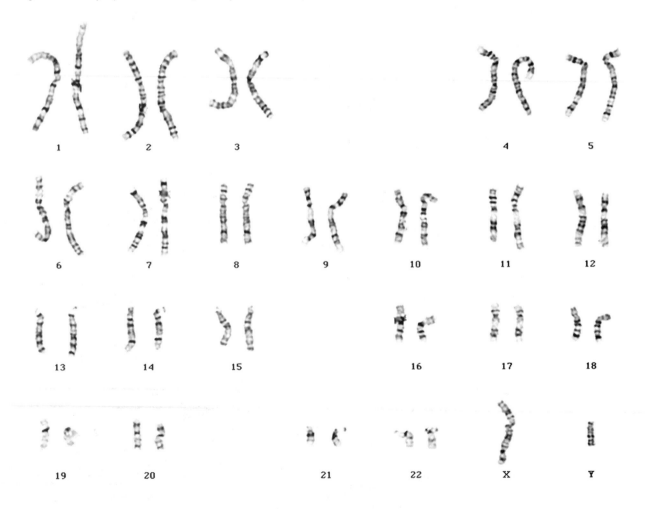

FIGURE 5-3. Diagram of the human karyotype according to the 1971 Paris Conference. (Based on Yunis, 1980.)

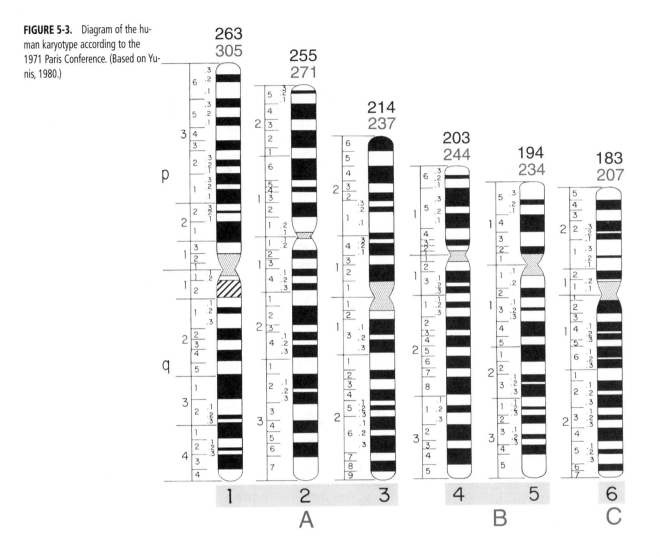

each chromosome are identified as *p-ter* and *q-ter*). Each chromosomal arm is divided into a number of regions that range between 1 and 4 in different chromosomes. The numbering systems always start at the centromere, and the numbers go up toward the telomeres.

Subdivision of these regions depends on the level of "resolution" being used—that is, whether the chromosomes are being studied at the 300 band level, the 500 band level, or the 900 band level. In the first case each region is divided into bands. In the second case, the bands are subdivided into subbands. In the third case, the subbands are themselves divided. The level of resolution depends on how stretched out the chromosomes are. The nomenclature is illustrated in Figure 5-4.

Q-dark bands seem to be structural and functional chromosomal elements. They

FIGURE 5-3. Continued

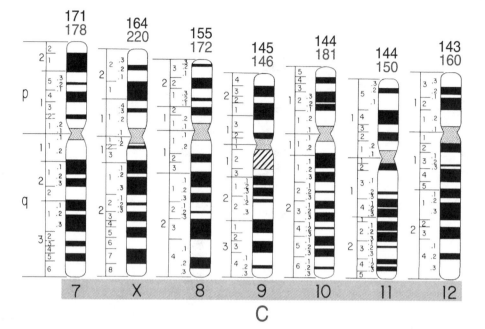

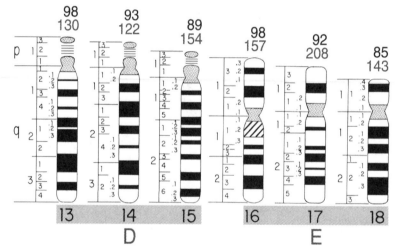

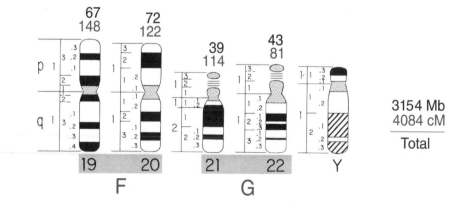

3154 Mb
4084 cM
───────
Total

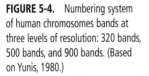

FIGURE 5-4. Numbering system of human chromosomes bands at three levels of resolution: 320 bands, 500 bands, and 900 bands. (Based on Yunis, 1980.)

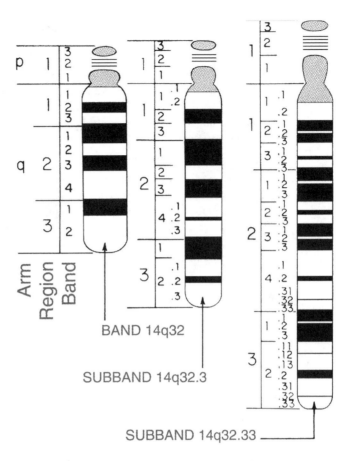

BAND 14q32

SUBBAND 14q32.3

SUBBAND 14q32.33

correspond to the *chromomeres* that characterize meiotic pachytene chromosomes, and all the DNA in a band replicates within a short period of time during the S phase. Q-dark bands replicate early in S, then Q-bright bands replicate, and finally C-bands (heterochromatin) do.

CHROMOSOMAL ABNORMALITIES

Soon after techniques permitted establishing a firm number of chromosomes for the human karyotype in the mid-1950s, the chromosomal defects associated with the most common viable abnormalities were characterized. These were all changes in chromosome number; thus, cases of Down syndrome showed an extra chromosome 21 (*trisomy* 21), Turner syndrome was associated with a single X-chromosome (*monosomy* X), and Klinefelter syndrome was associated with the karyotype 47,XXY. It would take the increased resolution of banding techniques to characterize in some detail structural rearrangements such as translocations and deletions.

Nomenclature for human karyotype abnormalities include the total number of chromosomes, the X/Y constitution, and an indication of any abnormalities; + signs indicate extra chromosomes and – signs indicate a missing chromosomes. Thus, 47,XY,+21 denotes a male with Down syndrome, or 45,X an individual with Turner syndrome. This system can also be applied to deletions or duplications of chromosomal arms or bands. For example, 46,XX,5p– denotes a female with a deletion in the short arm of chromosome 5. If the observation is more detailed, the specific bands or the region deleted is indicated. Similarly, for translocations, the nomenclature is 46,XY,t(8;14); if there were more detailed information and the breakpoints were known to be 8q22 and 14q31, then the designation would be expanded to include this information: 46,XY,t(8;14)(q22;14q31).

Abnormal Chromosome Numbers

The most common cases of abnormal chromosome numbers in human affect either whole chromosome sets (*polyploidy*) or single chromosomes (*aneuploidy*). Chromosomal anomalies are not defined by a single phenotypic characteristic, although some of them are recognizable by a combination of traits (e.g., Down syndrome). It is generally true, however, that all aneuploidies result in mental retardation, and most of them are associated with growth deficiencies and seizures of various types.

Polyploidy.   Triploidy and tetraploidy are the only cases of polyploidy found in humans (Table 5-1); they both result in spontaneous abortions or, in very rare cases, live births who die soon afterwards.

Triploidy can result from the fusion of a diploid gamete with a normal one, or from the fertilization of an oocyte by two sperms (see Plate 3). *Diploid gametes* result from failure of one of the meiotic divisions or from nuclear fusion of meiotic products (such as the oocyte nucleus and the second polar body).

Tetraploidy is most often the consequence of mitotic failure in an early zygotic division, which results in a cell with duplicated chromosome number. There are many cases of individuals who are *mosaics* for chromosomal abnormalities, meaning that some cells are normal and some carry the abnormality. In particular, tetraploid individuals who survive the longest are often mosaics of diploid and tetraploid tissues.

Aneuploidy.   We will deal here with autosomal aneuploidy; changes in sex chromosome numbers will be treated in the sections on the X- and Y-chromosomes. Autosomal monosomies are unknown, except for very rare 45,–21 spontaneous abortions. Trisomies are known for chromosomes 21 (the most frequent), 18, and 13 (for which most rare cases of live birth die in a few weeks). In 95% of Down syn-

TABLE 5-1.   Relative frequencies of chromosomal abnormalities, from a study of 1863 spontaneous abortions with chromosomal abnormalities.

Chromosome Abnormality	Percent
Trisomies (Down syndrome and others)	52
45,X (Turner syndrome)	18
Triplody	17
Tetraploidy	6
Other (mostly translocations)	7

SOURCE: Modified from Carr and Gedeon, 1977.

drome cases, the typical 47,+21 karyotype is found; 2% of Down syndrome cases are mosaics; and in the remaining 3% of cases a chromosome 21 is associated with another acrocentric by a translocation (see "Robertsonian translocations").

In approximately 80% of cases, trisomy 21 is due to chromosomal nondisjunction in female meiosis; the other 20% is due to nondisjunction in the male. The parental origin of the extra chromosome can be ascertained making use of chromosomal heteromorphisms. *Heteromorphisms* are visible differences in chromosome morphology (usually in the size of heterochromatic blocks) that have no obvious phenotypic consequence, but can be used to distinguish one homologous chromosome from another. The parental source of chromosomal abnormalities can also be determined by means of molecular markers, as we saw in earlier chapters.

Maternal age is a very significant factor in the frequency of trisomies (including 13, 18, and 21, as well as XXX and XXY) (Fig. 5-5). Trisomy 21 is 0.17% among 20-year-old mothers but rises 10-fold to 1.7% by age 45. Paternal age also has an effect, becoming significant after 55. In addition to parental age, genetic factors seem almost certain to affect the frequency of aneuploidy; but these have not yet been elucidated.

There is no good answer to the question of what proportion of human gametes carry chromosomal abnormalities. It is very difficult to study the frequency with which aneuploidies occur because most of them result in spontaneous abortions. Even studies of abortuses are incomplete because the most severe alterations may result in spontaneous abortions so early in development that they are not detected as pregnancies. For example, trisomies are known for all chromosomes except 1. Because the same mechanism that produces trisomy (nondisjunction) should also yield monosomy, there should be a sizable representation of monosomies among sponta-

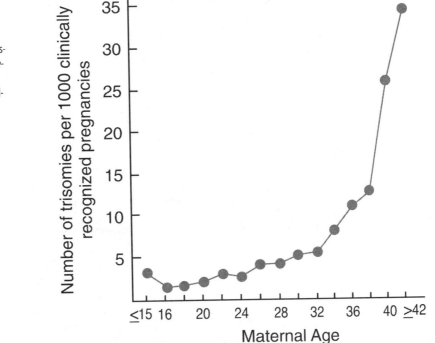

FIGURE 5-5. Incidence of trisomies among all clinically recognized pregnancies as a function of maternal age. (Redrawn with permission, from the *Annual Review of Genetics*, Volume 18, © 1984, by Annual Reviews http://www.AnnualReviews.org.)

neous abortions, yet this is not the case—only chromosome 21 monosomy has been detected. This suggests that monosomies for the other autosomes are so deleterious that they result in interrupted development before they can be studied.

There is one procedure that allows the direct detection of mutations in individual sperm. By removing the zona pellucida of the egg, it is possible to fertilize hamster oocytes with human sperm. The resulting cell is viable for a very limited period, but chromosomal condensation permits identification of the human karyotype and any abnormalities. In a sample of over 1500 sperm, 10.4% showed chromosomal abnormalities: 4.2% were aneuploidies and 6.2% were structural aberrations.

Abnormal Chromosome Structure

Chromosomal aberrations—deletions, duplications, inversions, and translocations—are the consequence of mutational DNA damage accompanied by improper repair, as if chromosomes had been cut and the wrong ends joined together. *Duplications* and, especially, *deletions* are aberrations that alter the amount of genetic material, and they are highly deleterious if the amount of material is large enough that it can be detected cytologically. *Inversions* and *translocations,* on the other hand, may be harmful or innocuous to the carriers, depending on whether the breakpoints involve specific genes or occur in intergenic regions. As we will see, even if harmless to the carriers, those rearrangements can have severe consequences for offspring.

All discussions of chromosome rearrangements in this section refer to their occurrence in whole individuals or in mosaic individuals in whom a significant portion of tissues is involved. We will not consider chromosomal aberrations in cultured cells or malignancies; these are often cell types with a great deal of chromosome damage and with very unstable karyotypes. The chromosomal rearrangements presented in this chapter are those large enough to be visible in chromosomal preparations. In Chapter 7 we will see smaller mutations, not visible at the microscope.

Deletions and duplications. The simplest form of chromosomal aberration would be a single chromosome break that results in the loss of a terminal, acentric fragment and a chromosome with a *terminal deletion*. In the absence of a telomere to ensure proper replication of the end of the DNA molecule (see Chapter 8, "Cellular Immortality and Telomeres"), the chromosome should be unstable and the deletion-bearing chromosome should be eventually lost, resulting in an inviable monosomy.

There are, however, a significant number of terminal deletions that have been described, and it is not clear that all of these can be explained by the occurrence of two breaks that leave a very small, telomeric fragment. It is possible that these are terminal deletions in which chromosomal shortening proceeds until the new end reaches a DNA sequence that can function as a de novo telomere, thus stabilizing the deleted chromosome.

Deletions involving all chromosomal arms and most regions have been observed. The most common are 4p–, 5p–, 9p–, 11p–, 11q–, 13q–, 18p–, and 18q–. Cri-du-chat syndrome, with an incidence of approximately 1 in 45,000, is the most frequently observed deletion. Associated with 5p, this syndrome is characterized by numerous abnormalities including a cat-like cry in infants. The 5p deletions may be very small or extend for as much as half the arm, with little correlation between severity of syndromes and size of the deletion. Careful study of many deletions with different breakpoints led to the identification of a region in band 5p15 as being responsible for most symptoms when it is deleted.

A special type of deletion leads to *ring chromosomes.* This occurs when breaks separate both telomeres and the new ends of the chromosomal arms link with one another to close the chromosome in a loop. Very frequently ring chromosomes are accompanied by serious developmental problems; their nature and severity depend on the specific chromosome involved and the size of the deleted fragments.

The occurrence of chromosome duplications in the form of the tandem repetition of a section of chromosome, large enough to be seen at the microscope, is very rare. When it occurs, the consequences are those of a partial trisomy, and their severity depends on the location and size of the duplication. Duplications and deletions also occur as a consequence of the unbalanced transmission of the members of a reciprocal translocation (see below).

Inversions.   *Inversions* result from two breaks on the same chromosome followed by reunion with the middle fragment rotated 180 degrees. There is no gain or loss of genetic material and usually any genetic consequences are due to the occurrence of genes at the breakpoints. If the breaks are in different arms, the inversion is said to be *pericentric.* If they involve only one arm, it is called *paracentric.*

Pericentric inversions have been observed in most chromosomes but they are not uniformly distributed. Thus, 40% involve chromosome 9, and breakpoints are concentrated in certain bands of other chromosomes—2p13, 2q21, 5q31, 6q21, 10q22, 12q13. Paracentric inversions are much less frequently described; in part this may be because they do not alter the position of the centromere and thus are harder to detect. It is frequent to find inversions carriers suffering no adverse consequences themselves, but having children with serious genetic defects much more often than the norm. This is due to the behavior of inversions in meiosis.

If the inversion is small, pairing of the inverted segments in meiosis does not take place, nor does crossing over within the inversion. With longer inversions, one of two arrangements is possible in meiotic prophase: either 1) one of the homologues forms a loop to accommodate pairing of the inverted segment so that the entire chromosome is more or less in synapsis, or 2) the inverted segments pair without forming a loop and the chromosomal ends are left unpaired (Fig. 5-6). In the case of pericentric inversions, if crossing over took place, within the inverted region, the chromatids involved would result in chromosomes with deletions and duplications for distal segments (see Fig. 5-6). In the case of paracentric inversions, the resulting products would include dicentric and acentric fragments with the concomitant irregularities in subsequent divisions.

A surprising number of inversions are quite benign both to carriers and their offspring. This is especially true of small inversions—those involving less than a third of the chromosome—probably because crossing-over in them is so rare.

Translocations.   *Reciprocal translocations* are the most common type of translocation and result from the exchange of terminal fragments of arms between two non-homologous chromosomes. Rarer, because they require three simultaneous breaks, are *insertional translocations,* in which a fragment from one chromosome becomes inserted into another. There are also complex rearrangements involving more than two chromosomes, which will not be considered here.

In the case of reciprocal translocations, if the two rearranged chromosomes are joined such that each has a centromere, and the breakpoints do not create dominant mutations in critical genes, the translocation can be quite innocuous. As in the case of inversions, in which no gain or loss or genetic material takes place, the majority of individuals with reciprocal translocations are phenotypically normal. Rare homozy-

FIGURE 5-6. **(A)** Schematic diagram of a pericentric inversion in chromosome 14. **(B)** Possible pairing figures in a heterozygote for this inversion. **(C)** The consequences of crossing over within the inverted segment. A woman heterozygote for such an inversion had an affected daughter carrying the 14q3 duplication. (Based on Trunka and Opitz, 1977.)

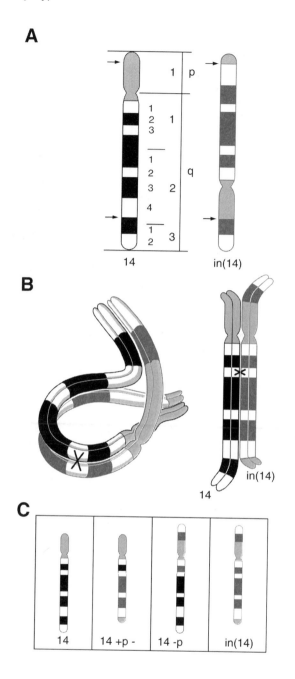

gotes, offspring of related parents, may be affected due to homozygosis of null mutations created at the breakpoints. Figure 5-7 shows the pedigree of a family in which the inheritance of neurofibromatosis type I (a dominant trait) is associated with t(17;19) (q11.2;q13.2). Southern analysis showed that the chromosome 17 break occurred in intron 23 of the NF1 gene (see Fig. 4-10).

All chromosomal arms have been seen involved in translocations, and exchanges are more or less at random, except for a few cases of great bias such as that of 11q23

FIGURE 5-7. Pedigree of a family with neurofibromatosis type I and t(17;19) (q11.2;q13.2) (see Fig. 4-10). Both traits appear in the shaded individuals. (Reproduced by permission from Fahsold R, Habash T, Trautmann U, Haustein A, and Pfeiffer RA. Familial reciprocal translocation t(17;19) (q11.2;q13.2) associated with neurofibromatosis type 1, including one patient with non-Hodgkin lymphoma and an additional t(14;20) in B lymphocytes. Hum Genet 96:65–69 [1995], Fig 1.)

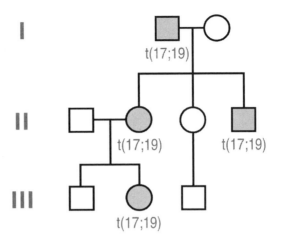

and 22q11. Nearly 80% of translocations involving 11q also involve chromosome 22. This strong preference could be due to the existence of short regions of homology that lead to ectopic pairing between chromosomes 11 and 22. (This is similar to the ectopic exchanges that produce deletions and duplications, as discussed in Chapter 7, "Errors in Recombination.")

In translocation carriers, during meiosis the chromosomes involved in a reciprocal translocation associate in groups of four—the two members of the reciprocal translocation and the two normal homologous partners (Fig. 5-8). The three possible two-and-two segregations of these chromosomes are:

Alternate. The two normal chromosomes go to one pole and the two translocated ones to the other. In this case either normal chromosomes or the balanced translocation is transmitted to viable progeny.

Adjacent I. One normal chromosome and one translocated chromosome go to each pole. In this case the future gametes carry *unbalanced translocations*—deleted for one chromosomal fragment and duplicated for another. The resulting progeny would be inviable in this particular case.

Adjacent II. This segregation also produces unbalanced and lethal translocations. The main difference from the adjacent I products is that there is nondisjunction of homologous centromeres. In general, adjacent II segregation is rare, and whether the aneuploidies are more or less severe than those produced by adjacent I depend on the position of the breakpoints. What proportion of divisions follow one or another alignment seems to depend on the specific translocation.

Robertsonian translocations. The term *Robertsonian translocation* (RT) or *Robertsonian fusion* is used to describe a whole-arm translocation between acrocentric chromosomes, in which the resulting large chromosome carries all the genetic information of the two acrocentrics and the small element with the two short arms is usually lost without consequence (Fig. 5-10). In humans, RTs are the most common type of translocation—three-fourths of them involve chromosomes 13 and 14. Their existence provides indirect evidence of the paucity of genetic information (except for the ribosomal genes) in the short arms of human acrocentric chromosomes. RTs may occur by ectopic recombination near the centromeric region. DNA sequence homologies

FIGURE 5-8.   Diagrammatic representation of the expected chromosomal association (in an equatorial plane perpendicular to the page) during meiotic prophase of t(17;19) (q11.2;q13.2) (see Fig. 5-7), and the three forms of two-and-two segregation: **(A)** alternate, **(B)** adjacent I, and **(C)** adjacent II. The boxes at the bottom represent the two products of the first meiotic division for each type of segregation. (Based on Fahsold et al., 1995.)

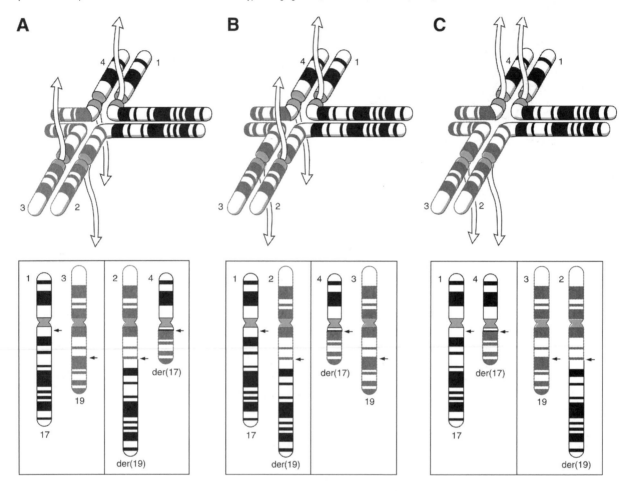

| EXAMPLE 5.1 | *Aneuploidy in a family with a reciprocal translocation* |

Figure 5-9 shows t(4q;11q) in which the fragment 11q22→ter became attached to 4q35 or 4qter. Chromosome 11 received a very small fragment (4q35→ter), if any, from chromosome 4 (Fig. 5-9A). From the synapsis of the four chromosomes in meiosis one can predict the possible outcomes (Fig. 5-9B). In the family in question (Fig. 5-9C), the mother (individual II-2) and first daughter (III-4) were normal carriers of the balanced translocation; the youngest daughter (III-7) suffered various congenital abnormalities and had the translocation chromosome der(4) and two normal chromosomes 11 (and, therefore, a duplication for 11q22→ter). Several spontaneous abortuses are suspected of having been carriers of 11q– [der(11)] (deletions for 11q22→ter). The occurrence of multiple spontaneous abortions in a family is an indicator of the possible existence of a translocation (Pihko et al., 1981).

FIGURE 5-9. Reciprocal translocation t(4q;11q). **(A)** Chromosomal constitution. **(B)** Proposed meiotic pairing. **(C)** Pedigree. (Based on Pihko et al., 1981.)

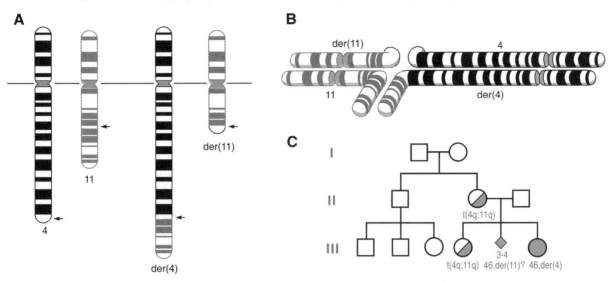

in the heterochromatic regions of acrocentric chromosomes may be important for these processes.

The second most common RTs are between chromosomes 14 and 21. The meiotic synapsis of chromosomes involved in this RT may include all three elements (see Fig. 5-10) or just the two larger chromosomes, with the small chromosome 21 free and going to one pole or another at random. In either case, aneuploidy resulting from adjacent segregation or random assortment of the unpaired chromosome results in trisomy or monosomy for the corresponding chromosome. Thus, while most cases of Down syndrome are sporadic, a significant minority of them are clustered in families with t(14;21). Another form of familial Down syndrome is due to an RT between the two chromosomes 21; such RTs give rise to an *isochromosome,* a chromosome whose two arms are homologous.

FIGURE 5-10. Robertsonian translocations/fusions. **(A)** Type of breaks (arrows) that may give rise to a RT between chromosomes 14 and 21. The breaks are indicated in the short arm of 21 and the long arm of 14. The translocation is, therefore, t(14q;21p). The centromere of chromosome 21 is preserved in the large element, 21 der, which is labeled 14q21q. **(B)** Meiotic synapsis in a heterozygote for t(14q;21p) in which the small element, 14 der, or 14p21p, has been lost.

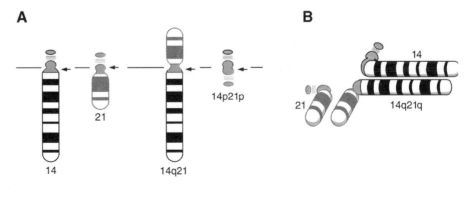

The Use of FISH to Identify Chromosomal Rearrangements

Fluorescence in situ hybridization (FISH) was discussed in Chapter 4. Plate 2 shows metaphase chromosomes from a peripheral blood study on an infant with birth defects. A routine G-banded karyotype had revealed an abnormal chromosome 3 with extra material attached to the end of the short arm. There was a suspicion that the extra material might come from chromosome 5. In this preparation, chromosomal DNA is stained red for contrast, and hybridized with a probe that is specific for chromosome 5, *chromosome painting* (yellow fluorescence). The two yellow chromosomes are two normal chromosomes 5, and the arrow points to the abnormal 3 with extra chromosome 5 material. Another preparation from the same patient was hybridized with a subtelomeric probe, a probe specific to the single sequence DNA found right next to the repeated sequences characteristic of telomeres (see Chapter 8).

As Plate 2 shows, the subtelomeric probe (green) hybridizes to both the normal and abnormal (arrow) chromosomes 3 (in this case DNA is stained blue). This indicates that there is very little if any material missing from chromosome 3, and the birth defects are caused by partial trisomy of chromosome 5. The parents of the child had normal chromosomal constitutions, indicating that this was a new mutation.

DiGeorge syndrome (DGS) is a complex birth defect caused by a small interstitial deletion in 22q that recurs with a certain frequency due to new mutations in the population. Plate 3 is of chromosomes from a DGS patient. The green fluorescent probe is used to identify chromosome 22, whereas the red probe is specific to the DGS critical region. Note that one of the chromosomes 22 (arrow) lacks red fluorescence. Notice also that the single red fluorescence spot can also be detected in the neighboring interphase cells.

SEX CHROMOSOMES

Sex determination in humans is controlled by the presence or absence of the Y-chromosome. This was made evident soon after human karyotypes could be accurately evaluated, when among the first chromosomal abnormalities detected were 45,X and 47,XXY individuals. The former, Turner syndrome, exhibits secondary sexual characteristics of a female, despite only one X; and the latter, Klinefelter syndrome, exhibits the secondary sexual characteristics of a male, despite two Xs. This shows that it is the presence of a Y rather than the number of Xs that determines sex.

Individuals with Turner syndrome have a set of distinguishing characteristics including small stature, typical facial features, and rudimentary gonadal development. Given the relative mildness of this syndrome, it is surprising that 99% of XO zygotes abort spontaneously. The incidence of 45,X children is independent of maternal age, and 80% of cases studied are the consequence of nondisjunction in male gametes.

Individuals with Klinefelter syndrome have slightly reduced IQ, tall stature, some feminization in hair distribution, and testicular atrophy. The majority of 47,XXY cases seems to be due to nondisjunction in the oocyte, and incidence is, therefore, dependent on maternal age.

The Y-Chromosome

The same size as the smallest autosomes, the Y-chromosome has a distinctive morphology; especially after quinacrine staining, it displays a very bright region in the distal half of the q arm. This heterochromatic segment may vary in length from almost absent to twice the size of the rest of the chromosome, apparently without phenotypic effect. The presence of one or more Y-chromosomes can usually be detected by Q staining in interphase cells, such as those obtained from buccal epithelium.

There seem to be very few genes on the Y. The occurrence of individuals who seemed to have Turner syndrome but had a constitution 46,X i(Yq) (i.e., an isochromosome for Yq) suggested that the male-determining factor (called TDF, testis-determining factor) is on Yp. Study of a number of XX males and XY females established that TDF is localized in the distal portion of Yp11.2. The 46,XX males usually have a small translocation of the tip of Yp to one of the Xs; they resemble individuals with Klinefelter syndrome, but they are shorter and with teeth in the size-range of females—which agrees with hypotheses that there are stature and tooth size genes on Yq. By correlating the extent of Y-translocations to phenotype—with the help of molecular and cytogenetic maps—the interval bearing the TDF was narrowed down to 35 kilobases (kb). Eventually *SRY* (sex-determining region on the Y), a gene with a testis-specific transcript, was identified as the TDF. Evidence includes the development of XX mice, transgenic for *Sry*—the homologous gene in the mouse—as males. *SRY* codes for a DNA-binding regulatory protein. Some deletions of Yp, next to the terminal heterochromatin, lead to male sterility, due to the arrest of spermatogenesis, without affecting secondary sexual characteristics. This indicates that in addition to *SRY*, there are one or more male fertility genes on the Y.

The tip of the short arm (Yp11.3) is homologous to the tip of the X-chromosome (Xp22.3). Genes such as those responsible for tooth enamel and a steroid sulfatase are present on both chromosomes, and pairing and crossing over during meiosis ensures the regular disjunction of the sex chromosomes in males. For those reasons, the segment of homology between X and Y is called the *pseudo-autosomal region* (PAR). *SRY* is located immediately proximal to the PAR.

The X-Chromosome and Dosage Compensation

There are two major aneuploidies that occur normally in mammals. One is the absence of a Y-chromosome in females; as we have seen, the effects are largely restricted to sexual differentiation because of the limited genetic content of the Y. Unlike the Y, however, the X-chromosome is quite similar to the autosomes—both in appearance and genetic content, so monosomy for any group C chromosome other than the X is totally inviable. To explain male viability in spite of such a major monosomy, geneticists postulated that a genetic mechanism had evolved to compensate for the difference in X-chromosome numbers between the sexes. It is characteristic of the Xs in females that one of them appears highly condensed and heteropyknotic during interphase (the Barr body); the Lyon hypothesis (named after Mary Lyon, the geneticist who proposed it in the early 1960s) suggested that the heterochromatic X was, in fact, inactive, so that both males and females have only one effective X.

That the Lyon hypothesis of dosage compensation is essentially correct has been demonstrated in many ways by Lyon herself and many others. The evidence includes the fact that female cats, heterozygous for coat color, show patches of the two allelic colors (calico cats). In humans, the demonstration requires more direct testing of gene products. For example, there are two alleles affecting the electrophoretic mobility of the enzyme glucose-6-phosphate dehydrogenase (G6PD); extracts from females heterozygous for them show both bands. However, cultures started from single fibroblasts of such females show either one band or the other, but not both. Studies combining such allelic detection with chromosomal rearrangements showed that the heterochromatic X is the inactive one.

EXAMPLE 5.2

Discordant monozygotic twins

The mosaic aspect of gene inactivation can be appreciated in the discordant phenotype of some MZ twins who are heterozygous for certain mutations. The family in Figure 5-11 is affected with the X-linked Fabry disease (OMIM 301500), a deficiency of α-galactosidase A that causes the lysosomal storage of certain sphingolipids in affected males. The two parents (I-1 and I-2) are genotypically normal, the first daughter (II-1) and one of the twins (II-3) are also normal, but the other twin (II-2) is symptomatic, as is the son (III-1) of the normal sister. That the twins are genetically identical was determined by testing the identity of 10 short tandem repeats in different chromosomes. Both twin daughters and the grandson carry an Asp231Asn amino acid substitution, which must have occurred de novo because all parental (generation I) alleles are normal.

It was proposed that the discordance between the twins was due to skewed X-inactivation. To test this hypothesis, the level of methylation of the androgen receptor gene was measured (see section "Identification of the inactive X"). It showed that the X present in the grandchild and the active X in the affected daughter is the paternal one (P, derived from I1), whereas the active X in the unaffected daughter is maternal (M, derived from I2). In skin fibroblasts, the ratio of activity of paternal:maternal (X-chromosome) was 100:0 in individual II-2 and 3:97 in II-3. The ratios in lymphocytes were not so skewed (Redonnet-Vernhet et al., 1996).

Many other cases of MZ female twins who are discordant because of asymmetric X-inactivation have been reported. They involve conditions such as Hunter disease, muscular dystrophy, color blindness, fragile X, and hemophilia. Actually, the frequency of symptomatic heterozygotes seems to be much higher among monozygotic twins than in the general female heterozygous population, which has led to suggestions that there may be a relationship between the processes that lead to X-inactivation and twinning.

X-inactivation occurs in the embryo in the first few days after fertilization when the embryo has a few hundred cells: in each cell one of the two Xs becomes heterochromatic and this inactivation is stably transmitted to all progeny cells. Inactivation occurs at random, so that, on average, in half the cells the active X is maternal and in half the active X is paternal—both Xs are active in gonial cells. Not all genes on the X are subject to inactivation; genes in the PAR are always active. Because of the irreversible nature of X inactivation and its clonal transmission, we can think of female

FIGURE 5-11. Family with Fabry disease, a deficiency in the enzyme α-galactosidase. The numbers represent the average of several measurements of α-galactosidase activity in arbitrary units (in hair follicles). In normal controls, averages range between four and six units. P and M indicates the source of the X-chromosome in generation I. The letter in color represents the predominantly active X. (Data from Redonnet-Vernhet et al., 1996.)

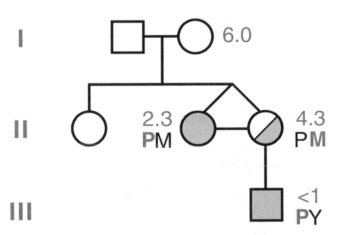

mammals as mosaics, with patches of tissue expressing the allele in one X and alternating patches expressing the allele in the other.

The expression of X-linked genes in heterozygotes. Mosaicism becomes phenotypically relevant in individuals heterozygous for deleterious mutations. First, we must note that, for X-linked genes, the notions of heterozygosity (and therefore, dominance) at the cellular level—such as the mixing of two types of subunits in a polymer (see Fig. 6-2)—do not apply, because each cell is functionally hemizygous. In the case of null mutations, the fate of cells in which the active X is the one with the mutant allele depends on the nature of the mutation. For example, in women heterozygous for hypoxanthine-guanine phosphoribosyl transferase (HPRT) deficiency—the mutation that causes Lesch–Nyhan syndrome in hemizygous males—fibroblasts are a mixture of $HPRT^+$ and $HPRT^-$ cells, depending on which X is inactivated. However, in these same women red blood cells are all $HPRT^+$, probably because of selection against $HPRT^-$ cells during hematopoiesis. At the level of the organism, the cells with enzymatic activity make up for the deficient cells, and the mutant allele usually appears recessive (as in the case of Lesch–Nyhan syndrome).

Although the average level of gene activity in heterozygotes for X-linked null mutations is pretty close to 50% of the normal level—as is the case for heterozygotes for autosomal null mutations (see Chapter 6)—the distribution around the mean value is very broad, with rare individuals showing close to the 0% and 100% levels. This heterogeneity is due to the existence of cellular patches of X-inactivation that can be large enough to affect most cells of the mutation's target organ; thus, heterozygous women may show mild symptoms of the disease they carry. Most women heterozygous for hemophilia are normal, but in a few cases the clotting process is slowed down so much as to be clinically significant. This great variability in the level of gene product in heterozygous females makes it very difficult to detect carriers of X-linked gene mutations by product measurements. See also Example 5.2.

Identification of the inactive X. The most practical methods to identify the heterochromatic X rely on the excess DNA methylation of this chromosome. One particular method makes use of sequences in the first exon of the androgen-receptor gene. There, within 100 base pairs (bp), there is a highly polymorphic CAG repeat and a cluster of *Hha*I and *Hpa*II sites. These enzymes digest GCGC and CCGG sites, respectively, but only if the Cs are not methylated. By means of flanking primers and polymerase chain reaction (PCR, Fig. 5-12), the allele in each X can be identified as maternal (M) or paternal (P) by the length of the CAG repeat; in a DNA sample from a woman, two bands would appear—one for each X. If prior to PCR the extracted DNA were digested with *Hha*I or *Hpa*II, however, only the band from the inactive X would be seen. In the absence of methylation, DNA from the active X would be cut between primer sites and PCR would fail.

The results obtained depend on the source of DNA. If the DNA were from a biopsy or blood sample, most often two bands would be seen, because one X would be inactivated in some cells and the other X in other cells. If instead the DNA were from a clonal cell line (started from a single cell), a tumor, or a patch of tissue within which all the cells have the same inactive X (as in the twins in Example 5.2), then only one band would occur. In this case it would be possible to identify whether the active X is maternal or paternal.

X-inactivation center (XIC). Studies of X-chromosome translocations and deletions showed that inactivation was possible only in those chromosomes carrying the

FIGURE 5-12. **(A)** Segment of the androgen-receptor gene indicating methylation-sensitive restriction sites, CAG repeat, and primers sites. **(B)** Example of monoclonal DNA in which we assume the paternal X (P) is inactivated—therefore methylated (Me) and resistant to digestion—and gives rise to a PCR fragment. The maternal X (M) is active, unmethylated, and unable to serve as PCR template because it is digested. M and P Xs are distinguished by the length of the CAG repeat, which affects the size of the PCR fragment. (Based on Allen et al., 1992.)

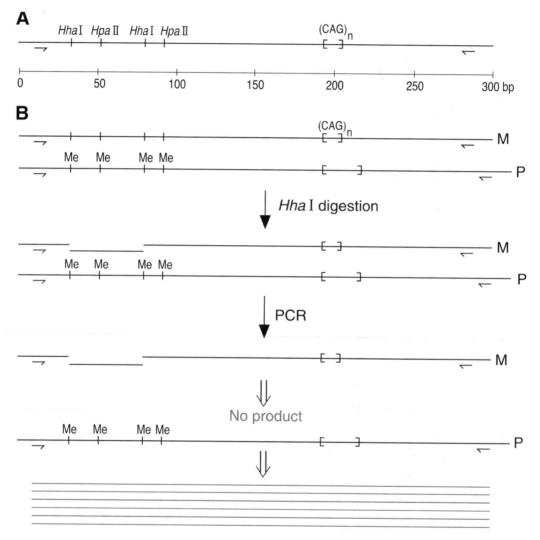

distal portion of segment Xp13. The *XIST* gene (X-inactive specific transcript) is located within the X-inactivation center (XIC) and has the unique properties of 1) being transcribed only on the heterochromatic X and 2) not coding for a protein product.

Mice X-chromosomes were engineered with a deletion of *Xist* (the mouse version of *XIST*), and it was observed that, in cells heterozygous for the deletion, only the normal X was inactivated. Transcription of *Xist* is thus required for inactivation to spread from the XIC, and it is known that the *Xist* transcript remains associated with the inactive X. Conversely, *Xist* seems sufficient for the *cis* control of inactivation because insertion of this gene into an autosome led to its behaving as a heterochromatic X.

The molecular mechanisms that bring inactivation about have not yet been elucidated, but it seems clear that inactivation is maintained by profound changes in the structure of the chromatin, including increased methylation of the DNA.

X-automosome translocations. Because X-inactivation is not autonomous for all points on the X-chromosome but rather it spreads from the XIC, the reciprocal translocation of fragments of X and autosomes has unexpected consequences. In a heterozygote for such a translocation, cells in which the normal X is active and the t(X;A) is silenced are usually selected against because of functional aneuploidy. In the X fragment with the XIC, inactivation spreads from this center out toward the fragment of autosome so that, for that piece of autosome, the cell is monosomic. Conversely, the piece of X without the XIC, which is attached to the other segment of autosome, is not inactivated so that there is double the level of normal expression for the genes in that part of the X. As a consequence, cells in which the normal X is heterochromatic and the translocation chromosomes are active predominate. This situation has consequences if the translocation happened to inactivate a gene, because, in this case, the female becomes functionally hemizygous for a null mutation. Figure 5-13 shows the translocation in a woman with Duchenne muscular dystrophy (normally a recessive trait).

THE USE OF CYTOGENETICS TO LOCALIZE GENES

The study of chromosomes has become fully integrated with molecular biology in the new field of genomics. We saw in Chapter 4 several methods that can be used to associate genes to specific chromosomes and to place them on the cytogenetic map. Those methods included in situ hybridization, hybridization to chromosome-specific libraries, and genetic linkage to previously mapped markers. There are some other ap-

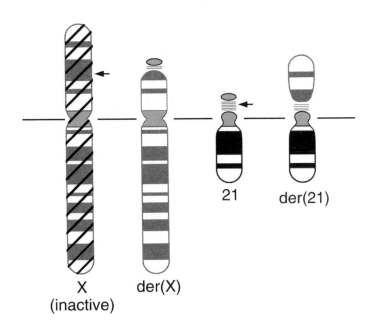

FIGURE 5-13. Partial karyotype of a woman heterozygous for t(X;21)(p21;q12). Arrows mark the position of the breakpoints. Inactivation of the normal X leads to Duchenne muscular dystrophy because the translocation disrupts *DMD*. Localization of the breakpoint on the X helped in the cytogenetic and molecular mapping of *DMD*. (Based on Bodrug et al., 1987.)

plications of chromosome studies to genome analysis, which are particularly helpful in identifying the target gene in the process of chromosome walks.

The localization of chromosomal rearrangements breakpoints was one of the methods that allowed the determination of the position of the Duchenne muscular dystrophy gene (*DMD*) on Xp21. DMD is normally a recessive condition affecting predominantly males; more than 20 female patients with DMD were observed to have translocations between the X and an autosome, and in each case the breakpoint on the X was at p21 while the autosomal breakpoint was variable. Cloning and sequencing of the breakpoint in a patient (see Fig. 5-13) showed that some of the ribosomal genes from 21q were attached to one of the *DMD* introns, thus separating several exons of *DMD* from the rest.

Deletion mapping is another method sometimes used to place a gene in the cytogenetic map. It consists of analyzing the extent of various deletions, all displaying the same phenotype; the section of the chromosome that is missing in all the deletions identifies the position of the gene in question. In the case of multiple genes located nearby, deletion mapping can be used to order the genes relative to one another.

Somatic Cell Genetics

Somatic cell genetics, in combination with molecular and cytological methods, has provided a powerful means to characterize human genes. Somatic cell genetics is based on the in vitro fusion of somatic cells, a rare event under normal culture conditions but one that can be promoted by treatment of the cells with inactivated Sendai virus or polyethylene glycol. After fusion of the plasma membranes, a transitory *heterokaryon* ensues; the hybrid cell proceeds with the cell cycle—usually taking cues from the most actively dividing of the two partners—and during the next mitotic division the two nuclei fuse to produce a *synkaryon*.

Cell fusion is a random event, so that, untreated, the culture would be a mixture of some *cell hybrids* and many unfused, *parental cells*. To obtain pure cultures of cell hybrids, selective systems have been devised that allow only the fused cells to propagate. A very common method to select against one parental type is to use fibroblasts carrying a mutation in the *HPRT* gene. The *HPRT*⁻ cells are unable to produce hypoxanthine-guanine phosphoribosyl transferase and therefore cannot use guanines recycled from the breakdown of nucleic acids (the *salvage route*). They must synthesize all nucleotides de novo to survive. De novo synthesis of nucleotides, in turn, is inhibited by the compound aminopterin. Thus, our *HPRT*⁻ fibroblasts are unable to prosper in medium with aminopterin because they have no source of nucleotides.

Another common method to select against unfused cells is to use lymphocytes as one of the parental types. In the absence of specific stimulation, lymphocytes survive for a time in culture but fail to divide. In the presence of aminopterin, a culture of *HPRT*⁻ fibroblasts and lymphocytes would have no dividing cells except for those that fused. Fused cells would have the impetus to divide from the fibroblast parent and HPRT from the lymphocyte parent. Such a culture would also need hypoxanthine (as a source of purines) and thymine (as a source of pyrimidines, also blocked by aminopterin); this is known as HAT medium.

In human genetic studies, the most frequent fusions are between human lymphocytes and mouse or Chinese hamster *HPRT*⁻ fibroblasts. Human/rodent cell hybrids are unstable in that they tend to lose human chromosomes more or less at random. This phenomenon can be used as a form of *segregation* to carry out genetic analysis. Thus, cell lines with single human chromosomes can be tested with gene-specific probes to establish linkage groups. As was discussed in Chapter 4, when the proper chromosome has been identified, those same lines can be treated with x-rays to break the chromosome in fragments and further localize the position of the gene (RH mapping).

It is not necessary to keep and test 23 hybrid cell lines, one with each human chromosome, to find the linkage group of any one probe. In theory, a panel of just five lines can be constructed that would serve the same purpose (Table 5-2). In practice more than five lines are usually necessary to cover all chromosomes. Table 5-3 shows the panel used to localize Huntington's disease to chromosome 4.

CONCLUSION

Just as methodological advances in the cloning and sequencing of genes bridged the gap between molecular genetics and Mendelian genetics, so too the chasm in the study of human inheritance at the cytogenetic and molecular levels is disappearing. Vectors with ever greater carrying capacity, more complete DNA physical maps, chromosome-specific probes used to "paint" chromosomes, and the development of fiber FISH, chromosomal purification, and other new techniques have coalesced to give rise to the field of genomics. Molecular tools have also greatly increased the power of traditional cytogenetic clinical studies, mainly through the application of FISH techniques.

EXERCISES

5-1. On average, how much DNA is contained in a chromosome G-band (assume a resolution of 900 bands)? How many β globin–size genes could fit in such a band? How many CFTR-size genes?

5-2. How can you explain an embryo in which approximately two-thirds of the cells are normal and one-third are tetraploid?

5-3. How would you explain an infant with cri-du-chat syndrome and a ring chromosome? Draw a diagram of a possible ring chromosome and identify specific bands, using Figure 5-3 as a guide.

TABLE 5-2. Imaginary panel of hybrid cells that can resolve the chromosome of any probe in a species with eight chromosomes. The + and − signs indicate the presence of any one chromosome in each line. To identify a linkage group, all lines are tested with the probe at hand. For example, if the signal is present in lines a, b, and c, the probe can only correspond to chromosome 1; if the signal is present in line b, but not in a or c, the probe can only correspond to chromosome 6. Simple inspection shows that the panel needs to be extended to four lines to resolve 16 chromosomes and to five lines to resolve 32 chromosomes.

Cell Lines	Chromosomes							
	1	2	3	4	5	6	7	8
a	+	−	+	−	+	−	+	−
b	+	+	−	−	+	+	−	−
c	+	+	+	+	−	−	−	−

SOURCE: Adapted from Ruddle and Creagan, 1975.

TABLE 5-3. Segregation of G-8 with human chromosomes in human mouse hybrids. This panel was used to establish that the Huntington's disease gene is linked to chromosome 4.

Chromosomes

Cell hybrid	G-8	1	2	3	4	5	6	7	8	9	10	11	12	13	14	15	16	17	18	19	20	21	22	X
1	■				■													▨			▨			
2								▨					▨									▨		
3			▨				▨				▨			▨			▨	▨						
4	▨				▨																			
5	▨				▨																			
6	▨																							
7		▨	▨			▨												▨						
8	▨				▨																			
9	▨				▨																			
10																	▨					▨		
11																								
12	▨				▨																▨			
13	▨	▨	▨		▨										▨									
14	▨				▨																			
15	▨				▨																			
16																								
17																								
18					▨																			

SOURCE: Based on Gusella et al., 1983.

5-4. Verify that the consequences of crossing over within a pericentric inversion are the same whether meiotic pairing involves a loop or not (see Fig. 5-6).

5-5. Do you think that the translocation t(17;19) (q11.2;q13.2), to which Figure 5-7 refers, is the cause of neurofibromatosis in that family, or is it just linked to *NF1*? Explain with a diagram of the translocation relative to the *NF1* gene.

5-6. The boxes at the bottom Figure 5-8 depict the possible chromosomal constitutions of the gametes produced by an individual who is a carrier of t(17;19) (q11.2;q13.2) (see Exercise 5-5). Indicate the phenotypes of individuals derived from each of those gamete types.

5-7. In Figure 5-9, can you tell what kind of disjunction led to the chromosomal imbalances of each of the affected family members?

5-8. Diagram the formation of an isochromosome 21 (21q21q) by means of a Robertsonian fusion and the meiotic behavior of such a chromosome. What type of progeny would a person have who carries a 21q21q chromosome?

5-9. Two sisters who have a brother with Lesch–Nyhan syndrome are heterozygous for electrophoretic variants at the G6PD locus (*Gd*). (Let us assume that their father carried *Gd*^{A} and their mother was homozygous for *Gd*^{B}.) Explain why, when blood samples are taken from the sisters, only one of the two alleles (*Gd*^{A}) shows up. (This question is loosely based on Nyhan et al., 1970.)

5-10. Active G6PD is a homodimer and the alleles A and B can be distinguished by electrophoresis. Diagram a gel with the G6PD bands produced by fibroblasts and blood samples of normal females heterozygous for the two alleles.

5-11. Contrast Example 5.2, "Discordant monozygotic twins," with the example in Exercise 5-9.

5-12. When the pedigree in Figure 1-2 was discussed in Chapter 1, female III-2 was presented as homozygous. Reconsider this interpretation in light of the phenomenon of X-inactivation.

5-13. If a human/rodent cell hybrid derived from *HPRT⁻* mouse fibroblasts and normal human lymphocytes was kept under constant selection in HAT medium, could all human chromosomes be lost? If not, what human chromosome(s) would remain in the hybrid?

REFERENCES

General references

Borgaonkar DS. (1984) Chromosomal variation in man: a catalog of chromosomal variants and anomalies, 4th ed. New York: Liss.

Therman E, and Susman M. (1993) Human chromosomes: structure, behavior and effects, 3rd ed. New York: Springer-Verlag.

Specific references

Allen CA, Zoghbi HY, Moseley AB, Rosenblatz HM, and Belmont JW. (1992) Methylation of *Hha*I and *Hpa*II sites near the polymorphic CAG repeat in the human androgen-receptor gene correlates with X chromosome inactivation. Am J Hum Genet 51:1229–1239.

Bodrug SE, Ray PN, Gonzalez IL, Schmickel RD, Sylvester JE, and Worton RG. (1987) Molecular analysis of a constitutional X-autosome translocation in a female with muscular dystrophy. Science 237:1620–1624.

Carr DH, and Gedeon M. (1977) Population cytogenetics of human abortuses. In: Hook EB and Porter IH, eds. Population cytogenetics. New York: Academic Press, 1–9.

Fahsold R, Habash T, Trautmann U, Haustein A, and Pfeiffer RA. (1995) Familial reciprocal translocation t(17;19) (q11.2;q13.2) associated with neurofibromatosis type 1, including one patient with non-Hodgkin lymphoma and an additional t(14;20) in B lymphocytes. Hum Genet 96:65–69.

Francke U. (1994) Digitized and differentially shaded human chromosome ideograms for genomic applications. Cytogenet Cell Genet 65:206–219.

Gusella JF, Wexler NS, Conneally PM, Naylor SL, Anderson MA, Tanzi RE, Watkins PC, Ottina K, Wallace MR, Sakaguchi AY, Young AB, Shoulson I, Bonilla E, and Martin JB. (1983) A polymorphic DNA marker genetically linked to Huntington's disease. Nature 306:234–238.

Hassold TJ, and Jacobs PA. Trisomy in man. (1984) Ann Rev Genet 18:69–97.

Herzing LBK, Romer JT, Horn JM, and Ashworth A. (1997) *Xist* has properties of the X-chromosome inactivation centre. Nature 386:272–275.

Jorgensen AL, Philip J, Raskind WH, et al. (1992) Different patterns of X inactivation in MZ twins discordant for red-green color vision deficiency. Am J Hum Genet 51:291–298.

Kleczkowska A, Fryns JP, and Van den Berghe H. (1987) Pericentric inversions in man: personal experience and review of the literature. Hum Genet 75:333–338.

Koopman P, Gubbay J, Vivian N, Goodfellow P, and Lovell-Badge R. (1991) Male development of chromosomally female mice transgenic for *Sry*. Nature 351:117–121.

Lyon M. (1996) X-chromosome inactivation: pinpointing the center. Nature 379:116–117.

Martin RH, Rademaker AW, Hildebrand K, et al. (1987) Variation in the frequency and type of sperm chromosomal abnormalities among normal men. Hum Genet 77:108–114.

Nyhan WL, Bakay B, Connor JD, Marks JF, and Keele DK. (1970) Hemizygous expression of glucose-6-phosphate dehydrogenase in erythrocytes of heterozygotes for the Lesch-Nyhan syndrome. Proc Natl Acad Sci USA 65:214.

Pihko H, Therman E, and Uchida IA. (1981) Partial 11q trisomy syndrome. Hum Genet 58:129–134.

Redonnet-Vernhet I, Ploos van Amstel JK, Jansen RPM, Wevers RA, Salvayre R, and Levade T. (1996) Uneven X inactivation in a female monozygotic twin pair with Fabry disease and discordant expression of a novel mutation in the α-galactosidase A gene. J Med Genet 33:682–688.

Ruddle FH, and Creagan RP. (1975) Parasexual approaches to the genetics of man. Ann Rev Genet 9:407–486.

Sinclair AH, Berta P, Palmer MS, Hawkins JR, Griffiths BL, Smith MJ, Foster JW, Frischauf A-M, Lovell-Badge R, and Goodfellow PN. (1990) A gene from the human sex-determining region encodes a protein with homology to a conserved DNA-binding motif. Nature 346:240–244.

Trisomy 21. (1990) Am J Med Genet Suppl 7.

Trunca C, and Opitz JM. (1977) Pericentric inversion of chromosome 14 and the risk of partial duplication of 14q (14q31→14qter). Am J Med Genet 1:217–228.

Yunis JJ. (1980) Nomenclature for high resolution human chromosomes. Cancer Genet Cytogenet 2:221–229.

6 How Mutant Alleles Affect the Phenotype

EFFECTS OF MUTATIONS ON THE QUALITY AND QUANTITY OF PROTEIN PRODUCTS

For supplemental information on this chapter, see Appendix Figures 4 and 5

How do mutations give rise to the corresponding phenotypes? Mutations lead to various syndromes because they alter either the quality or the quantity of specific proteins. Mutations can be classified on the basis of the type of change in the DNA molecule and the position of that change in the gene. With respect to the latter we will consider 1) mutations that affect the coding region, and 2) mutations that occur outside of the coding region. Mutations can also be classified according to their effect on the product into 1) loss-of-function mutations, and 2) gain-of-function mutations. The effect of a mutation on the phenotype—for example, whether it is dominant or recessive—is determined by the type of mutation it is, and by the type of gene prod-

---

www **Box 6.1** *Internet Sites*

The Online Mendelian Inheritance in Man (OMIM) database, courtesy of the Center for Medical Genetics (Johns Hopkins University, Baltimore) and the National Center for Biotechnology Information, National Library of Medicine (Bethesda, Maryland), is a catalog of over 10,000 human genes and genetic disorders, authored and edited by Dr. Victor A. McKusick and his colleagues at Johns Hopkins and elsewhere (1999). It was developed for the World Wide Web by the National Center for Biotechnology Information (NCBI). It can be searched by disease name or gene name. Each entry can also be accessed by a six-digit number that is provided here for the examples cited:

> http://www.ncbi.nlm.nih.gov/Omim/

The HUGO Mutation Database Initiative has a web site with links to specific gene databases where lists of all known mutations can be obtained:

> http://ariel.ucs.unimelb.edu.au:80/~cotton/mdi.htm

For lay people, there are two sources of information. One is the Directory of Genetic Support Group:

> http://members.aol.com/dnacutter/sgroup.htm

Another is the Alliance of Genetic Support Groups:

> http://info@geneticalliance.org/

---

uct affected. Thus, recessive mutations are usually loss-of-function alleles involving proteins that are not rate limiting (usually present in catalytic amounts, such as enzymes). Dominant alleles are often gain-of-function mutations in any type of gene product, or loss-of-function mutations in products needed in stoichiometric amounts (such as structural proteins), or a special type of loss-of-function allele called a dominant negative mutation.

Mutations in the Coding Region

Mutations that affect the coding region include:

1. Missense mutations that lead to amino acid substitutions. They usually result in a protein with altered properties, unless the substitution is a very conservative one (one charged amino acid for another, or one hydrophobic amino acid for another of similar size). When the substitution is nonconservative, the effect on protein function can be great or slight depending on the specific case. Amino acid substitutions are abbreviated by giving the designation of the residue in the normal protein, the position along the polypeptide chain, and the designation of the residue in the mutant allele. Thus, the sickle-cell mutation in β globin is $Glu^6 \rightarrow$

Val6, Glu6Val, or using the one-letter code for amino acids, E6V (see the table of amino acid abbreviations in the Appendix).

2. Small deletions or additions of three, or multiple of three, bases. As in the first group, the effect of the mutation is localized to one or a few amino acids.

3. Other mutations that cause more drastic changes in protein structure. These are usually truncations, such that the resulting proteins are unable to fold into stable structures and are often degraded by cellular proteases. Three types of mutations can lead to this outcome:

    A. Base substitutions resulting in stop codons that cause premature chain termination.

    B. Deletions or additions of a number of bases that are not a multiple of three. This causes a frameshift; the amino acid sequence becomes widely divergent from the normal product and with high probability very soon encounters a stop codon that results in protein termination.

    C. Mutations in the splicing signal sequences of one of the introns. Most often, donor site mutations result in the previous exon being excluded from the mature mRNA as if it were part of the neighboring intron (*exon skipping*) (Fig. 6-1B). This is thought to occur because, in the process of RNA splicing, exons seem to be recognized by the presence of an acceptor site at the 3' end and a donor site at the 5' end. If either site is mutant, the exon is not recognized as such, and splicing proceeds to the next well-defined exon (see Fig. 6-11 for specific examples, and see Chapter 3 on the limited size of mammalian exons). Acceptor site mutations also usually result in exon skipping (see Fig. 6-1C), although they may cause an intron to be retained as if it were an exon (see Fig. 6-1D). In some cases, mutations in splicing signals, or in their neighborhood, result in the activation of cryptic splicing sites, which leads to the production of defective mRNAs (see Fig. 6-1E and F). These mutant mRNAs may either include a stretch of intron or lack a segment of exon; in either case the coding triplets are usually thrown out of frame (see Figs. 6-12 and 6-13 for specific examples). Mutations that alter RNA processing often lead to prematurely degraded RNAs or mature mRNAs that fail to be properly exported into the cytoplasm. mRNA degradation is also often seen in connection with polypeptide chain-terminating mutations.

4. Another type of mutation that may lead to altered protein structure is a translocation in which two chromosomes are broken within specific genes in such a way that the amino terminal end of one protein becomes attached to the carboxy terminal end of another. Such chimeric proteins are very rare, but there are several cases of medical consequences in the development of cancer (see for example the Philadelphia chromosome in Chapter 8).

Mutations outside the Coding Region

Several types of mutations affect the level of mRNA and thus the level of protein that accumulates. These mutations occur often, but not always, *outside* of the coding region, and they usually lead to much reduced protein levels (although in some cases the phenotype is due to an increase in product levels). The protein made is usually normal.

1. Mutations in the transcribed region. Base substitutions in 5' untranslated regions (leader) or 3' untranslated regions of the mRNA may lead to reduced mRNA sta-

FIGURE 6-1. Schematic representation of the exons and introns of a gene. The asterisks represent mutations in splicing signals, and the consequences are shown by the lines that join exons indicating the splicing process.

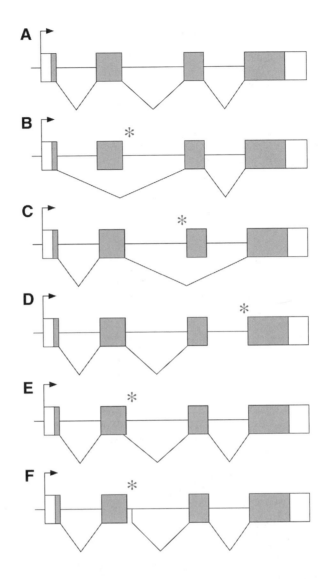

bility by facilitating the entry of mRNA into degradation pathways. Mutations in the coding region that cause premature termination of translation, and some mutations that interfere with intron splicing (as was described in the previous section) also result in very low levels of mRNA because of reduction in its half-life.

2. Mutations in the promoter region, proximal to the transcription initiation site, or in more remote enhancers. These may affect gene regulation and thus the adequate synthesis of mRNA.

Loss-of-Function Mutations

The phenotype is the result of a reduction in the amount of normal protein or the consequence of the presence of a nonfunctional protein. These *loss-of-function mutations* are sometimes called *amorphic* or "null" if they are associated with the total absence of a function. They are said to be *hypomorphic* or "leaky" if there is only a partial loss of function. They are the most common type of mutation and they are usually recessive, especially if they affect soluble enzymes. In the heterozygote there is 50% of the normal product encoded by the wild-type allele, and this is usually sufficient to produce a normal phenotype. This proportionality between the number of normal alleles and the amount of normal product is called *dosage effect*.

There are, however, also *dominant negative mutations* (also called *antimorphic mutations*). These are cases in which the mutant polypeptide lacks its normal function and, in heterozygotes, it also blocks or antagonizes the function of the normal polypeptide; examples include multimeric products in which the presence of any mutant subunit inactivates the whole. Another case of loss-of-function alleles with a dominant phenotype is found in *haplo-insufficient genes,* in which 50% of a product is not sufficient for normal function (usually found in mutations of abundant, structural proteins).

Gain-of-Function Mutations

Gain-of-function mutations, also known as *neomorphic,* are characterized by the ability of the mutant allele product to perform new functions. As would be expected, a single copy of a gain-of-function allele is sufficient to express the new function, and such mutations are typically dominant. The newly acquired function may be a consequence of structural changes in the protein product, or it may be the result of much increased levels of normal product brought about by greater mRNA or protein stability or by greater transcriptional activity. A special type of mutation that affects transcription is *translocations,* in which the enhancer of one gene is brought into proximity with the promoter of another, such that the second gene is overly activated. Such translocations have been detected repeatedly in Burkitt's lymphoma, a cancer of the lymphatic system. The translocation between chromosome 8 that contains the *Myc* oncogene and chromosomes 2, 14, or 22 that contain immunoglobulin genes brings *Myc* under the control of a powerful immunoglobulin gene enhancer, which leads to overexpression of *Myc*, one of the steps necessary for transformation of the cells in which the mutation occurred into cancerous cells. Note that these mutations occur in a somatic cell, and, although it has consequences for the individual affected, it is not a hereditary trait (see Chapter 8).

MUTATIONS IN SOLUBLE ENZYMES

Mutations in soluble enzymes were among the first in which mutant gene products could be studied. The ability to detect individual enzymes in a crude mixture of proteins through specific assays made it possible to identify genetic defects affecting soluble enzymes long before it was routine to isolate and characterize genes. These types of studies are the province of biochemical genetics.

Electrophoretic Variants

A very powerful tool that allowed the development of this type of analysis is *native* (non-denaturing) *gel electrophoresis*. This is carried out either on polyacrylamide or starch gels. A crude preparation of ground tissue is centrifuged and the supernatant is run electrophoretically on a gel under buffer and temperature conditions that prevent protein denaturation. Of the dozens of proteins on the gel, the enzyme of interest is then put in evidence by incubating the gel with appropriate substrates and precursors, which, under enzymatic catalysis, react to produce insoluble dyes at the site of migration of the enzyme under study. When a gel is thus "stained" for a given enzymatic activity (alcohol dehydrogenase, esterase, peptidase, lactate dehydrogenase, etc.), often multiple bands are visible—these are called *isozymes*.

The most obvious sources of isozyme differences are the existence of different genes whose products have different mobility but the same enzymatic activity, or the existence of two alleles of the same gene encoding variants of the same enzyme with slightly different electrophoretic mobility. Isozymes, however, can also be produced by various steps of post-translational modifications (phosphorylation, glycosylation, binding of cofactors, etc.) of only one gene product. Usually fairly simple genetic tests allow us to distinguish among these possibilities. Figure 1-13, for example, shows that individuals homozygous for any given allele of acid phosphatase show two bands, due to post-translational modifications, whereas the heterozygotes have four bands, as expected.

As was mentioned in Chapter 1, approximately 30% of proteins tested in this fashion show the existence of at least two alleles in the population (i.e., they are polymorphic) as a consequence of charged-amino-acid substitutions. In most cases these mobility alleles seem to be functionally equivalent. In heterozygotes, between two electrophoretic alleles the exact pattern of isozymes depends on the quaternary structure of the enzyme. If the enzyme is a monomer (such as the acid phosphatase seen in Fig. 1-13), then the pattern in a heterozygote is simply the sum of the patterns of the two homozygotes. If the enzyme is a homodimer, such as peptidase A, in a heterozygote with two alleles, one that produces a fast moving form of the enzyme and another that produces a slow form, the dimers can be made of two fast forms of the enzyme, two slow forms, or one fast and one slow form. This last "hybrid" dimer will have a mobility intermediate between the other two; because of the binomial distribution of probabilities, its concentration will be higher than the other two (Fig. 6-2). If the enzyme is a trimer, there are four possible subunit constitutions, and so on.

Mutations with Complete or Partial Loss of Enzyme Activity

Unlike electrophoretic variants, loss-of-function mutations usually have serious consequences for affected individuals. Alkaptonuria, a syndrome caused by the absence of a simple metabolic step, provided the first tractable human trait studied by geneticists. Enzyme deficiency syndromes affect the metabolism of amino acids, as in alkaptonuria and phenylketonuria; purines, as in the Lesch–Nyhan syndrome; sugars, as in galactosemia; and so on. These conditions were called *inborn errors of metabolism* by Garrod and have the following general characteristics:

1. They are caused by either null or leaky mutations.
2. The mutations are usually recessive, and homozygotes for the mutant allele have zero, or a small fraction of the normal level of, enzyme activity.
3. Heterozygotes display approximately 50% of the normal enzyme activity, indicating that in many of these enzymatic systems a significant reduction of enzyme activity has little or no consequence, although there are cases, such as porphyria, of haplo-insufficiency.
4. The medical consequences of these errors of metabolism are more often associated with the accumulation of high levels of toxic intermediaries than with the absence of a product, albinism and cretinism being contrary examples.

FIGURE 6-2. Electrophoretic migration of proteins synthesized in individuals heterozygous for mobility variants. **(A)** Monomeric protein showing two charges, at the origin before electrophoresis (left side) and separated into two bands, after electrophoresis. **(B)** Dimeric protein being assembled into three kinds of dimers in the cell (upper left side), and before and after electrophoresis. **(C)** Electrophoretic patterns of three individuals with different genetic constitution for peptidase A (an enzyme that hydrolyzes dipeptides into constituent amino acids). Two alleles are shown here, the fast-moving form in the first lane and the slow form in the last lane. Because the enzyme is a dimer, the sample of a heterozygote, in the middle lane, shows three bands. (Based on Harris, 1975, Fig.2.3.)

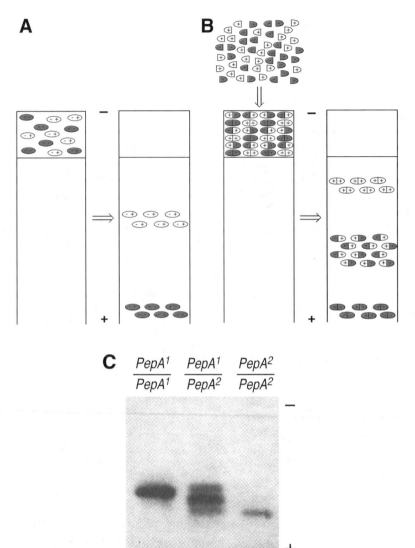

Numerous examples of inborn errors of metabolism have been described. Most of these are very rare diseases, but they illustrate that cellular processes involved in the manufacture, processing, storage, or breakdown of all the main cellular building blocks (carbohydrates, amino acids, lipids, and nucleotides) may be affected. Examples include enzymes of glycogen biosynthesis and breakdown, enzymes involved in the catabolism of amino acids and the urea cycle, and enzymes involved in the degradation of complex molecules in lysosomes.

The latter group are collectively known as *lysosomal storage diseases,* because the absence of specific hydrolases leads to the accumulation of partially processed compounds and a characteristic lysosomal morphology. Some of the substances that accumulate in various conditions include glycogen, various mucopolysaccharides and oligosaccharides, triglycerides, cholesterol derivatives, sphingolipids, and other lipids.

EXAMPLE **6.1**   *Phenylketonuria*

Phenylketonuria (PKU), an autosomal recessive disease, is caused by a deficiency in the liver enzyme phenylalanine hydroxylase (PAH), which converts phenylalanine to tyrosine. This is the normal route for the elimination of excess phenylalanine, as well as for the synthesis of tyrosine (Fig. 6-3). In the absence of PAH, the level of phenylalanine in serum and urine becomes elevated, concomitantly with the appearance of phenyl pyruvic acid and other ketonic derivatives not normally found.

In the absence of treatment, severe mental retardation is the most significant consequence of this metabolic defect. During the fetal development of PKU infants, maternal liver function keeps levels of phenylalanine down, but soon after birth they begin to rise. Newborn babies are routinely screened, a few days after birth, for abnormally high levels of blood serum phenylalanine; when the condition is encountered (approximately 1/15,000 births in European populations), the infant is placed on a restricted diet with controlled amount of phenylalanine. This treatment usually results in intellectual development within the normal range.

Note that the damage is caused by the excessive accumulation of phenylalanine derivatives rather than by a deficiency of tyrosine, which is found abundantly in the diet. Tyrosine, however, becomes an essential amino acid in patients with PKU, and it must be supplied in the low-protein diet of treated infants. See Chapter 9 for the possible risks to children of PKU mothers.

The introduction of large screening programs revealed the existence of a related condition, *mild hyperphenylalaninaemia,* in which there are slightly increased levels of serum phenylalanine but none of the severe effects of PKU, and does not require diet treatment. The gene for phenylalanine hydroxylase (*PAH*) was localized in chromosome 12, section 12q22–24; it is over 90,000 base pairs (90 kilobases) long and generates an mRNA of 2.4 kb. Over 50 different mutations have been identified in *PAH.* A few of them are amino acid substitutions that retain low levels of enzymatic activity and are associated with mild hyperphenylalaninaemia. The rest are mutations in intron splice sites and nonsense and missense mutations generally devoid of *PAH* activity and associated with PKU (OMIM 261600).

FIGURE 6-3. The amino acids phenylalanine and tyrosine, and some of their metabolic derivatives, including the pigment melanin and the thyroid hormone thyroxine. Steps blocked by color bars indicate mutations leading to several inborn errors of metabolism. Only in the absence of phenylalanine hydroxylase is excess phenylalanine oxidized to phenylpyruvic acid and other ketonic compounds. The normal elimination route for these amino acids is through the homogentisic acid pathway. In parenthesis are OMIM numbers.

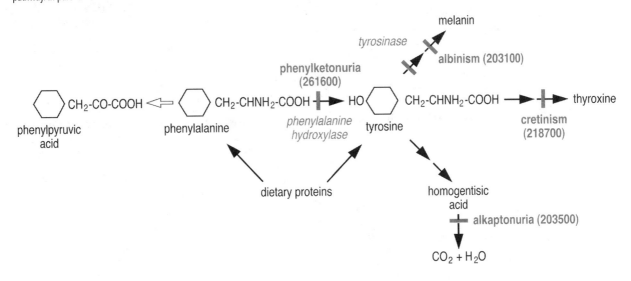

EXAMPLE 6.2

Galactosemia

Galactosemia, an autosomal recessive inborn error of carbohydrate metabolism, results from a deficiency in the enzyme galactose-1-phosphate uridyltransferase, which is necessary to convert galactose-1-phosphate to glucose-1-phosphate, a step in the metabolism of galactose. Galactose is one of the constituent monomers of the milk sugar lactose. Affected infants, if not treated, accumulate very high intracellular levels of galactose-1-phosphate; this interferes with normal carbohydrate metabolism and leads to liver hypertrophy, cataracts, and mental retardation. A strict galactose-free diet does not prevent all developmental problems, even when initiated early. It has been suggested that the residual problems might be due to deficiencies in galactose containing glycoproteins and glycolipids as a consequence of an overly strict diet.

The gene for galactose-1-phosphate uridyltransferase (*GALT*) has been localized to chromosome 9 (9p13), it is 3.9 kb in size and generates an mRNA of 1.3 kb. Over 30 mutant alleles have been described; most are amino acid substitutions that lead to enzymatic activities between 0% and 10% of normal. One variant in particular, known as Duarte, is the Asn314Asp substitution, which results in 50% activity in homozygotes and 75% in individuals heterozygous for the Duarte variant and normal allele. The Duarte variant can be quite frequent in some populations (4% to 5%), and homozygotes usually show very mild and transitory symptoms (OMIM 230400).

EXAMPLE 6.3

Porphyrias

Porphyrias are a group of diseases that result from mutations in any of the enzymes in the heme biosynthetic pathway. All porphyrias are accompanied by excretion of elevated levels of the porphyrin intermediates synthesized upstream of the step blocked by the mutation.

The acute porphyrias are characterized by sporadic episodes of severe intestinal colic and neurological disturbances. The nonacute porphyria cutanea tarda is characterized by light-sensitive dermatitis caused by circulating light-absorbing porphyrins—much remains to be learned about the immediate causes of these symptoms. The porphyrias are exceptional as enzyme deficiencies in that many of them are dominant. Affected individuals are heterozygotes for a mutation and typically have approximately 50% of the corresponding enzymatic activity (Fig. 6-4).

The penetrance of these conditions is very low, approximately 10%, and symptoms manifest themselves in adulthood and sporadically. Under normal conditions, 50% of enzyme activity in these patients appears to maintain an adequate stream of heme synthesis without the accumulation of intermediates. The appearance of symptoms is usually associated with extraneous factors, such as the administration of certain drugs or some stresses that stimulate porphyrin biosynthesis. Under conditions of increased synthesis, the partially deficient enzyme fails to keep up and the accumulation of intermediates results. Heme acts as a feedback repressor and, in some cases, symptom relief is derived from treatment with heme derivatives that inhibit synthesis of the enzyme responsible for the first step in the pathway, aminolevulinic acid synthetase, thus slowing down the biosynthetic pathway.

Obviously null mutations in the biosynthesis of heme would be recessive lethals. Several apparently homozygous patients have been observed, however; they present earlier and more severe symptoms than heterozygotes, and they have low but significant levels of enzymatic activity (see Fig. 6-4). The residual activity in homozygous patients may be the result of leaky mutations or may be due to the

FIGURE 6-4. Porphyrias. Heme biosynthetic pathway, the enzymes involved, the approximate activity level measured in patients of hereditary porphyrias, and whether the patients were homozygous or heterozygous for the mutation. In parenthesis are OMIM numbers.

Metabolic Pathway	Enzyme	Enzyme Levels in Mutants		Genetic Disease
GLYCINE + SUCCINYL CoA ↓	*ALA Synthetase*			
AMINOLEVULINIC ACID (ALA) ↓	*ALA Dehydratase*	Homozygous	12%	ALA Dehydratase Deficiency 125270
PORPHOBILINOGEN ↓	*Porphobilinogen Deaminase*	Heterozygous	50%	Acute Intermittent Porphyria 176000
HYDROXY-METHYLBILANE ↓	*Uroporphyrinogen III Synthetase*	Homozygous	15%	Congenital Erythropoietic Porphyria 263700
UROPORPHYRINOGEN III ↓	*Uroporphyrinogen III Decarboxylase*	Heterozygous Homozygous	50% 25%	Porphyria Cutanea Tarda Hepato-erythropoietic P. 176100
COPROPORPHYRINOGEN III ↓	*Coproporphyrinogen Oxidase*	Heterozygous	50%	Hereditary Coproporphyria 121300
HARDEROPORPHYRINOGEN ↓	*Coproporphyrinogen Oxidase*	Homozygous	10%	Harderoporphyria 121300
PROTOPORPHYRINOGEN IX ↓	*Protoporphyrinogen Oxidase*	Heterozygous Homozygous	50% V. low	Porphyria Variegata Early onset and fatal form 176200
PROTOPORPHYRIN (+ Fe^{++}) ↓	*Ferrochelatase*	Heterozygous Homozygous	50% 25%	Erythropoietic Protoporphyria Early onset and fatal form 177000
HEME				

combination of two different mutant alleles that happen to partially complement each other (*intragenic complementation*). For example, the most common form of variegated porphyria among South Africans of Dutch descent (a case of founder effect) is the Arg59Trp mutation in the protoporphyrinogen oxidase gene. This mutation is almost certainly homozygous inviable. However, a severely affected and apparently homozygous child was studied and, upon molecular analysis, proved to be heterozygous between Arg59Trp and another, milder mutation, Arg168Cys (see Fig. 6-4 for the OMIM references).

Gain-of-Function Mutations in Enzymes

Gain-of-function mutations are much rarer than loss-of-function alleles and involve amino acid substitutions that alter the spectrum of substrates, or perhaps change the allosteric response of an enzyme (see Example 6.4, "Familial amyotrophic lateral sclerosis").

EXAMPLE 6.4

Familial amyotrophic lateral sclerosis

Amyotrophic lateral sclerosis (ALS), also known as Lou Gehrig's disease, is a degenerative disease of motor neurons that affects 1 in 10,000 people and does not manifest itself until middle age. Approximately 10% of ALS cases are hereditary and they are due to mutations in *SOD1*, the gene for the cytosolic enzyme copper/zinc–dependent superoxide dismutase (CuZnSOD). Superoxide, the free radical $\bullet O_2^-$ with an unpaired electron, is a normal by-product of many energy exchange reactions. It is highly reactive and would be very dangerous inside the cell if not efficiently converted to H_2O_2 by the dismutase—the hydrogen peroxide is then broken down into H_2O and O_2 by catalase.

A deficiency of CuZnSOD might be expected to behave as a typical loss-of-function error of metabolism: accumulation of a precursor, superoxide, would lead to damage in the most sensitive tissues—in this case nerve cells. There were, however, several features that contradicted this simple picture. One feature was that approximately 25% of the cases of FALS are dominant; another, was that enzyme activity in patients was quite significant—

40% on average and as high as 70% in some. Sequence analysis of the mutant alleles demonstrated that all were amino-acid substitutions (and as could be inferred from activity levels, none were protein truncations or other null mutations). X-ray diffraction studies established that the amino acid substitutions were never in the active site (where they might lead to complete enzyme inactivation) but rather were in other regions of the protein, where they could still have a significant effect in protein conformation.

Finally, it was demonstrated that the mutant forms of SOD have enhanced ability to carry out peroxidation; these are reactions where, for example, the double bonds in membrane lipids are attacked by oxygen radicals. The mutant SOD also seems able to participate in the production of peroxynitrite, another very damaging radical. Thus, FALS, and the concomitant damage to motor neurons, seems to be caused not by the loss of SOD activity, but by the new function acquired by the enzyme as a result of these gain-of-function mutations (OMIM 147450).

MUTATIONS IN STRUCTURAL PROTEINS

The distinction between enzymes and structural proteins, although clear enough at the extremes of the spectrum, is quite arbitrary in many cases because there are so many structural proteins such as myosin that also have enzymatic activities. In the context of the phenotypic effect of mutations, the main point of the distinction is that enzymes occur in small amounts and mutations are significant only to the extent that they affect their catalytic activities; that is, it is the presence or absence of precursors and products of a certain metabolic reaction that defines the phenotype. Structural proteins exist in relatively large amounts, usually in multi-molecular aggregates, and the presence of significant amounts of mutant protein affects directly the organelle or molecular system of which they are part.

One group of structural proteins are those that participate in very large, quasi-crystalline aggregates of repeating units that reach macroscopic dimensions. Such is the case with the aggregation of procollagen into collagen fibers, keratin in skin, or actin and myosin in striated muscle fibers. As one might expect, mutations in these proteins are usually dominant, because if 50% of the protein participating in the aggregate is mutant (in the case of an amino acid substitution), the properties of the whole structure are affected, and if 50% of the protein is missing (in the case of a null mutation) there is likely to be haplo-insufficiency (see Example 6.5, "Osteogenesis imperfecta").

Another group of structural proteins are those that participate in the organization of more defined structural elements, the elements made up of only one or two molecules of each type. In this case, mutations are more likely to be recessive because in the heterozygote there are enough normal subunits to produce a significant number of structural elements without any mutant subunits (see Example 6.6, "Muscular dystrophies").

EXAMPLE 6.5	*Osteogenesis imperfecta*

Osteogenesis imperfecta (OI) or brittle bone disease is caused by dominant defects in collagen type I. Defective collagen leads to improper bone mineralization and easy fractures, in addition to abnormalities of skin and tendons.

Collagen fibers are aggregates of tropocollagen subunits (Fig. 6-5). Each tropocollagen subunit, in turn, consists of three polypeptide chains—two α1 chains and one α2 chain—entwined around each other to produce 300 nm by 1.5 nm cables. The helical path of the α chains brings every third residue toward the interior of the complex, and the very crowded conditions inside the triple helix allow only the smallest amino acid, glycine, to occupy that position. The α chains are synthesized as precursors, pro-α1 and pro-α2, and converted into the mature form by hydrolysis of an amino terminal peptide. The genes responsible are *COL1A1* and *COL1A2* on chromosomes 17 and 7, respectively.

OI is a syndrome with extremely variable expressivity. In its most severe form it is perinatally lethal, but its milder presentations permit full and nearly normal lives. Mutations in the procollagen gene *COL1A1* that result in substitution of one of the Gly with other residues (often Cys, Ser, or Arg) account for most cases of perinatally lethal OI, although some substitutions near the N-terminus seem to be less disruptive. Substitution of a single Gly with a bulky residue interferes with the assembly of the normal triple helix, and therefore with aggregation of the tropocollagen subunits into normal collagen fibers, even in the presence of normal polypeptide from the nonmutant allele. These are dominant negative mutations, because they prevent the formation of normal product in heterozygotes.

Fibroblasts from 23 patients with mild OI synthesized apparently normal pro-α1 peptides, but in amounts half of those in normal cells. Several such cases of OI are the result of mutations in splice donor sites that lead to exon skipping (see Fig. 6-1), sequestration of the mutant mRNA in the nucleus, and virtually no synthesis of mutant procollagen. These become effectively null mutations: patients are heterozygous, and have 50% as much pro-α1 mRNA as normal individuals. It is not clear whether the mild symptoms are due to small amounts of defective pro-α1 (a negative dominant effect) or to the imbalance in the amounts of the pro-α1 and pro-α2 polypeptides (haplo-insufficiency). Not all exon-skipping mutations are mild, as some lead to severe or lethal forms of OI, presumably because the level of mutant mRNA translated is elevated. Although there is no direct evidence on the matter, null or nearly null *COL1A1* mutations can be expected to be recessive lethals.

There are numerous mutable sites in the procollagen genes—338 sensitive Gly positions and over 100 splice sites (see *COL1A1* gene in Chapter 3)—and most cases of OI are due to de novo mutations. Thus, there is a great deal of allelic heterogeneity among OI cases and, as we have seen above, this explains most of

FIGURE 6-5. **(A)** Assembly of two α1 and one α2 molecules into a tropocollagen subunit. **(B)** The tropocollagen subunits assembled into extended collagen fibrils by cross-linking. (Based on Stryer, 1975.)

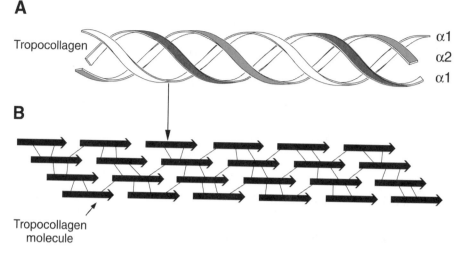

A

Tropocollagen

α1
α2
α1

B

Tropocollagen molecule

the variable expressivity observed. Probably because of this large number of mutable sites, OI mutations are relatively frequent.

The many cases of normal parents with two or more affected children had, in the past, been interpreted as instances of either recessive OI or incomplete penetrance. But molecular analyses of several such instances have shown them to be a result of one of the parents being a mosaic for one of the typical OI dominant mutations. That is to say, an OI mutation occurred during the embryonic development of the parent, such that he or she does not show the trait—or shows a very mild form. The majority of skin-, bone-, and tendon-producing cells carry the normal alleles, but the parent transmits it to the progeny if a significant fraction of the gametes happen to carry the mutant allele (OMIM 120150, 166200, and others).

EXAMPLE 6.6   *Muscular dystrophies*

Muscular dystrophies are a group of genetic diseases characterized by progressive degeneration of smooth, cardiac and skeletal muscle, and eventual complete loss of their function. The first one of the muscular dystrophies to be well understood was Duchenne muscular dystrophy (DMD), an X-linked, severe disease caused by mutations in the gene for dystrophin. Dystrophin is part of a complex that serves to transfer the mechanical force, generated by the actin–myosin complex, from the actin cytoskeleton to the basal lamina and the fibrillar matrix outside muscle cells through a rigid glycoprotein complex (Fig. 6-6). It has been proposed that in the absence of dystrophin the actin cytoskeleton is not properly anchored to the extracellular matrix and contraction of the muscle fibers leads to small tears in the plasma membrane.

Dystrophin (see Fig. 6-6) is a protein of 427 kilodalton (kd) with at least three functional regions:

1. A globular amino-terminal region that associates with cytoskeletal actin.
2. A large central region in the shape of a 125-nm rod made up of 26 repeats of a segment approximately 100 amino acids long. The repeats have only moderate similarity—10% to 25% amino acid identity—but their

FIGURE 6-6.   Structural role of the dystrophin molecule in muscle cells. The amino terminal end of dystrophin binds to actin filaments in the cortical region of muscle fibers while a region near the carboxy terminus associates with transmembrane glycoproteins. These in turn bind laminin, which is anchored in the extracellular filaments of the basal lamina. Thus, the contractile force of the actin–myosin complex (beyond the left edge of the figure) is transmitted through a rigid link to the extracellular matrix and the tendons. (Based on Koenig, et al., 1988, and Worton, 1995.)

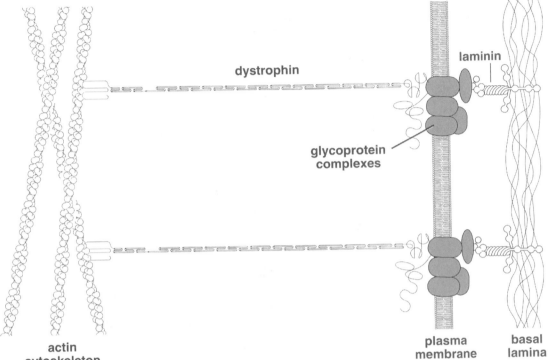

amino acid composition seems to favor the formation of α helices.

3. A carboxy-terminal globular region, a cysteine-rich portion of which is probably involved in binding to a transmembrane complex of glycoproteins, which, in turn, binds to laminin, a component of the extracellular matrix.

DMD is caused by the absence of dystrophin as a consequence of frameshift mutations, usually deletions. In-frame deletions produce partly functional protein and result in the milder form, Becker muscular dystrophy (BMD). A case was described in which more than half of the rod-like central domain was missing, yet the patient suffered a very mild form of the disease. Because the gene is X-linked, in heterozygous females only one of the two alleles is expressed (see Chapter 5), and the question of dominance at the

cellular level is further complicated by the fact that multinucleated striated muscle fibers result from the fusion of multiple myoblasts. Probably many fibers have enough dystrophin for normal function, and those that do not die and are replaced. At the phenotypic level DMD behaves as a recessive.

The dystrophin-associated glycoprotein complex (see Fig. 6-6) is composed of two groups of proteins, the dystroglycan (DG) complex (with subunits α and β) and the sarcoglycan (SG) complex (with subunits α, β, and γ). Mutations in the α, β, and γ sarcoglycan and the laminin genes lead to autosomal recessive forms of muscular dystrophy (ARMD). In general, amino acid substitutions and short in-frame deletions are responsible for milder, late-onset forms of the disease, whereas out-of-frame mutations are associated with the more severe forms. These mutations are all recessive, as would be expected given that the complex has only one subunit of each type; heterozygotes would have at least 50% normal complexes. This proportion is probably higher because many of the mutations lead to the absence of mutant subunits due to degradation of mutant mRNA (OMIM 310200, 128239, and others).

MUTATIONS IN TRANSPORT PROTEINS

The effect of mutations on transport proteins is somewhat intermediate between that in enzymes and in structural proteins. Whether that effect is closer to one or the other depends to some extent on how abundant the transport protein is. At one extreme is *hemoglobin* (Hb), the predominant protein in red blood cells. In hemoglobin, a particular amino acid substitution leads to crystalline aggregations that almost certainly would not occur if the intracellular hemoglobin concentration were lower. This mutation is β^S, which is responsible for sickle cell anemia in the homozygous condition. The mutation is partially dominant because in heterozygotes there is enough mutant protein for some aggregation to occur and cause the sickle cell trait (see Example 6.7, "Hemoglobin").

Hemoglobin also provides a good illustration of the effect of null mutations in genes for abundant proteins. Normal adult hemoglobin (Hb A) is a tetramer of the type $\alpha_2\beta_2$; each subunit bears an O_2-carrying heme group. There are several mutations that produce hemoglobin in which either the α or β subunits are unable to bind O_2; this is known as Hb M. Individuals heterozygous for Hb M are usually cyanotic due to a reduction in the O_2-transport capacity of blood, and homozygotes are inviable. Hereditary cyanosis thus is caused by an allele that is dominant for cyanosis (because of haplo-insufficiency) and a recessive lethal. Hb M mutations are null in the sense that they fail to function (carry O_2), even if the amount of protein present is normal.

Another group of null or leaky mutations are the *thalassemias,* diseases in which either the α or β chains fail to be made at all or are made in reduced quantities. Heterozygotes may have red blood cells that are smaller and with lower hemoglobin content and they may suffer mild anemia. In the case of complete absence of one of the subunits, the homozygous condition is lethal perinatally or in childhood.

In between those two extremes, there are many clinical forms of the thalassemias. This heterogeneity has two sources: one is that there are multiple copies of the globin genes and patients can have various numbers of functional genes; the other source of heterogeneity is the diversity of mutations. With respect to α thalassemias, there are two α globin genes (α1 and α2, see Figs. 3-15 and 3-16). Affected individuals can have 0, 1, 2, or 3 functional copies of this gene, and the severity of symptoms varies accordingly. As for β thalassemias, there are two β-type globin genes expressed in adults—β globin and δ globin—but β globin accounts for 97% of adult β-type globin (see Fig. 6-8). The symptoms of affected patients depend largely on the extent to which δ globin, or the fetal form γ globin, can compensate for the absence of β globin—as well as on the severity of the β globin gene mutation.

The defects in thalassemias are most often nonsense and frameshift mutations, or deletions (Fig. 6-7), all of which are null alleles, or defects in transcription or RNA processing that result in leaky alleles. The high frequency of some mutant alleles,

FIGURE 6-7. Extent of two common deletions in the α globin cluster. The color lines indicate the extent of the deletions; dashes indicate uncertainty in boundaries of the deletions. The $-\alpha^{3.7}$ is a 3.7 kb deletion that is missing one of the two α genes (it is not clear which) and the space between them; it is probably the result of unequal crossing over. Heterozygotes ($-\alpha/\alpha\alpha$) display α thalassemia 2 (Phillips, et al., 1980). $- -^{SEA}$ is a deletion of both α genes that is common in Southeast Asia; heterozygotes ($- -/\alpha\alpha$) display α thalassemia 1 (Pressley, et al., 1980). Several other deletions affecting one or more α globin genes have been described.

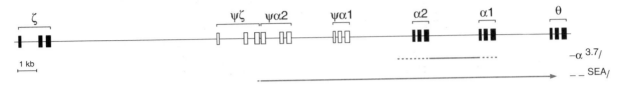

| EXAMPLE 6.7 | *Hemoglobin* |

The two subunits of hemoglobin are encoded by genes in two clusters. The α globin cluster is on chromosome 16 and the β globin cluster is on chromosome 11. The first includes three functional genes, α1 and α2, which are expressed during fetal development and in adults and ζ, which is expressed during embryogenesis (see Figs. 3-15 and 3-16). The β globin cluster consists of five functional genes: β and δ, which are expressed in adults (although δ is only 2.5% of total adult non-α globin); $^A\gamma$ and $^G\gamma$, which are expressed during fetal development, and ε, which is expressed during embryogenesis (Fig. 6-8).

The allele β^S, present in Hb S, results from an A → T base substitution that leads to the amino acid substitution Glu6Val. This residue occupies a position in the periphery of the folded protein, and Val creates a hydrophobic pocket that, in deoxygenated hemoglobin, facilitates aggregation of hemoglobin molecules into long fibers. In homozygotes for this mutation, those fibers are so prevalent that red blood cells are severely deformed, obstruct small capillaries, and break down at a faster rate than normal red blood cells. The consequence is sickle cell disease, characterized by anemia and complications resulting from the need to metabolize and dispose of the excess of heme-derivatives released by the lysed red blood cells. In heterozygotes, anemia is not evident under most conditions, but

FIGURE 6-8. Relative contribution of the various hemoglobin subunits during the course of development. Subunits of the α globin family are in color and subunits of the β globin family are in black. (Based on Bunn and Forget, 1986.)

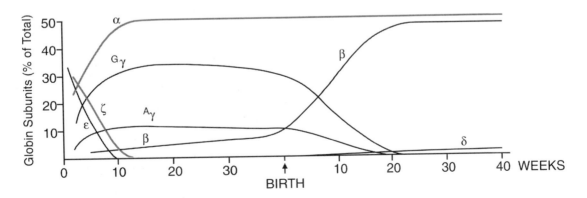

a certain amount of sickling can be observed—this is the *sickle cell trait*. Thus, the mutation in β^s is a gain-of-function mutation, although the new function, ability to polymerize, is quite deleterious.

Examples of loss-of-function mutations in hemoglobin are provided by the cases of Hb M. The heme group in each of the globin subunits is held in place by coordination between a His residue and the hemic Fe^{++}. That His, called F8, is at position 87 in α globin and at position 92 in β globin. If this His is substituted with Tyr,

the phenolic oxygen of Tyr binds to the iron atom which is then locked in the Fe^{+++} state—hence the name Hb M, for methemoglobin or ferrihemoglobin—and is thus unable to carry O_2 (Figs. 6-9 and 6-10). Mutations at other sites have a similar consequence, in particular α His58Tyr and β His63Tyr (see Fig. 6-9). Patients heterozygous for such amino acid substitutions in either the α or β chain display hereditary cyanosis, characterized by lavender-blue appearance and dark brown blood. The autosomal dominant

FIGURE 6-9. Amino acid sequence of the β chain of hemoglobin. The eight stretches of α helix are designated by the letters A through H. His92, which binds the heme group, and His63 are indicated. (Based on Vogel and Motulsky, 1997.)

BETA CHAIN

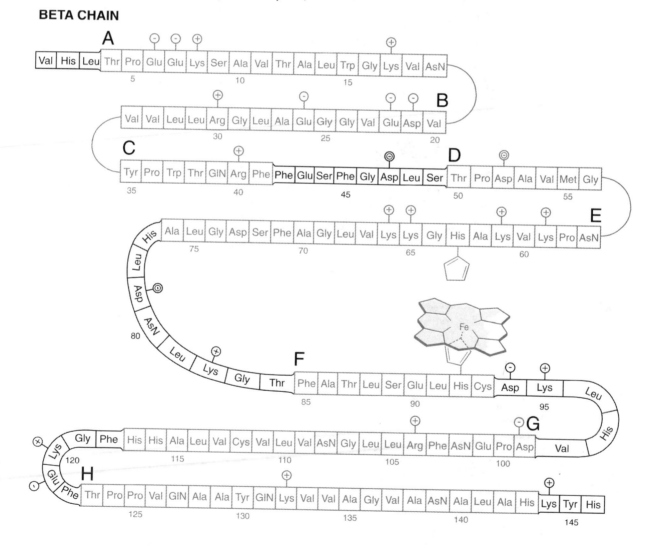

FIGURE 6-10. Tertiary structure of the β globin subunit of hemoglobin deduced from x-ray crystallography. The heme group is in color and the eight helical stretches from Figure 6-9 are indicated as straight cylinders. (Based on Perutz, 1964.)

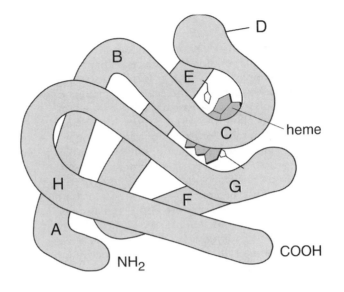

inheritance of cyanosis or *chikuro* (black blood) in a localized area of Honshu in Japan has been known since the 19th century. All these mutations are recessive lethals.

In the thalassemias, which are another group of loss-of-function mutations, the hemoglobin subunits made are normal, but there is a deficit in one or another of the subunits. Complete absence of α subunit is known as α° thalassemia; it comes about because of deletions that take out both adult α genes and is represented as (- -) (see Fig. 6-7). When homozygous (- -/- -), this condition, *hydrops fetalis,* is lethal in utero or at birth. Heterozygosis (αα/- -) is a mild condition detected usually on hematologic examination because of smaller and paler red blood cells (α *thalassemia 1*). The - - mutations are particularly prevalent in Asia. In African populations, on the other hand, the more common deletions remove only one gene (α–) (see Fig. 6-7). Heterozygous (α–/αα) are normal or only marginally detectable on hematologic examination (α *thalassemia 2*), while the homozygous (α–/α–) present the α thalassemia 1 phenotype as they have two copies of the α gene. Cases of α thalassemia are also found in which there is a combination of deletions so that only one functional α

gene remains (α–/- -). These individuals suffer chronic hemolytic anemia of variable severity, known as Hb H disease because, due to the great deficiency of α globin, the β subunits aggregate into $β_4$ or Hb H molecules. Hb H is not an efficient O_2 carrier and it tends to precipitate leading to red blood cell lysis.

The β° *thalassemias* are characterized by deficiency of the β chain. Over 97% of adult non-α subunit is the product of the single β gene, and the δ gene is responsible for the rest; thus, any null mutation of the β gene leads to virtual absence of non-α chain and severe anemia. The most common form of β° thalassemia in the Mediterranean region is the Ferrara-type, a nonsense mutation in codon 39 (CAG → UAG). In patients homozygous for this, and other nonsense mutations, bone marrow cells accumulate significant, but not normal, levels of β globin mRNA (5% to 30% of α globin mRNA) but there is practically no β globin chain synthesis. It is thought that the premature termination of translation of the mutant β globin mRNA leads to increased turnover and thus reduced levels of mRNA.

Other mutations causing β° thalassemia are those that affect RNA processing, such as G → A transitions at position 1 of introns 1 or 2, thus replacing the conserved donor site dinucleotide GT with AT. In the case of the globin gene, these donor site mutations result in two kinds of RNA products: either the skipping of exon 2, as is characteristic of other genes, or the activation of cryptic splice sites nearby that produce nonfunctional mRNA. When position 1 of intron 2 is mutant, for example, a cryptic donor site with GT at positions 48 and 49 of intron 2 is used so that the processed RNA produced includes 47 nucleotides of intron 2, and exon 3 is thrown out of frame (Fig. 6-11). There are also β° thalassemias caused by deletions.

Homozygosis for β° thalassemia is a severe condition, but unlike homozygous α thalassemia (- -/- -) anemia is not present at birth. This is because the γ genes rather than the β gene are expressed during fetal development (see Fig. 6-8). However, regular blood transfusions must be initiated within the first year. The severity of homozygous β° thalassemia is often ameliorated by elevated levels of Hb A$_2$ ($\delta_2\beta_2$) and Hb F ($\gamma_2\beta_2$). Heterozygotes for β° mutations (β thalassemia trait) display very mild or no symptoms of anemia; typically their red blood cells are smaller and paler.

There is also a milder form of β thalassemia known as β *thalassemia intermedia*. This is a heterogeneous entity that varies widely in severity. Its most common cause is the presence of leaky (β$^+$) rather than null (β°) mutations. The genotype of patients with thalassemia intermedia is often either β°/β$^+$ or β$^+$/β$^+$. The β$^+$ globin

FIGURE 6-11. Consequences of a G to A substitution in the donor site of the β globin gene intron 2. **(A)** Normal splicing. **(B)** Skipping of exon 2. **(C)** Splicing at a cryptic site 47 bp downstream from the legitimate donor site.

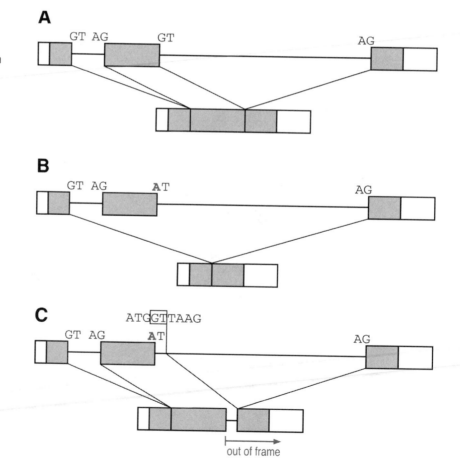

genes are characterized by the production of reduced levels of totally normal mRNA. There are numerous mutations that affect either the processing of mRNA or the regulation of transcription and result in leaky mutations. The first β⁺ mutation sequenced was a G → A base substitution at position 110 of intron 1. This generates an AG dinucleotide in a neighborhood that resembles an intron acceptor site, thus creating an alternative splicing site 19 nucleotides upstream of the normal site (Fig. 6-12). Splicing at the alternative site occurs preferentially such that only 10% of the transcript generates normal mRNA. The remaining 90% includes the extra segment and is degraded quickly, probably because of premature

termination sites encountered in the translation process.

Similarly, mutations have been described in the neighborhood of the donor site of intron 1. Figure 6-13, for example, shows mutations that generate alternative donor sites that are used in preference to the normal site. Thus, even though the normal donor site is left unaffected, only a minority of the RNA is spliced into functional mRNA.

Mutations affecting the polyadenylation signal or the promoter of the β globin gene have been described, as well as mutations that alter the expression of several genes in the β globin gene cluster and lead to the hereditary persistence of fetal hemoglobin syndrome.

FIGURE 6-12. Sequence near the 3' end of intron 1. **(A)** Normal sequence and splice acceptor site (↓). **(B)** Use of the normal splice site (↓) in the RNA bearing a mutation at position 110 of intron 1 (G → A). **(C)** Use of the newly generated splice site (↓) in the RNA bearing a mutation at position 110 of intron 1. Ninety percent of the transcripts use the new site. (Based on Westaway and Williamson, 1981.)

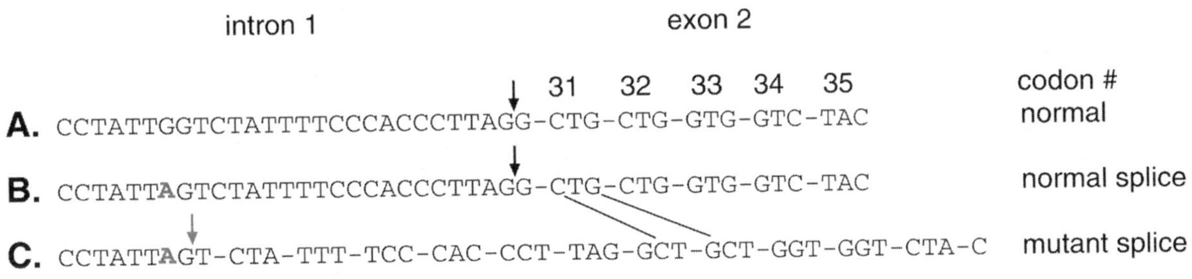

FIGURE 6-13. Three mutations toward the 3' end of exon 1 that create a favorable splicing donor site in codon 25 (↓). The splicing site in codon 30 is the normal one. (Based on Orkin, et al., 1984.) Other mutations at the 5' end of intron 1 have the similar effect of stimulating splicing at cryptic sites.

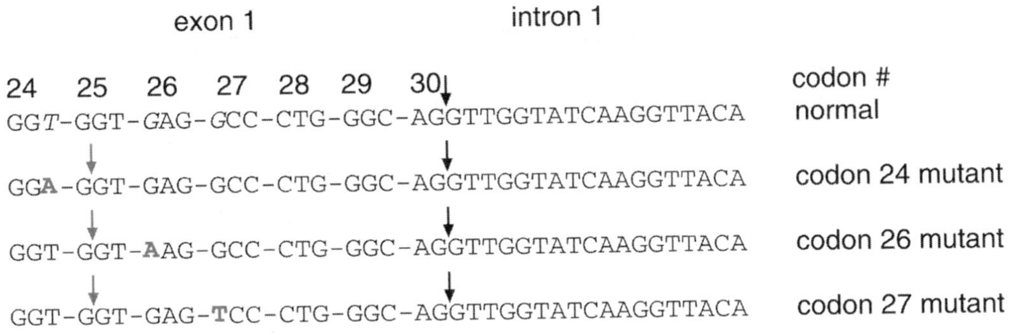

those for sickle cell anemia and thalassemias in particular, in certain human populations is due to the greater tolerance that heterozygotes for these alleles present to infection by the malaria parasite. Numerous other, rarer, mutations that affect heme-binding and other structural properties of globins have been described.

The large amounts of hemoglobin required are due to its role as a "long-distance" carrier, which necessitates that each molecule of hemoglobin be bound to an oxygen molecule for a relatively long period of time. Thus, hemoglobin is present in stoichiometric amounts. Other proteins, involved with transport over very short distances have much greater turnover numbers (the number of molecules they transport per unit time) and they are present in catalytic amounts only.

Examples of such short-distance carriers are proteins that take part in membrane transport systems, such as the cystic fibrosis transmembrane conductance regulator (CFTR). CFTR is a large protein (1480 aa) encoded by a gene on chromosome 7 and localized in the apical membrane of many exocrine cells—in sweat glands, tracheal and intestinal mucous glands, the pancreas, and so on. CFTR is a regulated chloride ion transporter: when phosphorylated, and in the presence of ATP it moves Cl⁻ across the plasma membrane (Fig. 6-14). Null mutations in this gene cause cystic fibrosis (CF), a disease characterized by abnormal mucous secretions and the formation of mucous obstructions. Exactly how a deficiency in Cl⁻ transport causes those symptoms is not clear, but water balance and defective transport of Na⁺ are thought to be involved. Heterozygotes for CF mutations are completely normal individuals; however, it has been suggested that minor changes in the properties of mucous secretions in these heterozygotes provide increased resistance to certain bacterial infections, and this would explain the high frequency of mutant alleles in many human populations (especially Northern European).

MUTATIONS IN REGULATORY PROTEINS

Regulatory proteins are involved in the transmission of a signal from some source to a final effector. They include hormones, hormone receptors, cytoplasmic kinases that

FIGURE 6-14. Schematic diagram of the CFTR protein imbedded in the lipid bilayer. The six pairs of transmembrane segments are shown in a hexagonal arrangement defining a pore. NBF, nucleotide-binding fold; Δ, site of mutation ΔF508; R, regulatory region. Also indicated are the amino (N), and carboxy (C) termini. (Based on Tsui, 1995.)

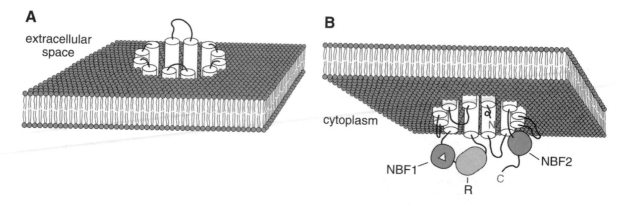

EXAMPLE 6.8

Cystic fibrosis transmembrane conductance regulator (CFTR)

The *CFTR* gene includes 27 exons that span 250 kb (see Fig. 3-14). Sequence comparisons allowed the identification of CFTR as one of the traffic ATPases, a family of prokaryotic and eukaryotic membrane proteins. CFTR seems to be organized in two halves, each with a membrane-spanning domain (which include six transmembrane segments) and a nucleotide-binding domain (where ATP is bound and, presumably, hydrolyzed). Separating the two halves there is a single regulatory domain, which blocks or permits Cl⁻ transport, depending on its phosphorylation status (see Fig. 6-14).

Analysis of the population of mRNA molecules produced by normal CFTR genes showed that a significant fraction of those molecules undergoes alternative splicing such that exons 4, 9, or 12 may be excluded from the final mRNA. In some individuals as little as 10% of the RNA seems to be correctly spliced, yet they present a normal phenotype. These results suggest that a very low level of CFTR activity is sufficient for a normal phenotype. All CF mutations are of the loss-of-function kind, and they are subdivided into four classes.

- *Class I mutations* are characterized by defective protein production. Examples of nonsense, frameshift, and splice-site mutations are all represented in this class. As usual, the defective proteins are quickly degraded and undetectable, there is no physiological function, and these are all null mutations.
- *Class II mutations* are characterized by defective processing. They are typically single amino acid deletions and missense mutations that lead to failure of proper glycosylation and localization of the protein on the cell membrane. The mutant protein is quickly degraded and there is no detectable function. One such mutation is ΔF508, the deletion of phenylalanine at position 508 that accounts for 70% of all CF mutations worldwide; in this case, the mutant protein is unable to transit out of the endoplasmic reticulum.
- *Class III mutations* are characterized by defective regulation. They are usually missense mutations in one of the nucleotide-binding domains and the resulting protein is defective in binding or processing ATP. Such mutant CFTR are unresponsive or weakly responsive to ATP, and these mutations maybe of the null or leaky category.
- *Class IV mutations* are characterized by defective conduction. They are usually leaky missense mutations. The mutant protein accumulates in sufficient amounts and is properly positioned on the membrane, but the resulting channel is not as efficient as the normal protein in Cl⁻ transport. As expected, several class IV mutations have been localized to the membrane-spanning domains, the region of the protein that is thought to form the pore lining.

Altogether, more than 300 mutant alleles of CFTR have been found and sequenced; mutations occur throughout the length of the protein and affect 20 of the 52 splice sites (OMIM 219700, 602421).

transmit a signal from the membrane to the interior of the cytoplasm or the nucleus, proteins that control gene expression by association to transcription factors, and the transcription factors themselves, which allow the RNA polymerases to recognize transcription initiation sites. At another level, the various proteases involved in blood clotting are also regulatory proteins, as they amplify the signal from a damaged blood vessel until a massive polymerization of fibrin forms a clot.

Regulatory proteins act in small amounts, and null and leaky mutations are usually recessive, as in the case of enzymes. The specific phenotype of homozygotes for such mutations depends on the physiological function missing. Thus, improper blood clotting results from mutations in clotting factors, deficient growth results from mutations in growth hormone or its associated signal transduction molecules, or poor assimilation of glucose results from mutations in insulin.

A key characteristic of regulatory proteins is that their activity must be transitory. Thus, clotting factors circulate in an inactive form, they are activated by proteolysis only when needed, and they quickly degrade when their function is completed.

Other regulatory proteins can be activated reversibly, by binding GTP, for example. After a short period of time, the same regulatory protein hydrolyzes the GTP to GDP and inorganic phosphate; this ADP-bound protein is inactive and it remains so until the GDP is again replaced by GTP (usually by action of another signal protein). This type of protein is thus multifunctional, usually with separate domains responsible for different functions. In this case, one protein domain binds and hydrolyzes GTP, while another transmits the signal when activated; thus, one domain controls the action of another through allosteric effects. Reversible activation provides an opportunity for gain-of-function, dominant mutations. In this example, a mutation that inactivates the GTPase activity of the protein would leave it activated for a much longer time period than normal, leading to a situation equivalent to the presence of excessive stimulatory signal. Some mutant oncogenes acting this way result in uncontrolled cell division and cancer (see Chapter 8 for specific examples).

CONCLUSION

In almost all cases understanding the biochemical effects of a mutation helps us to understand the causes for the phenotype produced, and very often mutations provide unique leads into the analysis of complex, multimolecular cellular processes by allowing us to study the effects of removing or modifying a single molecular component at a time. This approach has been called the *genetic dissection* of cellular processes. The enterprise is very laborious because the genes must be analyzed one at a time, and gross phenotypes often provide little guidance.

One of the goals that has been rather elusive so far is to understand the mechanistic causation of complex and quantitative traits (see Chapter 2). The best successes have been in explaining heterogeneous diseases—those in which the same or very similar symptoms are caused by different genes, as revealed by examples of complementation (see Fig. 2-3). Loss of hearing taken as a whole could be considered a complex hereditary trait because it can be congenital (present at birth) or progressive; it may occur by itself or be associated with loss of sight (Usher syndrome); in different families it can be sex-linked, autosomal, recessive, or dominant. In recent years 13 genes have been identified that can lead to this condition; their products include six membrane associated proteins, two unconventional myosins, and two transcription factors. In most cases each of these genes shows a characteristic phenotype. One of the genes (*DFNB1*) is for a *connexin*, a protein that in groups of six subunits forms small pores connecting neighboring cells (gap junctions); mutations in this gene can lead to either loss-of-function recessive alleles or dominant negative alleles. *DFNB3*, which we saw as an example in Chapter 2, encodes one of the unconventional myosins specific to cells in the inner ear. For more details, see The Hereditary Hearing Loss Homepage at http://dnalab-www.uia.ac.be/dnalab/hhh/.

EXERCISES

6-1. Review the examples described in this and previous chapters and place each mutation in the proper category in the table below. Indicate other characteristics of the mutations such as amino acid substitution, chain termination, exon-skipping, cryptic splice-site, recessive lethal, and so on. This classification is not perfect, and you may find that some mutations could fit more than one category. The term *dominant* is used to indicate that the heterozygote is distinguishable from homozygotes for the normal allele. Are there mutations or genetic variants that do not fit any of these categories? How would you describe them?

6-2. How would you apply the concept of *liability* (from Chapter 2) to the incomplete penetrance of some porphyrias?

REFERENCES

Bunn HF, and Forget BG. (1986) Hemoglobin: molecular, genetic and clinical aspects. Philadelphia: W.B. Saunders.

Cohn DH, Starman BJ, Blumberg B, and Byers PH (1990) Recurrence of lethal osteogenesis imperfecta due to parental mosaicism for a dominant mutation in a human type I collagen gene (*COL1A1*). Am J Hum Genet 46:591–601.

Cole WG, and Dalgleish R. (1995) Perinatal lethal osteogenesis imperfecta. J Med Genet 32:284–289.

Dalgleish R. (1998) The Human Collagen Mutation Database. Nucleic Acids Res 26:253–255.

Deng HX, Hentati A, Tainer JA, et al. (1993) Amyotrophic lateral sclerosis and structural defects in Cu,Zn superoxide dismutase. Science 261:1047–1051.

England SB, Nicholson LVB, Johnson MA, Forrest SM, Love DR, Zubrzycka-Gaarn EE, Bulman DE, Harris JB, and Davies KE. (1990) Very mild muscular dystrophy

TABLE 6-1. Exercise 1 table.

Types of Mutations	Examples
Loss of function	
Null (Amorphic)	_____
Recessive	_____
	_____
Dominant (Haplo-insufficient)	_____
	_____
Leaky (Hypomorphic)	_____
Recessive	_____
	_____
Dominant (Haplo-insufficient)	_____
Dominant (Antimorphic)	_____
	_____
Gain of function	_____
Recessive	_____
	_____
Dominant (Neomorphic)	_____

associated with the deletion of 46% of dystrophin. Nature 343:180–182.

Garrod AE. (1909) Inborn errors of metabolism. Oxford: Oxford University Press.

Harris H. (1975) The principles of human biochemical genetics. Amsterdam: North Holland.

Hindmarsh JT. (1993) Variable phenotypic expression of genotypic abnormalities in the porphyrias. Clin Chim Acta 217:29–38.

Ingram VM. (1957) Gene mutations in human hemoglobin: the chemical difference between normal and sickle cell hemoglobin. Nature 180:326.

Jenkins T. (1996) The South African malady. Nat Genet 13:7–9.

Koenig M, Monaco AP, and Kunkel LM. (1988) The complete sequence of dystrophin predicts a rod-shaped cytoskeletal protein. Cell 53:219–228.

Kunkel LM. (1986) Analysis of deletions in DNA from patients with Becker and Duchenne muscular dystrophy. Nature 322:73–77.

Leslie ND, Immerman EB, Flach JE, Florez M, Fridovich-Keil JL, and Elsas LJ. (1992) The human galactose-1-phosphate uridyltransferase gene. Genomics 14:474–480.

Matsumura K, Tomé FMS, Collin H, Azibi K, Chaouch M, Kaplan J-C, Fardeau M, and Campbell KP. (1992) Deficiency of the 50K dystrophin-associated glycoprotein in severe childhood autosomal recessive muscular dystrophy. Nature 359:320–322.

Moore MR. (1993) Biochemistry of porphyria. Int J Biochem 25:1353–1368.

Orkin SH, Antonarakis SE, and Loukopoulos D. (1984) Abnormal processing of beta (Knossos) RNA. Blood 64:311–313.

Orkin SH, Cheng T-C, Antonarakis SE, and Kazazian HH Jr. (1985) Thalassemia due to a mutation in the cleavage polyadenylation signal of the human beta-globin gene. EMBO J 4:453–456.

Perutz MF. (1964) The hemoglobin molecule. Sci Am 211:64–76.

Phillips JA III, Vik TA, Scott AF, Young KE, Kazazian HH Jr, Smith KD, Fairbanks VF, and Koenig HM. (1980) Unequal crossing-over: a common basis of single α-globin genes in Asians and American blacks with hemoglobin H disease. Blood 55:1066.

Pressley L, Higgs DR, Clegg JB, and Weatherall DJ. (1980) Gene deletions in α thalassemias prove that the 5'-ζ locus is functional. Proc Natl Acad Sci USA 77:3586.

Redford-Badwal DA, Stover ML, Valli M, McKinstry MB, and Rowe DW. (1996) Nuclear retention of *COL1A1*

messenger RNA identifies null alleles causing mild osteogenesis imperfecta. J Clin Invest 97:1035–1040.

Rosen DR, Siddique T, Patterson D, et al. (1993) Mutations in Cu/Zn superoxide dismutase gene are associated with familial amyotrophic lateral sclerosis. Nature 362:59–62. Erratum, Nature 364:362 (1993).

Stryer L. (1975) Biochemistry. San Francisco: W.H. Freeman.

Sykes B. (1990) Bone disease cracks genetics. Nature 348:18–20.

Tennyson CN, Klamut HJ, and Worton RG. (1995) The human dystrophin gene requires 16 hours to be transcribed and is cotranscriptionally spliced. Nature Genetics 9:184–190.

Trecartin RF, Liebhaber SA, Chang JC, Lee KY, Kan YW, Furbetta M, Angius A, and Cao A. (1981) β thalassemia in Sardinia is caused by a nonsense mutation. J Clin Invest 68:1012.

Tsui L-C. (1995) The cystic fibrosis transmembrane conductance regulator gene. Am J Respir Crit Care Med 151:547–553.

Tsui L-C, and Buchwald M. (1991) Biochemical and molecular genetics of cystic fibrosis. Adv Hum Genet 20: 153–266.

Vogel F, and Motulsky AG. (1997) Human genetics: problems and approaches. Berlin: Springer.

Weatherall DJ, Clegg JB, Higgs DR, and Wood WG. (1995) The hemoglobinopathies. In: The metabolic and molecular bases of inherited disease, vol 3. Scriver CR, et al., eds. New York: McGraw-Hill, 3417–3484.

Welsh MJ, and Smith AE. (1995) Cystic fibrosis. Sci Am 273:52–59.

Welsh MJ, Tsui L-C, Boat TF, and Beaudet AL. (1995) Cystic fibrosis. In: The metabolic and molecular bases of inherited disease, vol 3. Scriver CR, et al., eds. New York: McGraw-Hill, 3799–3876.

Westaway D, and Williamson R. (1981) An intron nucleotide sequence variant in a cloned β^+ thalassemia globin gene. Nucleic Acids Res 9:1777.

Wiedau-Pazos M, Goto JJ, Rabizadeh S, Gralla EB, Roe JA, Lee MK, Valentine JS, and Bredesen DE. (1996) Altered reactivity of superoxide dismutase in familial amyotrophic lateral sclerosis. Science 271:515–518.

Willing MC, Pruchno CJ, and Byers PH. (1993) Molecular heterogeneity in osteogenesis imperfecta type I. Am J Med Genet 45:223–227.

Worton R. (1995) Muscular dystrophies: diseases of the dystrophin–glycoprotein complex. Science 270:755–756.

7 | Mutations: Damage and Repair of DNA

Mutations, hereditary alterations of the DNA sequence, develop in a two-step process. In the first step, sometimes called *DNA damage*, *DNA lesion*, or *premutation*, the composition or structure of the DNA is modified. The second step, *fixation* of the mutation, is the reestablishment of the normal DNA structure but with a new DNA sequence that all progeny cells inherit. (Note that the expression "a mutation is fixed," used in this sense, means that a mutation will be transmitted, and not that one has been corrected.) Once a premutation occurs, whether it results in a mutation depends on which of two cellular processes gets to the damage site first, the next round of replication or a DNA repair system. However, some types of damage act as replication blocks, such that DNA synthesis does not proceed through them until the damage is repaired. Any one of several repair processes may be involved, and the usual outcome is that the normal chemical structure of the DNA is restored and the original nucleotide sequence is recovered. In the repair of some forms of damage, such as multiple double strand breaks, the chemical integrity of the DNA may be restored, but the original sequence is not.

Mutations may occur in the germ cells and be transmitted to the next generation or they may occur in a somatic cell and affect only a patch of tissue. DNA damage may have exogenous causes—mutagenic chemicals, ultraviolet (UV) light, x-rays—or endogenous causes, occurring "spontaneously" in the absence of an external insult. It is useful to consider these two sources separately.

ENDOGENOUS (OR SPONTANEOUS) DNA DAMAGE

There are three main sources of DNA damage in the normal course of biological processes: 1) instability of the DNA molecule, 2) errors in replication (DNA synthesis), and 3) errors in recombination.

Instability of the DNA Molecule

In some respects, DNA is a remarkably stable compound. It is very resistant to hydrolysis of the phosphodiester backbone—thanks to the absence of a 2' hydroxyl in deoxyribose—and it is also resistant to denaturation. But several bonds, especially in the nitrogen bases, are quite reactive. The chemical processes that are particularly relevant to human genetics are tautomeric shifts, deamination and oxidation of bases, and hydrolysis of glycosidic bonds with the concomitant loss of bases and creation of apurinic/apyrimidinic (AP) sites (Table 7-1 and Fig. 7-1). We will consider each type of damage in turn.

Tautomeric shifts. The bases T (at position C_4) and G (at C_6) have oxygen substitutions on their rings. The favored state of these oxygen atoms is in the form of carbonyl groups (C—O), which act as acceptors of protons to form H bonds (Fig. 7-2). A small fraction of molecules can be found in the unstable hydroxyl form (OH)—

TABLE 7-1. Some forms of DNA damage caused by inherent reactivities of the DNA.

Lesion	Incidence	Main Repair Mechanism
C deamination → U	100–500	uracil-DNA glycosylase
G oxidation → 8-OH-G	100–500	glycosylase
A methylation → 3-Me-A	600	3-methyladenine-DNA glycosylase
Base loss → AP	2000–10,000	AP endonuclease

Incidence is the estimated number of nucleotides damaged per day per mammalian cell. Note that these numbers are subject to varying, and large, uncertainties. They are quoted to emphasize the high frequency with which potentially mutagenic events occur.
SOURCE: Data extracted from Lindhal, 1993.

FIGURE 7-1. Schematic representation of **(A)** a segment of DNA, and **(B)** some types of lesions that may lead to mutations.

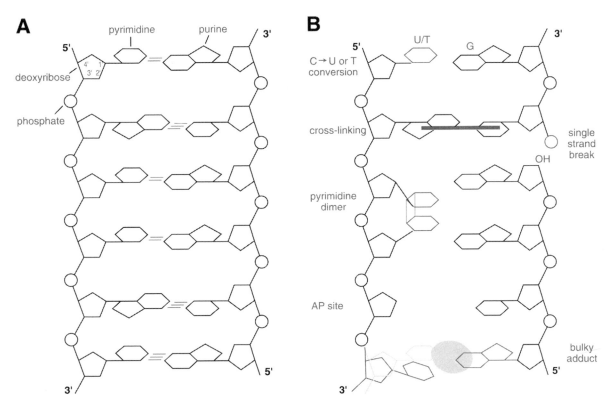

called a *tautomeric shift*—for which the proton is supplied by the neighboring N. If this transition occurs during the replication process, the shifted base will tend to mis-pair (*T•G and *G•T, where * marks the shifted base, and the symbol "C•G" is used to indicate nucleotides on complementary strands attached to each other by H-bonds). Similarly, C and A have amino groups ($-NH_2$) that may be shifted to the imino form (=NH) and thus tend to pair improperly (*C•A and *A•C). If left uncorrected these mispairs would give rise to transitions. A *transition* is a mutation in which a purine is replaced by another purine, whereas in a *transvertion* a purine is replaced by a pyrimidine, or vice versa.

Deamination. Bases subject to spontaneous hydrolytic deamination are A, G, C, and the derivative 5-methyl-cytosine (5mC). The most important, as sources of human mutations, are the deamination of C to give U, and methyl-C to give T (Figs. 7-1, 7-2, and 7-3). If left uncorrected by the repair systems, both give rise to C•G → T•A transitions (Fig. 7-4). Deamination of A and G to hypoxanthine and xanthine, respectively, occur at rates that are only 2% to 3% of the de-amination of C. Both hypoxanthine and xanthine base-pair preferentially, but not exclusively, with C so that, uncorrected, the deamination of purines may lead to transitions.

FIGURE 7-2. Schematic representation of the four DNA bases. Hydrogens attached directly to the carbon backbone have been omitted. The glycosidic bond to C1' of deoxyribose is through N1 of pyrimidines and N9 of purines.

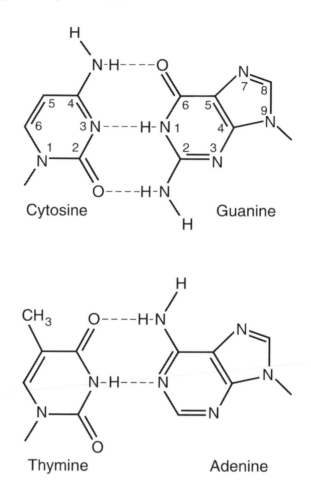

Cytosine   Guanine

Thymine   Adenine

EXAMPLE 7.1

Deamination of cytosine and 5-methylcytosine

The hydrolytic deamination of C leads to the insertion of a OH group in position 6 and its conversion to U (see Fig. 7-3), with the potentially mutagenic formation of a U•G base pair. The presence of deoxyuridine in the DNA molecule is detected by an abundant glycosylase (uracil-DNA glycosylase) which hydrolyzes the glycosyl bond, releasing U and leaving behind an AP site. The AP site is then corrected by the corresponding repair system. If DNA synthesis were to proceed before action by the glycosylase, the U•G pair would lead to a C → T transition. At the mRNA level, there would be a C → U transition if C were on the coding (nontemplate) strand or to a G → A transition if C were on the noncoding (template) strand (see Fig. 7-4).

Approximately 3% of C in mammalian DNA exists in methylated form (5mC), and deamination of 5mC leads to T (see Fig. 7-3). The T•G base pair formed is also mutagenic, but as T is a normal component of DNA there is no specific glycosylase, and repair depends on other, less efficient, repair systems. Thus, although overall damage to 5mC is less frequent than damage to C (because 5mC is rarer), 5mC sites are many times more likely to undergo mutation (mutation rate per nucleotide) than other nucleotides. As in the case of the U•G pairs, T•G pairs lead to a C → T transition or to a G → A transition (see Fig. 7-

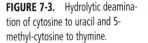

FIGURE 7-3. Hydrolytic deamination of cytosine to uracil and 5-methyl-cytosine to thymine.

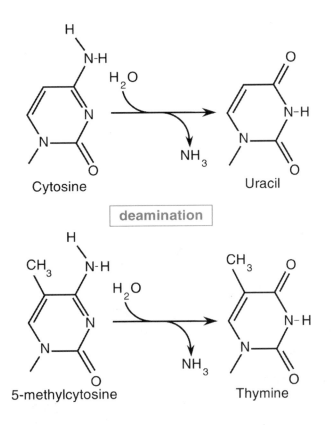

4). Most methylated C occurs in CpG dinucleotides as shown in Figure 7-4 (The symbol "CpG," is used to indicate neighboring nucleotides on the same strand and linked by a phosphate bond in the 5' → 3' direction.) That C in general and CpG dinucleotides in particular are mutational hot spots can be verified by molecular analysis of disease-causing base substitutions:

1. Of the 12 possible base substitutions, two (C → T and G → A) account for more than 50% of over 3500 mutations studied.
2. Of those base substitutions, 27% involved CpG dinucleotides and C → T or G → A transitions.

Most CpG dinucleotides are methylated, and the high mutation rate has led to their attrition in the mammalian genome; indeed less than 1% of dinucleotides are CpG, when we would expect 4% based on GC content. Two groups of CpG dinucleotides seem to be selectively preserved in the genome: one group is that of the isolated cases in coding regions that happen to encode conserved amino acids; the other group is the CpG clusters or islands found near transcription initiation sites (see Chapter 4). CpG islands at other locations in the genome are thought to be long gone during evolution, as intensive genome methylation seems to have started with the evolution of vertebrates.

Oxidation and methylation of bases. The oxidation of G to 8-hydroxyguanine by hydroxyl and superoxide radicals is the most important oxidative premutagenic lesion. 8-Hydroxyguanine pairs preferentially with A, with the potential to result in

FIGURE 7-4. Possible mutations by deamination at a methylated CpG pair.

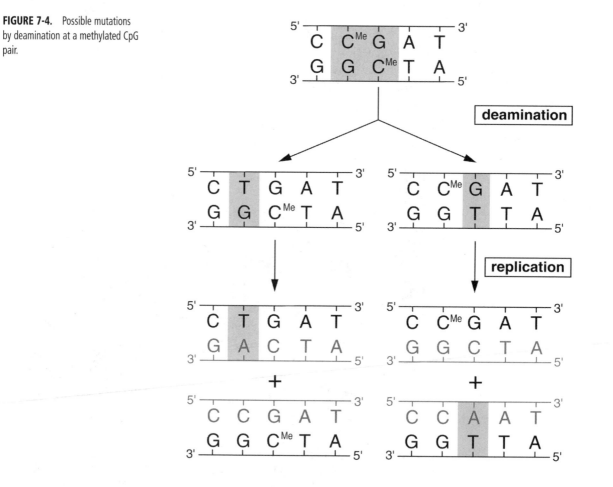

G•C → T•A transversions. Other forms of oxidative damage include the saturation of the C5–C6 double bond in pyrimidines, giving rise to a number of derivatives, including pyrimidine dimers (covalently linked neighboring bases).

Another endogenous source of DNA damage is the enzymatic cofactor S-adenosyl-methionine (SAM), which is capable of nonenzymatic methylation of bases. Of the products of SAM methylation, 3-methyladenine is the one with more serious consequences as it leads to a block in DNA synthesis.

Loss of bases. Hydrolysis of the glycosyl bond between carbon 1' and the nitrogen base occurs in native DNA much more frequently in purine than pyrimidine nucleotides, and for this reason these lesions are sometimes called *apurinic sites*. The abbreviation AP, however, stands for *apurinic/apyrimidinic site*. The consequence of unrepaired AP sites is a block in DNA replication.

Errors in Replication

DNA replication is involved in mutagenesis in two ways. One is the fixation of premutation lesions (of the type described in the previous section, and others) into base substitutions when the replication fork reaches a damaged base before repair occurs (see Fig. 7-4). Replication is also involved in creating premutation DNA damage: errors of replication include misincorporation of nucleotides leading to base substitutions, and *strand slippage* leading to small deletions and duplications. If the accuracy of base pairing were entirely dependent on the H-bonding abilities of the four bases, in the absence of the specificity provided by DNA polymerase and postreplication repair mechanisms, the misincorporation of nucleotides might be as high as 1% to 10%.

Strand slippage. Strand slippage is a type of misalignment that occurs in clusters of repeating sequences with several types of mutations as possible consequences. Figure 7-5 illustrates the production of a one-base deletion as a result of slippage in the template strand. Similar slippage of the priming strand results in base additions. If the repeating unit is longer than a single base, the possibility of intrastrand base pairing and the formation of a hairpin may stabilize the misalignment and therefore increase the chance of the deletion, or addition, of longer segments (Fig. 7-6). The best documentation of this type of mutation in humans is provided by the trinucleotide repeat expansion diseases.

FIGURE 7-5. Slippage of a single base during replication. Notice that, in the absence of a repeating sequence, a one-base displacement would be very unlikely because the two strands would be unable to form base pairs downstream of the displacement. The newly synthesized strand (color) contains 9 As in this stretch instead of 10, as in the parental strand. Another round of replication would fix this deletion.

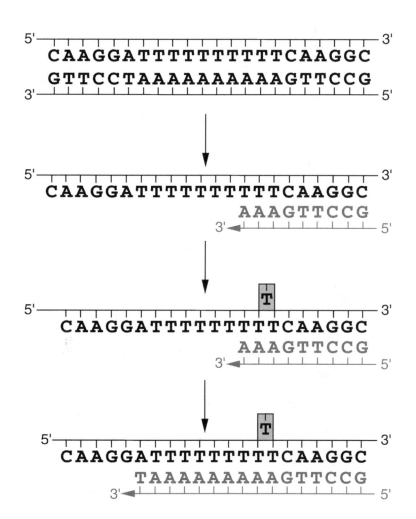

FIGURE 7-6. Slippage of a trinucleotide repeat. In this case the newly synthesized strand is the one displaced, and this leads to an expansion of the number of repeats. Although the possible formation of a hairpin is highlighted in this figure, many deletions and expansions involve only one or two repeating units, too few to form a hairpin.

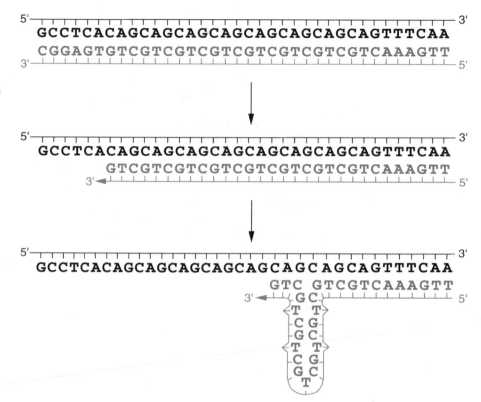

Trinucleotide repeat diseases. More than 10 loci have been identified in which expansion of trinucleotides repeats interfered with normal gene function. The location of the repeat within the gene is not constant in the various loci (Fig. 7-7 and Table 7-2). In most of these genes the number of repeats in the normal population is highly polymorphic, which suggests that mutations changing the number of repeats—both increases and decreases caused by slippage during replication (see Fig. 7-6)—occur rather frequently. Nevertheless, the mutation rate is not so high as to prevent the stable transmission of alleles within a family.

FIGURE 7-7. Ten loci in which trinucleotide expansions cause disease. The position occupied by the repeats within their genes is indicated. Other expansion sites away from genes, and without phenotype, are known. (Adapted from Warren, 1996.)

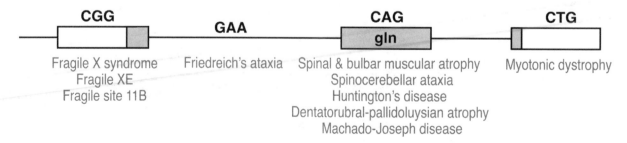

TABLE 7-2. Trinucleotide expansion sites and number of repeats.

Syndrome	Locus	NNN	Number of Repeats	
			Normal	Mutant
	In noncoding regions			
Fragile XA	FMR1 (FRAXA)	CGG	6–52	60–200 (premutation)
				230–1000 (full)
Fragile XE	FRAXE[2]	GCC	7–35	130–150 (premutation)
				230–750 (full)
None	FRAXF[3]	GCC	6–29	300–1000
None	FRA16A[4]	CCG	16–49	1000–1900
Fragile 11B	FRA11B[5] (CBL2)	CGG	11	80 (premutation)
				100–1000 (full)
Myotonic dystrophy	DMPK[6]	CTG	5–37	50–3000
Friedreich's ataxia	FRDA[7]	GAA	7–22	200–900
	In coding regions			
SBMA	AR[8]	CAG	11–33	38–66
Spinocerebellar ataxia	SCA1[9]	CAG	6–39	41–81
Huntington's disease	HDH[10]	CAG	10–35	36–121
DRPLA	B37[11] (DRPLA)	CAG	7–25	49–75
MJDI	MJDI[12] (SCA3)	CAG	12–37	61–84

Phenotypes: 1, fragile X mental retardation; 2, fragile XE mental retardation; 3,4, none; 5, predisposition toward Jacobsen (11q–) syndrome in offspring; 6, myotonic dystrophy protein kinase; 7, Friedreich's ataxia; 8, spinal and bulbar muscular atrophy (androgen receptor gene); 9, spinocerebellar ataxia type 1; 10, Huntington's disease; 11, dentatorubral-pallidoluysian atrophy; 12, Machado-Joseph disease.
1–5 display fragile sites. CAG repeats in the coding region are read in the frame encoding polyglutamine.
SOURCE: Based on Ashley and Warren, 1995.

Although increases and decreases in the number of repeats have been detected, there seems to be a slight overall tendency toward expansion. When repeat expansion reaches a certain level, however, the element becomes unstable, expansion accelerates, and it reaches the threshold at which it causes a phenotype, usually in one generation (Fig. 7-8). Those unstable intermediates are sometimes called "premutations." (Note that this use of the word should not be confused with the one employed earlier in the chapter to designate damaged DNA; the term is found in the literature with both meanings.)

FIGURE 7-8. Emergence of new TRD mutations from normal parents. Numbers below each symbol represent the number of units, in each allele, of the corresponding trinucleotide repeat. Haplotype analysis confirmed that for each of these genes the final mutational expansion always occurred in the intermediate, premutational allele (bold numbers) of the carrier parent. Note that the number of units in alleles below the premutational threshold are stable within each family. Half-shaded symbols indicate asymptomatic carriers of premutations; fully shaded symbols indicate affected individuals. **(A)** Extended family affected with fragile X syndrome. Note that expansion is greater in female carriers than in male carriers. **(B)** Family affected with myotonic dystrophy. **(C)** Family affected with Huntington's disease. The sex of individuals in generations II and III are not disclosed to ensure anonymity. (Parts A and B adapted from Caskey, et al., 1992; part C adapted from Goldberg, et al., 1993.)

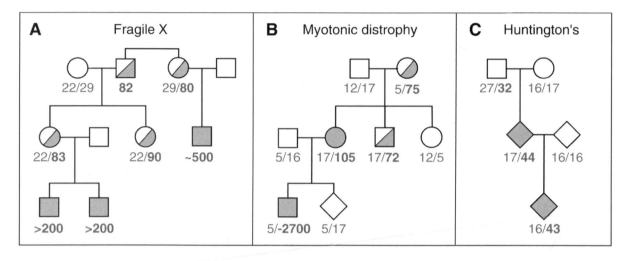

In some cases, if the number of repeats is only slightly over the threshold, symptoms may be less severe, or appear later in life. But because of the inexorable expansion of such unstable repeats, any offspring of such a patient would carry significantly larger repeats, accompanied by more severe symptoms, or earlier appearance of the disease. This peculiar aggravation of a genetic disease from one generation to the next—*genetic anticipation* or *Sherman's paradox*—was quite puzzling until the molecular basis of these mutations was discovered.

The largest group among the trinucleotide repeat diseases is that of the neuropathies associated with polyglutamine stretches (see Fig. 7-7). Two- to threefold expansion of these polyglutamine tracts seems to produce a gain-of-function, and therefore dominant, allele (see Chapter 3).

Other trinucleotide repeat mutations include 10^2- to 10^3-fold increases that cause alterations in chromosome morphology, visible gaps called *fragile sites*. The most notorious is fragile X syndrome (*FRXA*), the most frequent cause of hereditary mental retardation. Fragile sites due to expansions have been observed at other loci, some with a phenotype (*FRXE, FRA11B*) but others without a phenotype, probably located away from any genes (*FRXF, FRA16A*) (see Table 7-2).

Myotonic dystrophy (DM) is caused by a dominant mutation that consists of a trinucleotide expansion in the 3' untranslated region of a protein kinase gene. DM is a neuromuscular degenerative disease, but it may be accompanied by a host of other symptoms including balding and diabetes. The expansion interferes with processing of the mRNA from the mutant *and* the normal alleles, which explains its dominant negative characteristic. This mutation also seems to interfere with the expression of other genes, such as the insulin receptor gene.

EXAMPLE 7.2

Huntington's disease

Huntington's disease (HD) is an autosomal dominant, fully penetrant, neurodegenerative disease. This is one of the rare genetic conditions in which the heterozygous and homozygous mutant phenotypes are indistinguishable (see Fig. 1-3). HD symptoms usually start in the 30s and 40s and culminate with death 10 to 20 years later; they include progressively severe involuntary movements and mental disorders caused by neuronal attrition (see the Huntington's disease Examples in Chapters 1 and 4).

HDH, the gene responsible, has 67 exons that extend through 180 kilobases (kb); huntingtin, the corresponding protein, is 3144-amino-acids long and has no obvious sequence similarity to other proteins of known function. Seventeen codons downstream from the AUG site there is a short stretch of 10 to 30 CAG repeats encoding polyglutamine, with most alleles carrying less than 24 repeats. Expansion beyond 40 repeats is responsible for the majority of HD cases; no HD cases have fewer than 30 repeats.

There is an inverse relationship between the age of onset of the disease and the number of repeats. Individuals with 30 to 40 repeats are usually asymptomatic but are very likely to transmit further expansion to the progeny (see Fig. 7-8 C). In alleles of this intermediate range, changes in repeat lengths from one generation to the next are much greater in males than females; cell divisions in the male germ line can lead to a doubling of number of repeats, whereas in the maternal line changes are usually limited to increases or decreases of less than four repeats.

HDH has a high level of expression in neurons and several other tissues. Heterozygous for null mutations do not present any of HD symptoms, but no homozygotes for such nulls are known. Expansion of the polyglutamine tract is thought to be a gain-of-function mutation; perhaps by increasing the interaction ("stickiness") of huntingtins with other cellular components. The end result, at the cellular level, is the stimulation of apoptosis, or programmed cell death in affected neurons.

Friedreich's ataxia is caused by trinucleotide expansion in an intron that behaves as a recessive, loss-of-function, mutation. It is thought that the massive repetition of splice-site signals, $(GAA)_n$, may interfere with normal RNA processing.

Thus, as with other types of mutations, the effect of trinucleotide expansions on gene expression depends on the specific circumstances of the gene. All that seems to be in common among the trinucleotide expansion diseases is the mechanism that generates the mutations. Although most expansion mutations derive from a repeat of trinucleotides, such repeats are not the only possible substrates. Thus, expansion of a dodecamer repeat was found associated with a recessive mutation in the gene for cystatin B, which is responsible for a rare form of hereditary epilepsy, and amplification of a 33-bp repeat has been identified in a fragile site on chromosome 16 with no corresponding phenotype.

Errors in Recombination

Errors in recombination include homologous ectopic exchanges, also called unequal crossing over—resulting from pairing of repeated sequence elements that occupy nonallelic positions on the chromosome—and nonhomologous recombination. The premutation in this case corresponds to mispairing, which almost certainly leads to the formation of temporary heteroduplexes (DNA double helices in which one strand

EXAMPLE 7.3

Fragile X mental retardation

Fragile X mental retardation (FMR) is a dominant trait with incomplete penetrance (in a group of some 200 families, males showed 80% penetrance and females 30%), although, as in other cases of genetic anticipation, later generations of the pedigrees showed greater penetrance. Fragile X syndrome is due to expansion of a CGG repeat in the 5' untranslated region of *FMR1* (synonym, *FRAXA*), a gene encoding an RNA-binding protein with high levels of expression in the brain. The normal number of repeats is between 6 and 52 (most chromosomes carrying low numbers); when expansion reaches 60 repeats, it becomes irreversible, and during female meiosis expands into the mutant range producing affected males in the next generation. Males and females with expansions in the 60 to 200–copy range may show the chromosomal gap, but are asymptomatic (they are said to carry a premutation) (Figs. 7-9 and 7-10). Expansion beyond 230 copies is accompanied by methylation of the repeat and a neighboring CpG island, which seems to lead to deficient transcription.

Individuals with full mutations display mitotic instability in the number of repeats, thus clones of cultured fibroblasts from such patients often differ from one another in the number of copies. In some patients in the 200 to 300–copy range hypermethylation does not occur, but secondary structures in the mRNA, created by the repeats, lead to a block in translation and the consequent reduction of protein product.

derives from one chromosome and the other from the pairing partner) that are resolved and result in a fixed mutation when recombination-specific DNA synthesis completes the process.

Homologous ectopic exchanges seem to be the cause of many deletions and duplications. Deletions and duplications produced by unequal crossing over are usually larger than those produced by slippage during replication. *Alu* elements, ubiquitous throughout the genome, are responsible for many such ectopic exchanges. For example, most deletions in the low-density lipoprotein receptor gene (*LDLR*)—responsible for familial hypercholesterolemia (FH) in homozygotes—have breakpoints within *Alu* repeats. Figure 7-11 illustrates four such deletions and Figure 7-12 shows the proposed ectopic pairing that would lead to a deletion (or duplication) in case of crossing over.

Although *Alu* sequences are responsible for several known deletions in some genes, they seem to be less involved in others. For example, the growth hormone gene (*GH1*) has 48 *Alu* elements and deletions are frequent; most deletions, however, occur at almost identical direct repeats that flank *GH1* and not at *Alu* sites. Analysis of 10 patients showed that all carried 6.7-kb deletions, but because of haplotypic differences and diverse ethnic backgrounds, the deletions were postulated to be of independent origin. In all cases, the deletions extended to blocks of repeated sequences that flank the gene. These blocks are approximately 900 base pairs (bp) long and better than 98% identical.

Mispairing of tandemly duplicated genes, such as members of the globin and visual pigment gene families, may result in fusion genes, in addition to deletions and duplications. The hemoglobins *Lepore*, for example, are the result of unequal crossing over between the β and δ globin genes (Figs. 3-16 and 7-13). The recombination occurred in such a way that the β globin gene is replaced with a fusion gene composed of the 5' end of the δ globin gene and the 3' end of the β globin gene. Three different exam-

FIGURE 7-9. Study of DNA methylation and CGG expansion at the *FMR1* site in a family in which a son expressed a mild form of fragile X mental retardation. **(A)** Restriction map of the 5' UTR of *FMR1* (transcription is from left to right). The *Bss*HII and *Eag*I sites are at a CpG island 250-bp upstream of the CGG repeat. When the repeat is expanded, the *Bss*HII and *Eag*I sites become methylated so that the enzymes fail to cut. The arrows flanking the CGG repeat indicate PCR primers used to obtain the results shown in Figure 7-10. **(B)** Southern analysis of DNA from all family members, digested with *Eco*RI and the methylation-sensitive enzymes *Bss*HII (first seven lanes) or *Eag*I (last lane). In the absence of methylation, the *Eco*RI fragment is cut into two fragments of 2.4 kb and 2.8 kb, respectively. The pattern is shown by the normal males I-1 and II-1. In a normal female (III-1) the inactivated X is methylated (see Chapter 5), so half the sample is cut at the *Bss*HII site, yielding the 2.4-kb and 2.8-kb fragments, but the other half is not, producing a 5.2-kb fragment. The carrier females I-2, II-2, and II-3 have a similar pattern, but the fragments containing the triplet repeat (those corresponding to 2.8-kb and 5.2-kb fragments in the normal chromosomes) are slightly longer in one of the chromosomes, approximately 3.0 kb and 6.0 kb. In the affected male, the repeat-containing fragment has shifted in size to 3.6 kb. What is unusual about this patient is that so much of the DNA is digested by *Bss*HII or *Eag*I, which indicates that most of it is not methylated. Approximately 10% of III-2's DNA is methylated at this locus, and it appears as a 6.0-kb band representing the *Eco*RI fragment uncut by *Bss*HII or EagI. The low level of methylation probably explains the mildness of symptoms. (Redrawn with permission from Fig.1A in Feng, Y., Zhang, .F, Lokey, L.K., et al. Translational suppression by trinucleotide repeat expansion at *FMR1*. Science 268:731–734 [1995]. Copyright 1995 American Association for the Advancement of Science.)

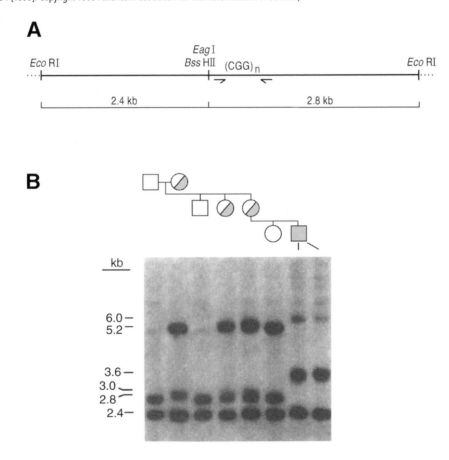

ples are known in which various amounts of each gene are present, indicating the position within the gene where crossing over took place.

Similarly, unequal crossing over may occur among the visual pigment genes (see Fig. 3-19) in such a way that the single red pigment gene is replaced with a fusion gene including part of the red pigment gene and part of a green pigment gene. The severity of the defect depends on how much of the red pigment gene is replaced.

FIGURE 7-10. The same family as in Figure 7-9, autoradiogram of PCR products obtained with primers flanking the CGG repeat. (The original image was foreshortened for clarity.) The size of fragments and number of repeats are indicated. The affected male III-2 displays a 10-fold expansion in number of copies, whereas the carrier females have intermediate numbers. In all lanes, there are multiple bands at each position due to artifacts of amplification, the polymerase failing to complete some runs. In male III-2 the heterogeneity is much greater, extending from 150 to over 400 repeats; this is due to mosaicism caused by mitotic instability. (Redrawn with permission from Fig.1B in Feng, Y., Zhang, F., Lokey, L.K., et al. Translational suppression by trinucleotide repeat expansion at *FMR1*. Science 268:731–734 [1995]. Copyright 1995 American Association for the Advancement of Science.)

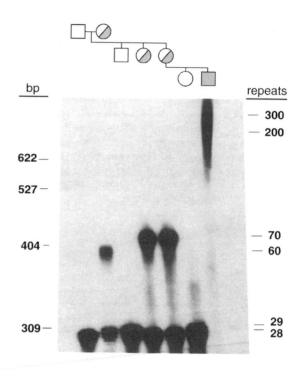

FIGURE 7-11. Deletions in the low-density lipoprotein receptor gene in two regions of the gene (exons 4–6 and 13–18), showing the position of exons and *Alu* sites (arrows) and the extent of four deletions. Note that some deletions involve elements in opposite orientation, which suggests some type of intrastrand recombination. Some deletions involve only one *Alu* element. (Adapted from Lehrman, et al., 1987.)

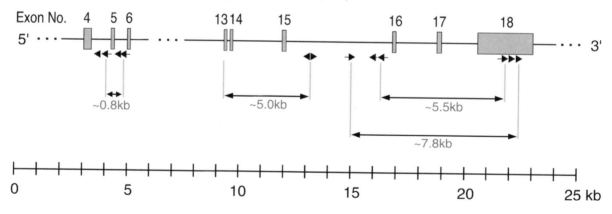

Unequal crossing over also explains some duplications, such as that of the γ globin genes (see Fig. 3-16), which occurred before the evolutionary split of the Old and New World monkeys. Crossing over in long interspersed repeats, *L1*, situated approximately 2 kb upstream and 2.5 kb downstream of the γ globin genes, gave rise to this duplication.

Ectopic pairing can occur even when the repeated sequences are quite distant from each other. Charcot-Marie-Tooth (CMT) disease type 1A is a hereditary peripheral neuropathy caused in most patients by a duplication of the peripheral myelin protein-22 gene (*PMP22*). Flanking the *PMP22* gene are two repeats of 24 kb in size, separated

FIGURE 7-12. Proposed pairing in the *LDLR* region. **(A)** Region of exons 13–18 of *LDLR*. Some restriction sites are indicated to identify segments of homology. For simplicity only two of the *Alu* sites are shown. **(B)** The same region showing ectopic pairing between two *Alu* elements in nonhomologous positions. The intervening segments loop out, as they are matched with nonhomologous sequences, thus allowing the rest of the chromosome to pair properly. A crossing over at the synapsed *Alu* sites would result in the deletion of 7.8 kb shown in Figure 7-11 or a corresponding duplication. (Based on data from Lehrman, et al., 1987.)

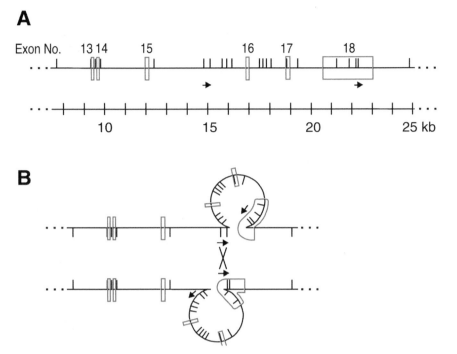

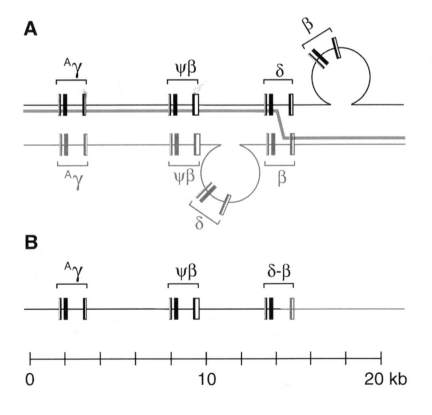

FIGURE 7-13. Unequal crossing over in the β globin cluster that produces Lepore hemoglobin.

by 1.5 megabase of DNA. The existence of this duplication, as well as the reciprocal deletion, responsible for a milder neuropathy (hereditary neuropathy with liability to pressure palsies, HNPP), was elegantly demonstrated by fiber FISH (see Chapter 4). Plate 1 shows chromatin fibers of the chromosome 17p11.2–12 region hybridized to a *PMP22* probe (red) and a probe to the flanking repeats (green). Plate 1A is a sample from a normal individual, 1B is from a CMT patient (two *PMP22* genes and three copies of the repeats), and 1C is from a HNPP patient (with only one repeat and no *PMP22* gene).

Insertional Mutagenesis

As was described in Chapter 3, another source of DNA damage that results in mutations is the insertion of transposable elements into structural genes. Several examples were described in that chapter. The same transposition process first creates a discontinuity in the DNA molecule and then mends it, fixing the mutation. Insertional mutagenesis is most often endogenous, in that the agents are the ubiquitous *Alu* and *L1* elements, but it could also be exogenous if it were caused by an infecting virus.

EXOGENOUS (OR ENVIRONMENTAL) DNA DAMAGE

Mutagenic agents are usually divided into physical and chemical. Some environmental agents increase the incidence of damage that occurs spontaneously, whereas others have unique and specific effects.

Ionizing Radiation

X-rays and γ-rays represent the very short end of the wavelength spectrum of electromagnetic radiation. They interact with water and biological molecules to create reactive ions and free radicals. It is estimated that one-third of the damage caused by ionizing radiation is due to direct interaction with the DNA molecule; the rest seems mostly due to indirect damage caused by hydroxyl radical attacks on the DNA. Ionizing radiation generally causes oxidative damage. Damage to bases may lead to base substitutions or a blockage of the replication process. Ionizing radiation damage to the sugar-phosphate backbone of DNA is less frequent than damage to bases, but it is still highly significant because it leads to single- and double-strand breaks and gross chromosomal aberrations. The Earth has always been exposed to ionizing radiation, but modern diagnostic and therapeutic methods also have become a significant source of potential exposure.

Ultraviolet Radiation

UV light is electromagnetic radiation, with wavelength between 100 and 400 nm. It is nonionizing and is significant as a mutagen mainly in the wavelength around 254 nm, the absorption peak of DNA nitrogen bases. Sunlight is the major source of UV radiation; although absorption of wavelengths below 320 nm by stratospheric ozone obviates the vast majority of the problems that sunlight might pose for humans, its premutagenic action is quite significant. Damage is probably due to a combination of the small amount of 254-nm light that reaches the surface of the Earth, and the attenuated but non-zero efficiency of longer wavelengths in causing DNA damage.

FIGURE 7-14. Cyclobutane pyrimidine dimer. Adjacent pyrimidines are covalently joined through their carbons C5 and C6. The extent of helix deformation and the ability to maintain H-bonds depend on the specific bases involved.

Thymine dimer

As opposed to the damage caused by ionizing radiation, UV light's effects result from direct interaction with DNA. UV light approaching 254 nm causes adjacent pyrimidines in DNA to form covalent bonds (Figs. 7-1 and 7-14). At least some of these pyrimidine dimers interfere with the secondary structure of DNA and the template function of the corresponding strand. DNA synthesis seems to be able to proceed across unrepaired pyrimidine dimers, but in an error-prone process that has a preference for inserting A on the complementary strand, thus resulting in frequent $G \bullet C \rightarrow A \bullet T$ transitions.

Although pyrimidine dimers are the main premutation lesions in UV-irradiated DNA, other photoproducts are found, including the cross-linking of DNA to proteins, and strand breakage.

Chemical Agents

Numerous chemicals are capable of reacting with DNA. We will discuss the main groups.

Alkylating agents. Alkylating agents are reagents that can modify most N and O atoms in the four bases by addition of alkyl groups. Alkylated bases induce a weakening of the glycosyl bond and often result in AP sites (see Fig. 7-1). Two alkylated bases in particular, O^6-methylguanine and O^4-methylthymine, have a tendency to form $O^6mG \bullet T$ and $O^4mT \bullet G$ pairs, thus leading to transition mutations.

Cross-linking agents and bulky adducts. Cross-linking agents and bulky adducts are chemicals that cause significant distortion of the secondary structure of DNA (see Fig. 7-1). Cross-linking can be intrastrand or interstrand. Interstrand links effectively block DNA replication; agents that promote them, such as nitrogen mustard, mitomycin, platinum derivatives, and psoralen derivatives, have been used in cancer chemotherapy.

Inactive chemicals metabolized to reactive mutagens. Some inactive chemicals have been recognized for their carcinogenic effect in the tissues where the conversion occurs; extensive studies have shown that carcinogenic compounds are also mutagenic. Polycyclic aromatic hydrocarbons, which are produced during coal combustion, and aflatoxins, which are produced by some fungi, are examples of such metabolically activated agents.

DNA REPAIR SYSTEMS

There are many enzymes whose function is to restore the molecular integrity of the DNA molecule and to preserve its sequence. Some of them are highly specific and correct only one type of damage, such as removal of methyl groups from O^6-mG. Others are more general and can repair an entire category of damages, such as a deformed double helix due to cross-linking or bulky modification of bases. Many repair enzymes are remarkably conserved in function, and to a certain extent in amino acid sequence, in all life forms. Although prokaryotes and yeast systems are the best known, we will, for the most part, limit our discussion to those functions that have been documented in mammals, because significant differences are known between repair in humans and in prokaryotes. For example, the *Escherichia coli* DNA polymerase involved in replication is known to have a proofreading function that allows it to detect a newly added nucleotide that is mispaired, remove it, and replace it with the correct one. Such a function has not been demonstrated for the mammalian replication enzyme.

The various repair processes are not yet fully characterized. Nor are they totally independent of one another, as several of them share certain common steps. Thus, the following classification is rather arbitrary, but is a useful way to approach a very complex subject.

Reversal of Damage

Reactions that neatly reverse the premutation chemical change are one repair process. One example is the very efficient removal of methyl groups from O^6-mG by the enzyme *O^6-methylguanine-DNA methyltransferase* (O^6-MGT) also known as the *Ada protein*. O^6-MGT may also act on O^4-methylthymine, but with much less affinity.

Another case of direct damage reversal is the ligation of single-strand breaks (see Fig. 7-1) in double-stranded DNA by ligase. This is possible when the phosphodiester bond hydrolysis leaves in position adjacent nucleotides with 3' OH and 5' phosphate ends. In practice, ligation is not always possible because single-strand breaks produced by ionizing radiation are often accompanied by damage to the sugar or base left behind, and repair must include removal of the damaged nucleotides (see below).

The classic example of damage reversal—and the first form of DNA repair found, in the 1940s—is *photoreactivation* (PR): the light-dependent monomerization of pyrimidine dimers by *photolyase* or PR enzyme. PR is a widespread process among prokaryotes and eukaryotes that does not seem to occur in placental mammals.

Base Excision Repair

Base excision repair (BER) represents a group of repair pathways characterized by specific enzymes that recognize some of the most common forms of endogenous DNA damage. The process is initiated by *glycosylases* that hydrolyze the N-glycosyl bond between certain improper bases and d-ribose in DNA. The most abundant of these enzymes is *uracil-DNA glycosylase*, which removes the U produced in DNA by deamination of C, as we saw in the previous section. Other glycosylases found in human cells include some with a broader spectrum of specificity, such as *3-alkyladenine-DNA glycosylase*, which acts on many alkylated purines. Glycosylases remove a single base, leaving behind an AP site with the sugar-phosphate backbone intact (Fig. 7-15). Subsequent steps in the pathway repair these AP sites, as well as those generated by spontaneous base loss.

The first step in AP site repair is carried out by a specific *AP endonuclease,* which hydrolyzes the phosphodiester bond on the 5' side of the AP site. The 5' end of this gap is recognized by *DNA polymerase* β, a two-domain protein that removes the deoxyribose phosphate with one domain, and replaces the missing nucleotide with the other domain, using the 3' OH as a primer. *DNA ligase I,* which is found in a multimolecular complex with polymerase β, then forms the phosphodiester bond (see Fig.

FIGURE 7-15. Base excision repair and repair of AP sites.

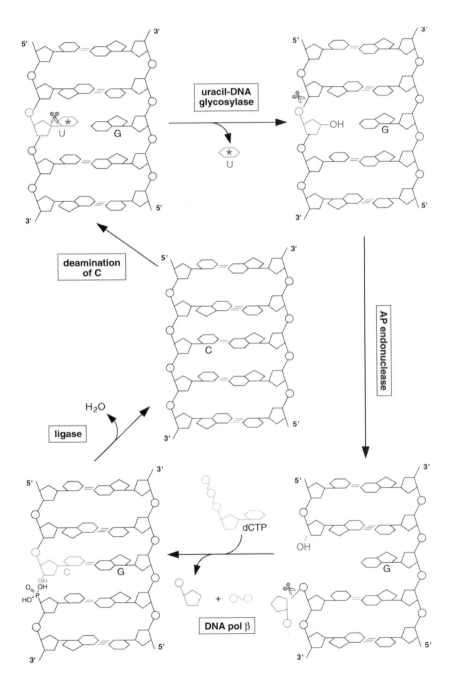

7-15). DNA polymerase β and ligase may also be associated with the AP endonuclease; thus, the entire set of reactions is carried out very efficiently.

The form of BER shown in Figure 7-15 is called *short-patch* or single-nucleotide BER. There is also *long-patch* BER, in which multiple nucleotides are replaced. Additional proteins are involved in long-patch BER. If the gap is between 2 and 6 nucleotides, repair is carried out by polymerase β; but if it is between 7 and 13, polymerase δ or ε seem to be involved.

There are no known human mutations in enzymes involved in BER. A null allele of β polymerase in the mouse is a recessive embryonic lethal. Cell lines obtained from very early, homozygous mutant mouse embryos can be maintained in the lab. They are very sensitive to alkylating agents, but not to x-rays or UV light, which gives an indication of the range of damage repaired by this enzyme.

Nucleotide Excision Repair

The nucleotide excision repair (NER) process corresponds to the *dark repair* (as opposed to light repair or PR) of pyrimidine dimers originally described in *E. coli*. NER exists in all organisms; it is not as specific as the glycosylase-initiated reactions, but rather it is a more generic repair system. NER is the primary repair system involved in the correction of damage that distorts the double helix, such as pyrimidine dimers or bulky adducts; these would include benzo[a]pyrene-guanine from smoke, or thymine-psoralen and guanine-cisplatin, from chemotherapeutic drugs. But it can also act on nondistorting damage such as methylated, oxidized, and even mismatched bases.

NER is carried out by a complex multienzyme aggregate. It includes proteins that recognize and bind to damaged sites, specific excision nucleases that introduce single-strand breaks on either sides of the damaged nucleotide(s), and helicases that unwind the DNA. A 29-np oligonucleotide, including the damaged nucleotide is thus removed, and DNA synthesis and ligation then repair the gap (Fig. 7-16).

In the absence of photoreactivation in humans, the brunt of thymine-dimer repair falls on NER. This is illustrated by three genetic diseases: xeroderma pigmentosum (XP), Cockayne syndrome (CS), and trichothiodystrophy (TTD), which are all conditions caused by deficiencies in NER. XP is a very rare disease characterized by the occurrence of dermatitis, excessive freckling, and numerous skin tumors on parts exposed to sunlight. CS patients suffer physical and mental underdevelopment and some light sensitivity. TTD patients have brittle hair, physical and mental retardation, and some of them are also UV sensitive. Seven genes have been identified that can cause the XP phenotype, and two more are associated with CS.

NER is, however, a very complex process known from biochemical studies to involve at least 12 proteins. NER seems to proceed by two alternate pathways, one that deals with transcribed DNA strands and one that acts on the genome in general. Damage in transcribed strands is repaired faster than damage in nontranscribed strands.

Mismatch Repair

As is BER, mismatch repair is another relatively specific DNA repair system that acts on endogenous damage. It detects and corrects mismatched bases and looped-out single strands, especially those that occur in connection with DNA replication and meiotic recombination. Mismatches during recombination occur when complementary strands of homologous chromosomes form a double helix at a locus for which the or-

FIGURE 7-16. Nucleotide excision repair. Excision nucleases hydrolyze the 5th phosphodiester bond 3' of the damaged site, in this case a pyrimidine dimer, and the 24th on the 5' side. The position of these incisions, three helical turns from each other, is probably determined by the size of the multienzyme complex that carries out NER. The single-stranded oligonucleotide is then removed with the help of helicases, and the gap is filled by DNA polymerase and ligase.

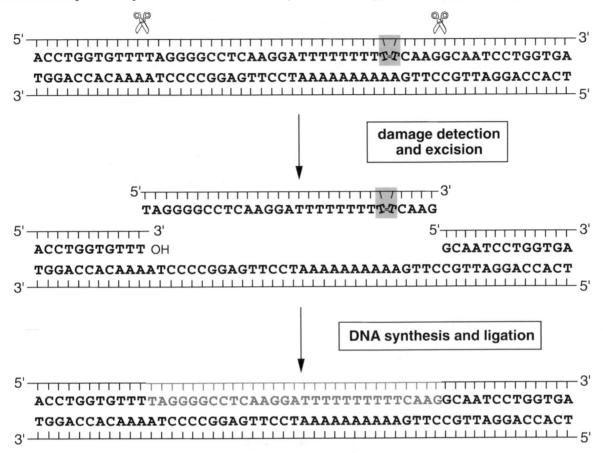

ganism is heterozygous. The best-studied function of mismatch repair in mammals, however, is in dealing with damage produced during DNA synthesis, namely the incorporation of mismatched bases and the single-stranded loops created by slippage along stretches of simple sequence DNA.

The type of damage repaired by mismatch repair is unique in that all the elements present—such as T•G mispairs in newly synthesized DNA—are normal components of the DNA molecule. There is, therefore, ambiguity concerning the original information: should the T be replaced with a C, or the G replaced with an A? Both replacements would correct the damage, but although the proper correction would recover the original sequence, the improper one would fix a mutation. Establishing which is the right replacement requires that the repair system be able to identify which is the original template strand and which is the priming or newly synthesized one. In *E. coli,* DNA is methylated at palindromic GATC sites some time after replication. Thus, newly synthesized DNA is hemimethylated, the priming strand being as yet unmethylated. This asymmetry allows the repair system to recognize the strand with the error and replace it. It appears, however, that methylation is not the

EXAMPLE 7.4 *Xeroderma pigmentosum and nucleotide excision repair*

XP behaves as an autosomal recessive condition, but it is heterogeneous in origin; that is to say, it may be caused by mutations in any one of several genes. When it was discovered that cultured fibroblasts from XP patients showed increased sensitivity to UV radiation, the suspicion that these patients were deficient in some form of DNA repair was reinforced. That observation also led to a convenient in vitro assay for genetic studies. One group of studies consisted of fusing cells from many patients in various pairwise combinations. The surprising result was that many of the cell hybrids had normal tolerance to UV light. This *genetic complementation* indicated that the same clinical condition arose from mutations in different genes. Cell hybrids that were as sensitive as the parental cells, showing *failure to complement,* pointed to mutations that were in the same gene. In this fashion, seven complementation groups were defined *XPA* through *XPG*.

Independently, by treating them with mutagens and selecting for UV-sensitivity, it was possible to identify several rodent cell lines that also appeared deficient in UV-damage repair. Transfecting the mutant rodent lines with human DNA and looking for functional restoration, it was possible to isolate many of the corresponding human genes. Those repair genes were designated *ERCC* (for *excision repair cross complementing*). Most *ERCC* genes turned out to be *XP* genes, but one in particular, *ERCC1*, does not have a counterpart in a human disease gene—it

has been argued that *ERCC1* is involved in other cellular functions because homozygous mutant mice die soon after birth.

The protein XPA recognizes damaged sites on DNA and, together with XPE, ERCC1, and the singles-stranded DNA-binding protein HSSB, binds to the lesion (Fig. 7-17). The transcription/repair factor TFIIH—which includes the helicases XPB and XPD—joins in and XPC and XPG are then recruited. The ERCC1/XPF heterodimer then makes the 5' incision and XPG makes the 3' incision, and both the oligonucleotide and at least some of the proteins in the excision aggregate are released. Finally, DNA polymerase δ or ε and ligase close the gap (see Fig. 7-16).

NER seems to occur by one of two pathways: the transcription-dependent pathway, and the transcription-independent pathway (see Fig. 7-17). Some of the NER proteins, such as the helicases XPB and XPD, are part of the transcription factor TFIIH, so that when DNA damage interferes with the transcription process, repair can be effected quickly. It has been suggested that this process permits focusing the immediate repair response on the small fraction of the genome that matters most to any given cell. Most CS patients have mutations in one of two genes, *CSA* and *CSB*, whose products participate only in the transcription-dependent pathway, and are less sensitive to light (see Fig. 7-17). Some *XPB*, *XPD*, and *XPG* alleles are responsible for TTD; probably because they interfere with transcription itself, they cause symptoms far

signal used in mammalian genomes to distinguish between newly synthesized and template strands.

In mammals, recognition of the newly synthesized strand may occur through its free 3' end, or perhaps by the association of repair proteins with the replication machinery, such that the two DNA strands can be distinguished by their position relative to these proteins. In any case, repair of DNA damage involves the degradation and replacement of several hundred nucleotides in the nascent strand, from the position used to distinguish between the strands to some distance past the damaged site. The gap left is then repaired by DNA polymerase and ligase (Fig. 7-18). Because all com-

FIGURE 7-17. Known genes involved in NER. (Based on Cleaver, 1994.)

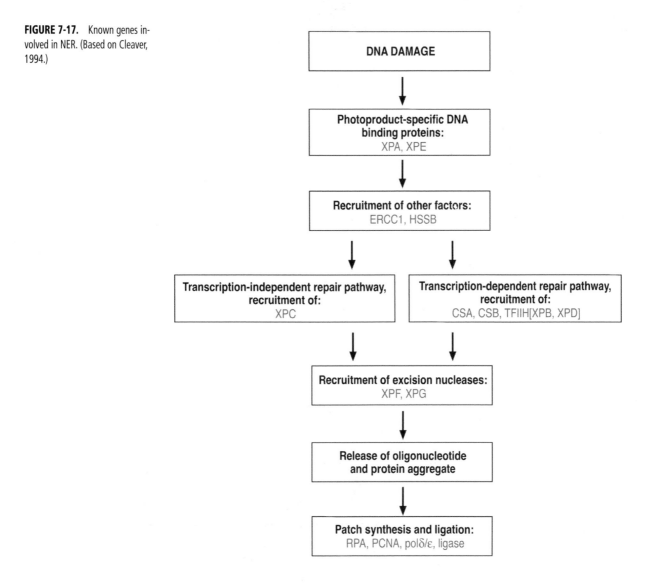

ponents in the damaged site are normal—the damage consists in the mismatching of the two strands—removal of the nascent strand allows resynthesis of an undamaged stretch of DNA, even when the original damage consisted of the looping out of nucleotides on the template strand (see Figs. 7-5 and 7-6).

Elucidation of the molecular processes and genes involved was possible thanks to a fortunate combination of the discovery of genetic diseases that disable the mismatch repair process, work from various labs using mutations in yeast and *E. coli*, and in vitro assays with human cell cultures. The players in this process are designated with reference to the three main components of mismatch repair in bacteria (MutS, MutL, and MutH), although in mammals there are often multiple proteins that combine to play the role of each bacterial enzyme.

Human mismatch repair can rectify all eight base mismatches and small insertion/

FIGURE 7-18. Proposed mechanism of mismatch repair. The DNA molecule is represented by the double line, with the 3' arrowhead indicating the growing point of the nascent strand. The damaged site is indicated by the wiggle in the lines; this may represent mispaired bases or a single-stranded loop either on the template or the priming strand. From the incision point, the newly synthesized strand is unwound by helicases and degraded exonucleolytically past the damaged site; the gap is then filled in. The incision point is arbitrarily positioned at some distance from the 3' end. In this example, the incision is 3' of the damage site, but repair seems to proceed just as effectively from a 5' nick, as one might find in the replication of the lagging strand.

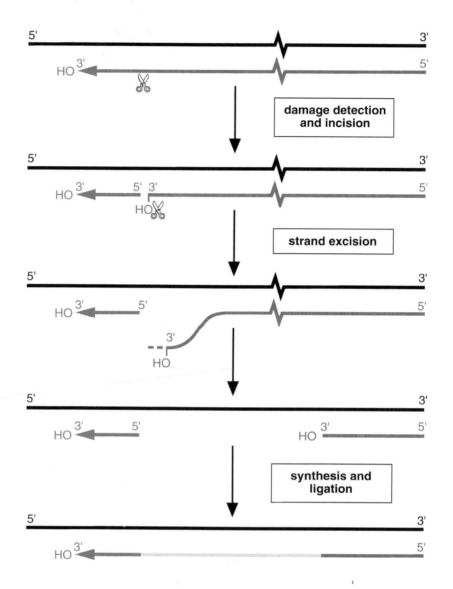

deletion loops that have a maximum size probably between 4 and 14 nucleotides. On the basis of what is known from mammalian cells, and by analogy to bacterial systems, it is likely that the heterodimer hMutSα (human MutS) binds to the mismatch site, with hMutLα associating to the complex. This binding stimulates an endonucleolytic activity (MutH in bacteria), which causes the single-strand incision. With the participation of a helicase and exonucleases, the damaged strand is unwound and hydrolyzed to a point 100 to 150 bases beyond the damage site. The resulting gap is then filled by, probably, one of the replication DNA polymerases α, δ, or ε, and ligase.

The role of mismatch repair is dramatically illustrated by the consequence of mutations in protein components of hMutSα (MSH2) and hMutLα (MLH1, PMS1, and PMS2). Mutations in these genes are responsible for hereditary nonpolyposis colon cancer (HNPCC). The connection between mutations and cancer is discussed in Chapter 8, but it should be noted here that tumor cells from these patients are deficient in the in vitro assays of mismatch repair, and display a great deal of genomic in-stability. The instability manifests itself in increased mutation rates of specific genes, and in extensive changes in number of repeats in microsatellites—$(A)_n$, $(GGC)_n$, $(CA)_n$. Microsatellite instability is precisely the phenotype one would expect of cells unable to correct insertion/deletion loops caused by strand slippage in tandemly re-peated sequences.

Other Repair Processes

There are other, less well-defined, repair processes that may be functionally analogous to bacterial *error-prone* or *SOS repair* and *post-replication repair.* In general, these are systems that restore the structural integrity of the DNA even when the informational content is compromised. Post-replication repair is defined by some mutant rodent cell lines that display slightly increased sensitivity to UV light, but reduced UV mutability. That is to say, normal cells seem able to complete repair of UV damage by a process that is prone to cause mutations; a deficiency in some step in this process renders the cells more sensitive to UV radiation (because, unrepaired, the lesions are cytotoxic) but less mutable. Similar characteristics are found in cells of Fanconi's anemia patients, an autosomal recessive disease characterized by increased predisposition to cancer.

UV radiation—as well as other DNA-damaging treatments—leads to a rapid, but temporary, inhibition of DNA synthesis. It is not clear whether this is due to a regu-latory response or to the blockage of replication forks that encounter pyrimidine dimers. The inhibitory period is time that the cell takes to repair the damage. If, after UV radiation, DNA synthesis is inhibited experimentally, and sufficient time is al-lowed to elapse before releasing the inhibition, there is practically no mutagenic effect from UV radiation.

Under normal conditions, DNA synthesis restarts before all damage has been re-paired, and there is evidence for bypass or *translesion synthesis* of DNA: elongation of the new strand bypassing a pyrimidine dimer on the template strand. The template ca-pacity of such dimers is at least partly, but not totally, impaired; translesion synthesis often results in the incorporation of A with the consequent tendency for $C \bullet G \rightarrow T \bullet A$ transitions. Translesion synthesis is not carried out by the normal DNA replication complex, but rather by one of two translesion synthesis systems. One system is error-free and implemented by DNA polymerase η, and the other is error-prone and in-volves DNA polymerase ζ. Patients suffering from a variant form of XP (XP-V) are deficient in pol η and error-free translesion synthesis. XP-V patients have the same high predisposition to skin cancer as XP patients, although they are not deficient in NER, which indicates that error-free translesion synthesis is a common process in cells exposed to UV damage.

MUTATION RATES

Mutation rate in a given human population is defined as the number of gametes in a sample that carry a *new mutation* divided by the total number of gametes in that sam-ple. The number of gametes can be obtained from the total number of individuals in the sample ($\times$2), or estimated by studying a sample of consecutive births; allowances must be made, however, for spontaneous abortions that may go clinically undetected (estimated to be 15% of pregnancies).

Counting mutations is less straightforward. In the case of chromosomal aberrations, it is possible to analyze all newborn and aborted fetuses in a sample and obtain the number of *detectable* mutations, meaning those visible with the techniques used, again taking into account that mutations causing miscarriages will be lost. This type of study provides an *incidence rate;* to obtain a mutation rate, it is necessary to establish what fraction of those mutations occurred de novo—as opposed to being the transmission of preexisting mutations—by testing the parents. For chromosomal aberrations, most numerical abnormalities are de novo mutations, whereas 70% to 80% of structural rearrangements are preexisting in one of the parents. Studies involving over 50,000 live births and spontaneous abortions gave mutation rates for all chromosomal rearrangements combined on the order of 1×10^{-3}. Numerical errors occur at a slightly higher rate.

With respect to gene mutations, the direct method described above can be applied to dominant mutations detectable at birth. It is possible to use indirect methods to estimate mutation rates for dominant traits not detectable at birth, and for X-linked recessive traits. These methods make use of the incidence rate of *sporadic cases,* and assume that the human population is in equilibrium for such mutations. That is to say, new mutations balance the loss due to the death of individuals carrying the trait. Table 7-3 shows estimates of mutation rates for some specific genes. Both human and model-organism studies have shown that mutation rates in different genes, or even various sites in the same gene, are highly heterogeneous. Thus, estimates such as those in Table 7-3 are likely to be very skewed by genes that show high mutation rates, and they are not adequate to calculate median values of mutation rates for all genes.

For autosomal recessive genes, it is not possible to apply either of those methods,

TABLE 7-3. Mutation rates for several dominant and sex-linked traits.

Trait	Mutation Rate
Autosomal Dominant Traits	
Achondroplasia	$0.7 - 1.4 \times 10^{-5}$
Aniridia 1, 2	$0.3 - .5 \times 10^{-5}$
Myotonic dystrophy	$0.3 - 1.1 \times 10^{-5}$
Retinoblastoma	$0.6 - 1.2 \times 10^{-5}$
Osteogenesis imperfecta (I, II, IV)	$0.7 - 1.3 \times 10^{-5}$
Neurofibromatosis 1	$4 - 10 \times 10^{-5}$
Marfan syndrome	$0.4 - 0.6 \times 10^{-5}$
X-Linked Traits	
Hemophilia A, B	$2 - 3 \times 10^{-5}$
Duchenne muscular dystrophy	$4 - 10 \times 10^{-5}$
Incontinentia pigmenti, Mainz type 2	$0.6 - 2 \times 10^{-5}$

SOURCE: Summary of data collected by Vogel and Motulsky, 1996.

because the vast majority of affected individuals are offspring of heterozygous parents. Identifying de novo mutations would be laborious.

Extrapolating from dominant mutation rates and from work in experimental animals, a median mutation rate of 1×10^{-6} to 1×10^{-7} (mutations per gene per cell cycle) is estimated. On the other hand, estimates of DNA damage are of the order of 1×10^{-3} (premutations per gene per cell cycle). The three or four orders of magnitude difference between the frequency with which damage occurs to a gene and the frequency with which that damage is fixed in a mutation is due to the effectiveness of the various repair processes.

SOMATIC MUTATIONS AND MOSAICISM

In the traditional view of mutations, these can be either somatic or germinal. *Somatic mutations* give rise to patches of mutant tissue in individuals, which are called *mosaics*. Unless the mutation affects skin morphology or pigmentation, somatic mutations usually go undetected. The size and shape of a mutant patch depends on the stage in development when the mutation occurred (Fig. 7-19). Genetic mosaicism was first

FIGURE 7-19. Distribution of EH nevi along Blaschko lines in a patient whose son suffered from uniform EH. (Based on Nazzaro, et al., 1990.)

detected for chromosomal abnormalities, and, as we saw in Chapter 5, it is found quite often.

Germline mutations, on the other hand are hereditary and transmitted to the progeny in Mendelian fashion. This dichotomy was a very formalistic view by necessity, because not much evidence was available to shed light on the origin of mutations, or on the cells and individuals in which they occurred. Advances in cellular and molecular techniques have made possible studies in this very area. Foremost among those techniques is PCR, which allows detection of certain mutations in samples of very few cells. Also, there is the ability to study human chromosomes of single sperm cells by fertilization of Chinese hamster eggs (see "Aneuploidy" in Chapter 5).

Those studies have blurred the distinction between somatic and germinal mutations. The notion of a single, ill-starred, mutant gamete in a pool of normal congeners is probably largely incorrect—except for nondisjunction and other meiotic errors that are only received by the gamete(s) that result from one particular meiocyte. Many hereditary mutations seem to start out in mosaic individuals who carry the mutation in a significant fraction of germline cells or sometimes in mixed patches of germline and somatic cells.

Examples of mosaicism are available in de novo cases of several genetic diseases, as in the case of an unaffected man who transmitted the same partial deletion of the Duchenne muscular dystrophy gene (an X-linked recessive mutation) to two daughters but not to three others. There are also cases in which the mutation can be detected in the normal parent of an affected child. For example, in a family in which two normal parents had two children with neurofibromatosis type 1 (an autosomal dominant mutation), a 12-kb deletion of *NF1* was detected in 10% of paternal sperm. (See also Example 7.5, "Epidermolytic hyperkeratosis and keratins K1 and K10," and Example

---

EXAMPLE 7.5

Epidermolytic hyperkeratosis and keratins K1 and K10

Epidermolytic hyperkeratosis (EH) is characterized by skin thickening and blistering due to the clumping of keratin filaments. Keratins K1 and K10 form heteropolymers in epidermal skin cells engaged in terminal differentiation, and some amino acid substitutions (such as His156Arg in *K10*) behave as typical dominant negative mutations (see Chapter 6) so that the presence of a single mutant allele of either gene is sufficient to interfere with the proper assembly of filaments. In several cases it had been noticed that one parent of children affected with EH had a type of nevus with the same skin characteristics of EH. *Nevi* are sectors of skin that differentiate from the rest by pigmentation, thickening, and so forth, often taking the form of stripes and swirls called lines of Blaschko (see Fig. 7-19). The lines of Blaschko are made up of clonally related cells that are put in evidence when a mutation or X-inactivation differentiates cells in neighboring sectors.

The presence, in parents, of clonally derived cells showing the same phenotype presented, uniformly, by affected children suggested that such parents might be mosaic for the EH mutation. Indeed, molecular analysis of skin biopsies from nevi and normal tissue of each parent uncovered the same amino acid substitution found in all cells of the affected child. Assay of DNA from parental fibroblasts (from nevi as well as from normal skin) showed that some of them also carried the mutation. This indicates that the mutation occurred in these parents very early in development, before differentiation of the ectodermal and mesodermal layers and before segregation of the germline and somatic tissues (Paller et al., 1994).

FIGURE 7-21.   Family transmitting a small deletion within the muscular dystrophy gene. All members were genotyped with probes that detect polymorphic sites on the short arm of the X in region Xp21–Xp22. Five of those sites cover two-thirds of the *DMD* gene, while the other four lie outside of this gene. Each probe detects the presence or absence of a particular restriction site, and the two alleles are indicated by lowercase versus uppercase symbols. In addition, probe J-Bir detected the presence of a MD-causing deletion of marker J (indicated by Δ and a discontinuity). This deletion was observed only in the black haplotype, and inspection of the females in generation II suggests that individual I-2 was a germline mosaic for it. Note that, in the absence of genotype information, this pedigree (which shows all the characteristics of a sex-linked recessive trait) gives no indication that this is a new mutation. (Redrawn with permission from Fig. 1 in Darras BT and Francke U. A partial deletion of the muscular dystrophy gene transmitted twice by an unaffected male. Nature 329:556–558 [1987]. Copyright 1987 Macmillan Magazines Limited.)

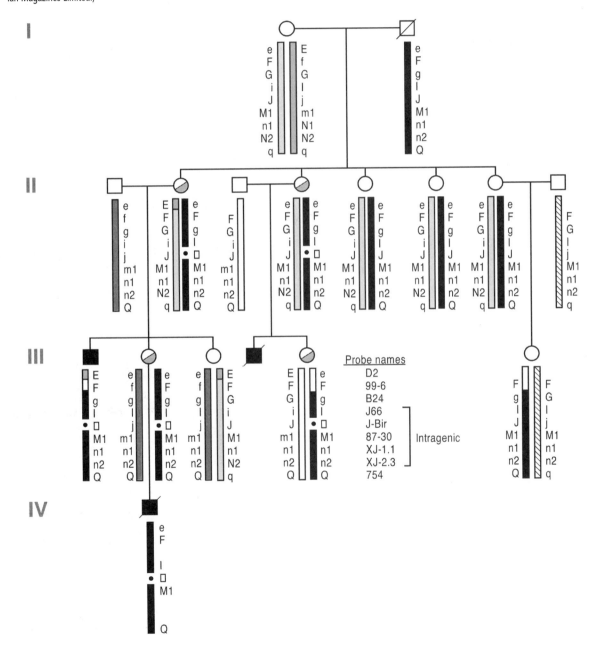

7-1. Show how the shifted forms of T and G would form hydrogen bonds with G and T, respectively, rather than with A and C.

7-2. Make a table with all possible dinucleotides (CpG, etc.). If all four bases occurred with equal frequency, what would be the frequency with which each dinucleotide is found? Given that human DNA is approximately 42% G+C, with what frequency would you expect to find each dinucleotide in the absence of any bias?

7-3. Compare the sequence of the trinucleotides that give rise to trinucleotide expansions. Is there a property in most of them that might explain their apparent tendency for slippage?

7-4. Write the genotypes (in terms of number of CGG repeats) for all the members of the family in Figure 7-9.

7-5. Diagram the chromosomal synapses that would explain the 6.7-kb deletions in the growth hormone gene region and the duplication of γ globin genes from one copy to two.

7-6. Diagram the mispairing of the flanking repeats of gene *PMP22* and the consequences of crossing over as visualized in Plate 1. The repeats are 24 kb in size and the duplication is 1.5 megabases. Indicate in your diagram the position(s) from which the probes were derived.

7-7. On the basis of Figure 3-18, diagram the chromosomal synapsis that would explain some forms of color blindness by unequal crossing over.

7-8. Design a study that would measure the rate of de novo mutations for one or more autosomal recessive traits.

7-9. In the absence of any bias in the transmission of homologous chromosomes in the family of Figure 7-20, what is the a priori probability that all seven offspring in generation III would be affected?

7-10. Redraw the pedigree in Figure 7-21 leaving out all genotype information. Shade the affected individuals. Shade half of the circle of females who are *obligatory heterozygotes* (i.e., those who transmitted Duchenne muscular dystrophy to their progeny). Interpret this family as a classical pedigree and, for as many individuals as possible, indicate their genotypes (*DMD*+ = normal allele; *DMD* = mutant allele). How do your genotypes compare with those in Figure 7-21 obtained by RFLP analysis? What implications would the two versions of the pedigree have for the descendants of a sister of I-1?

7-11. Add the mutations described in this chapter to the classification table in Exercise 1 of Chapter 6.

REFERENCES

General references

Caskey CT, Pizzuli A, Fu Y-H, Fenwick RG Jr, and Nelson DL. (1992) Triplet repeat mutations in human disease. Science 256:784–789.

Cleaver JE. (1994) It was a very good year for DNA repair. Cell 76:1–4.

Cooper DN, and Krawczak M. (1993) Human gene mutation. Oxford: BIOS.

Daniel A. (1988) The cytogenetics of mammalian autosomal rearrangements. New York:Alan R. Liss.

Friedberg EC, Walker GC, and Siede W. (1995) DNA repair and mutagenesis. Washington, DC: ASM Press.

Krawczak M, and Cooper DN. (1996) Mutational processes in pathology and evolution. In: Human genome evolution. Jackson M, Strachan T, and Dover G, eds. Oxford: BIOS, 1–33.

Warren ST. (1996) The expanding world of trinucleotide repeats. Science 271:1374–1375.

Specific references

Albin RL, and Tagle DA. (1995) Genetics and molecular biology of Huntington's disease. Trends Neurosci 18:11–14.

Ashley CT Jr, and Warren ST. (1995) Trinucleotide repeat expansion and human disease. Annu Rev Genet 29:703–728.

Darras BT, and Francke U. (1987) A partial deletion of the muscular dystrophy gene transmitted twice by an unaffected male. Nature 329:556–558.

Feng Y, Zhang F, Lokey LK, Chastain JL, Lakkis L, Eberhart D, and Warren ST. (1995) Translational suppression by trinucleotide repeat expansion at FMR1. Science 268:731–734.

Fitch DHA, Bailey WJ, Tagle DA, Goodman M, Sieu L, and Slightom JL. (1991) Duplication of the γ-globin gene mediated by the L1 long interspersed repetitive element in an early ancestor of simian primates. Proc Natl Acad Sci USA 88:7396–7400.

Fu Y-H, Kuhl DPA, Pizzuti A, Pieretti M, Sutcliffe JS, Richards S, Verkerk AJMH, Holden JJA, Fenwick RG Jr, Warren ST, Oostra BA, Nelson DL, and Caskey CT. (1991) Variation of the CGG repeat at the fragile X site results in genetic instability: resolution of the Sherman paradox. Cell 67:1047–1058.

Goldberg YP, Kremer B, Andrew SE, Theilmann J, Graham RK, Squitieri F, Telenius H, Adam S, Sajoo A, Starr E, Heiberg A, Wolff G, and Hayden MR. (1993) Molecular analysis of new mutations for Huntington's disease: intermediate alleles and sex of origin effects. Nat Genet 5:174–179.

Jacobs PA. (1981) Mutation rates of structural chromosome rearrangements in man. Am J Hum Genet 33:44–45.

Krawczak M, and Cooper DN. (1996) Mutational processes in pathology and evolution. In: Human genome evolution. Jackson M, Strachan T, and Dover G, eds. Oxford: BIOS, 1–33.

Lázaro C, Ravella A, Gaona A, Volpini V, and Estivill X. (1994) Neurofibromatosis type 1 due to germ-line mosaicism in a clinically normal father. N Engl J Med 331:1403–1407.

Lehrman MA, Russell DW, Goldstein JL, and Brown MS. (1987) Alu–Alu recombination deletes splice acceptor sites and produces secreted low density lipoprotein receptor in a subject with familial hypercholesterolemia. J Biol Chem 262:3354–3361.

Lindhal T. (1993) Instability and decay of the primary structure of DNA. Nature 362:709–715.

Masutani C, Kusumoto R, Yamada A, Dohmase N, Yokoi M, Yuasa M, Araki M, Iwai S, Takios K, and Hanaoka F. (1999) The XPV (xeroderma pigmentosum variant) gene encodes human DNA polymerase η. Nature 399:700–704.

Modrich P, and Lahue R. (1996) Mismatch repair in replication fidelity, genetic recombination, and cancer biology. Annu Rev Biochem 65:101–133.

Nazzaro V, Ermacora E, Santucci B, and Caputo R. (1990) Epidermolytic hyperkeratosis: generalized form in children from parents with systematized linear form. Br J Dermatol 122:417–422.

Paller AS, Syder AJ, Chan Y-M, Yu Q-C, Hutton E, Tadini G, and Fuchs E. (1994) Genetic and clinical mosaicism in a type of epidermal nevus. N Engl J Med 331:1408–1415.

Rautenstrauss B, Fuchs C, Liehr T, Grehl H, Murakami T, and Lupski JR. (1997) Visualization of the CMT1A duplication and HNPP deletion by FISH on stretched chromosome fibers. J Peripheral Nervous Sys 2:319–322.

Verkerk A, and Oostra B. (1992) Segregation of the fragile X mutation from an affected male to his normal daughter. Hum Mol Genet 1:511–515.

Vnencak-Jones CL, and Phillips JA III. (1990) Hot spots for growth hormone deletions in homologous regions outside of Alu repeats. Science 250:1745–1747.

Vogel F, and Motulsky AG. (1996) Human genetics, 3rd ed. Berlin: Springer.

Willems PJ, Van Roy B, De Boulle K, Vits L, Reyniers E, Beck O, Dumon JE, and Wood RD. (1996) DNA repair in eukaryotes. Annu Rev Biochem 65:135–167.

Cancer: A Genetic Disease

8

Cancer, a disease known since ancient times, is characterized by excessive proliferation of cells that, if they are constituents of solid tissues, form tumors. These tumors with time become invasive and lead to dysfunction of the organ or system in which they reside. Advances in genetics and molecular biology over the last 20 years have finally provided an understanding of how cellular processes are disrupted in this disease, or more accurately, in some forms of this disease. To provide points of reference to the many abnormal molecular activities found in cancer cells, we will begin the chapter with a brief review of the normal cell cycle. This is not a comprehensive review, but rather one that highlights precisely those regulatory steps in the cell cycle that are disrupted in cancer.

THE CELL CYCLE UNDER CONTROL

Cells divide and proliferate by orderly progression through the four stages of the cell cycle: mitosis and cytokinesis (M), gap 1 (G1), DNA synthesis (S), and gap 2 (G2) (Fig. 8-1). After mitosis, cells go into G1 and depending on environmental signals—growth or differentiation factors, mitogens, antiproliferative cytokines—they complete G1 and reenter S, or they proceed to G0, a resting state. Most nondividing cells in the organism—including differentiated cells, which under normal conditions will rarely divide again—are in G0. The "decision-making" stage in G1 that determines whether cells will proceed immediately to another cycle of cell division is sometimes called the late-G1 *restriction point.*

Once cells enter S, they become committed to divide, and proceed through the remaining stages independently of outside cues. Control of cell transit through the cycle is effected in part through protein kinases—*cyclin-dependent kinases* (CDKs)—which are in turn activated by protein cofactors that occur in sharp peaks at appropriate times in the cell cycle (*cyclins*). These are, however, internal controls, checkpoints, ensuring that a new stage is not entered until the previous one is finished. Thus, if DNA synthesis is not complete, cells are kept from chromosome condensation and mitosis, regardless of the time elapsed since S began. Similarly, until all chromosomes are properly lined up in the metaphase plate, chromatid separation and anaphase movements are held in check.

Nuclear Events

As we have seen, it is by regulating the transit through the late-G1 restriction point that organisms control cell proliferation, and cancer is almost invariably associated with a breakdown in the regulation of this transit. One of the main players in the G1 restriction point is the transcription factor E2F. E2F stimulates transcription of several genes (*MYC, MYB* B, thymidine kinase, etc.) required for synthesis of nucleotides and

FIGURE 8-1. Main stages of the mammalian cell cycle. Most of the variability in cycle length in various cell types comes from G1, which can be very short, for a total cell cycle of less than 15 hours or extended indefinitely (G0). The color asterisk indicates the point in the cycle illustrated by Figure 8-2.

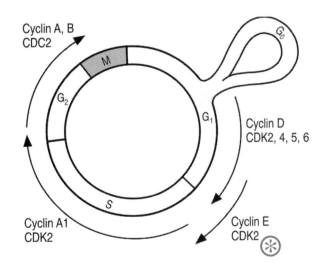

DNA polymerase. E2F also stimulates the synthesis of cyclins A and E, which in conjunction with CDK2, directly activate the replication machinery that is bound to DNA near origin-of-replication sites (see Fig. 8-1).

From the time of the previous S phase, and through the early part of G1, E2F is kept inactivated by the *RB* protein (pRB, product of the retinoblastoma gene)(Fig. 8-2). Production of D cyclins in response to *growth factors* signals progression through the G1 restriction point: D1 cyclin binds CDK4 and this complex phosphorylates pRB. Phosphorylated pRB releases E2F, which is then able to play its role as transcription activator (see Fig. 8-2).

The system also responds to negative signals. CDK4 is inhibited by the protein p21^{CIP1}, whose production, in turn, is stimulated by p53 when conditions are not appropriate for DNA synthesis. Thus, when there is DNA damage, p53 can hold cells in G1 until the repair systems have a chance to eliminate premutations. Similarly low oxygen supply induces p53 to halt the cell cycle (see Fig. 8-2). Other inhibitors of pRB phosphorylation, such as p15 and p16, respond to transforming growth factor β (TGF-β), which, despite its name, acts to repress cell division. Contact inhibition also blocks cell cycle progression.

FIGURE 8-2. Control of the transition between G1 and S by pRB and p53. This is a highly simplified version of the regulatory processes involved. Over two dozen components have been identified. Many of the proteins shown here are actually families of proteins, with individual members playing slightly different roles or being expressed in different tissues. Many intermediate steps and all feedback regulatory loops have been left out for the sake of simplicity. Finally, there must be other controls of the G1 to S transition, because cells that are deficient in pRB still need some growth factors to divide. The color asterisk indicates the point in the cycle illustrated by Figure 8-3.

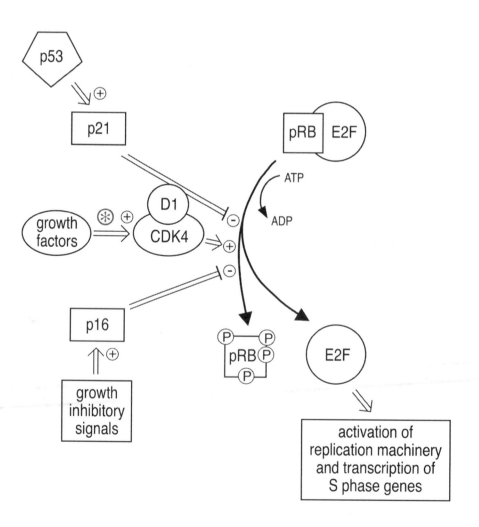

Thus, it is only under the proper combination of extracellular signals that pRB is phosphorylated and releases E2F. As the cell cycle progresses, phosphatases dephosphorylate pRB, which regains the ability to bind E2F in time to block the next G1 to S transition.

Cancer is often said to be not one, but many diseases. This is no doubt true in medical terms, but a common underlying process in all cancers is excessive cellular proliferation. Probably in all cases, the loss of control of the cell cycle comes about because of breakdown in the regulation of the G1 to S transition. In cancer cells, the mechanisms that restrain the cell cycle in G1, or those that shunt it into G0 and differentiation, fail, and cells are moved through an abbreviated G1 and onto S as soon as the previous cell division is complete.

It was through their failure or absence in specific cancers, that many of the components of G1 regulation were identified. Thus, proteins such as p53 and pRB, which act to reign in cell division, are examples of *tumor suppressor* gene products. Tumor suppressor genes are those that contribute to tumorigenesis when deleted or mutated to null alleles. On the other side, *oncogenes* are those whose products normally promote cell division, but under regulatory control. They contribute to cancer when they are hyperactivated, either by overexpression of the normal product or by gain-of-function mutations in the structural gene that make the protein product irrepressible. The wild-type allele of division-promoting genes are called *proto-oncogenes,* with the term "oncogene" reserved for the mutant, oncogenic allele. Cyclin D1 is one such proto-oncogene, and it is frequently overexpressed (oncogenic) in cancer cells.

The Cytoplasm: Receiving and Transducing Signals

The regulatory events presented in Figure 8-2 are centered on the action of transcription factors in the nucleus, and they represent the last stages of a regulatory pathway that starts at the cell surface with the binding of growth factors and hormones to specific receptors. More than 50 extracellular growth factors have been identified from studies of the growth requirements of cells in culture. They include polypeptides such as the platelet-derived growth factor (PDGF), which is released by platelets during blood clotting to stimulate proliferation of fibroblasts, smooth muscle cells, and other cells important in wound healing. Other examples include epidermal growth factor (EGF), which stimulates epithelial and some nonepithelial cells, and fibroblast growth factor (FGF). Related functions are performed by polypeptides such as insulin-like growth factor (IGF), which stimulates cell metabolism and, in concert with other growth factors, cell proliferation. Many growth factors occur as families of proteins, each member of the family expressed or detected by specific cell types. Cells usually respond to particular combinations of growth factors, so that a few different species of growth factors can be combined into a large number of signals for specific cell types.

As mentioned above, target cells must display specific transmembrane *receptors* on their surface in order to respond to growth factors (Fig. 8-3). Growth factor receptors are proteins with a ligand-binding domain on the outer surface of the plasma membrane and an intracellular domain. In most cases, the intracellular domain is a protein tyrosine kinase, or is associated with a protein tyrosine kinase. Binding of the ligand on the outer surface activates the cytoplasmic protein kinase (often by promoting the formation of dimers), which undergoes self-phosphorylation and, in turn, activates a protein of the Ras family. Activated Ras initiates a phosphorylation cascade of *mitogen-activated protein (MAP) kinases.* Ras binds and activates a protein serine/threonine ki-

FIGURE 8-3. Regulatory cascade initiated by binding of mitogen to a target cell. This illustration is also highly simplified. At any given time a cell receives a multitude of signals; some signal paths coalesce and others split. Thus, rather than responding to a simple, unidirectional informational cascade, cells must integrate a web of signals to produce a unique, reproducible outcome. This simplified version of cell-cycle control serves to illustrate the role played by various cancer-causing mutations. Although there are several known MAP kinase signaling pathways, each with its specific kinases, we will treat them here in a generic fashion and refer to them as follows: 1) MAPK, the enzyme that activates specific transcription factors by phosphorylation, and which is, in turn, activated by double phosphorylation of Ser and Tyr residues. 2) MAPKK, the dual specificity kinases that activate MAPK. 3) MAPKKK protein Ser/Thr kinases, which activate MAPKK. The color asterisk indicates the point in the cycle illustrated by Figure 8-7.

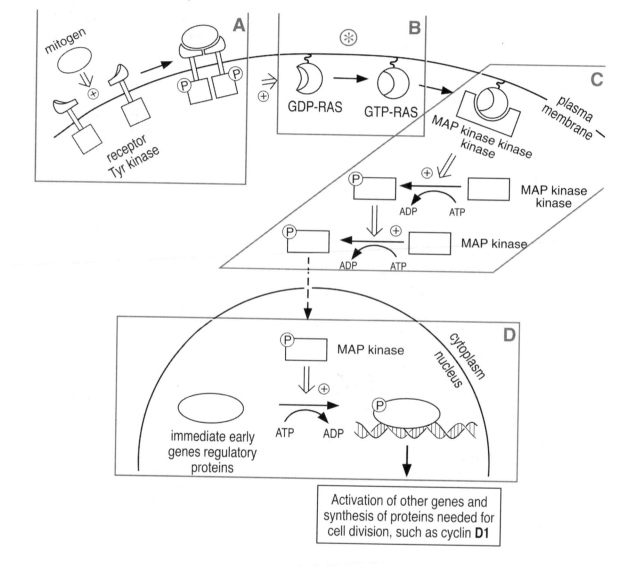

nase, MAP kinase kinase kinase (MAPKKK), which in turn activates MAP kinase kinase (MAPKK), which activates MAP kinase (MAPK). MAP kinase phosphorylates a group of proteins involved in activation of the *immediate early genes* of mitogen response. Products of these genes, in turn, stimulate transcription of another set of genes, one of which codes for the cyclin D1 that participates in the regulation of pRB (see Fig. 8-2).

Key factors in this type of regulatory cascade are the multiplicity of steps and the short half-life of the activated state of all intermediates. Because each step activates an enzyme, there is an exponential amplification of the signal as it travels down the pathway. On the other hand, quick inactivation of the intermediates ensures that the signal, if strong, is very brief, so the cell can respond nimbly to new environmental cues.

CANCER CELLS

Progression from Normal Cells to Cancer

Cancers are classified according to the cell type from which they derive. *Carcinomas,* human cancers of epithelial origin, account for 90%. Other cancers include *sarcomas,* cancers of muscle or connective tissues; *leukemias,* of blood cell precursors in the bone marrow; and *lymphomas,* of blood cells outside of the bone marrow.

Cancers occur mostly in tissues in which there is a certain level of replacement through cell division. Normally cell proliferation is kept under very strict control by molecular mechanisms of the type described in the previous section. Cell replacement usually occurs by a process in which relatively undifferentiated stem cells divide into two: one daughter cell remains undifferentiated and capable of subsequent growth, while the other is launched into a program in which it divides only a finite number of times and then all progeny cells differentiate. In this fashion, population growth follows a linear, rather than a geometric, function, and it can be adjusted to a level appropriate for replacement of cell loss.

For normal cells to become cancerous, or *neoplastic,* they must undergo several changes, each usually accompanied by a corresponding mutation. The first change might involve alterations in stem cell development that cause both daughter cells to remain undifferentiated; although cell divisions might still be under restraint, cells would now begin to accumulate geometrically. A second possible mutation, reducing the effectiveness of a tumor suppressor gene, might allow some cells to grow more vigorously and outpace more normal cells.

In this fashion, a system of natural selection and hopeless evolution is established. By now there may be a population of 10^5 or 10^6 cells, the probability of new mutations becoming greater as the population increases. These early stages are very slow and may take 5 or 10 or 20 years from the time of the first mutation. In solid tumors, these mutations often result in a *benign tumor* (adenomas or polyps, in epithelial tissues), in which cells are proliferating excessively, but they are confined in a well-defined mass. Not all benign tumors develop into malignancies, most of them remain benign or are reabsorbed.

Other changes undergone by neoplastic cells include the ability to grow under conditions of reduced oxygen (hypoxia), to induce blood vessel formation, and to grow indefinitely (we will return to this). Finally, in the case of *adenomas,* cells become detached from one another (probably through the loss of cell-adhesion molecules like E-cadherin) and they become invasive (i.e., traverse the basal membrane and penetrate the surrounding connective tissue).

The invasive last stage defines a *malignant tumor* (carcinoma in epithelial tissues). By this stage cells have been selected to grow very fast, the process is much accelerated, and in most cases death ensues relatively shortly if the disease is left untreated. Another property of malignant tissue is the ability to disperse through the vascular system and form *metastases* at remote sites.

To reiterate the point, an entire benign tumor does not undergo all the changes listed above and become malignant. Rather, at each stage a single cell mutates to acquire a new selective advantage and founds a clone that outgrows the others. Sampling and molecular analysis of cells from a single tumor show that it comprises several cell populations carrying different mutations (Fig. 8-4).

FIGURE 8-4. Schematic representation of a normal epithelium in which one cell acquires a mutation that increases its proliferative capacity. New mutations, indicated by the various shadings, lead to faster growing clones, which crowd out the slower cells. Cell death due to hypoxia, until mutants arise that have lost p53, is indicated by crosses.

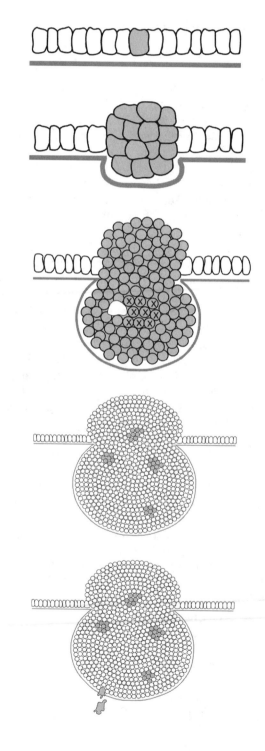

The mutation process is exemplified by colon cancer, in which, because of the accessibility of the tissue, it is possible to sample abnormal growths in various stages—from very small, benign polyps through larger adenomas and carcinomas. Analysis of genetic changes at each stage gives some clues of the number of steps involved (maybe six or more) and the genes involved. Figure 8-5 represents a hypothesis of colorectal tumorigenesis with some of the genes that have been identified as being often associated with transitions between stages. It should be emphasized that, ultimately, the outcome is dictated by the aggregate of mutations that occur rather than by the order in which they occur.

Cells in Culture

Much research into cancer cells has been carried out using cells cultured in vitro. Valuable comparisons can be made between the properties of normal (or almost normal) and malignant cells in culture. When a biopsy sample of normal tissue is placed in a culture dish with a nutrient medium that provides the basic requirements—carbohydrates, amino acids, vitamins, salts, and other appropriate precursors to complex biological molecules—the cells can be kept alive for a period of time, but they will not, as a rule, divide. All cells will reach G1 and stall there. It was discovered many years ago that the key to stimulate cell division in these explants was the addition of fetal calf serum to the medium. And it has since been demonstrated that it is the presence of specific growth factors in the fetal serum that promotes cells into an active mitotic cycle.

The treated, dividing cells slowly spread from the edges of the explanted mass, giving rise to a *primary culture*. After a few days, those dividing cells can be collected,

FIGURE 8-5. Progression of colon cancer. Mutations and altered processes that have been observed in colonic polyps at various stages.

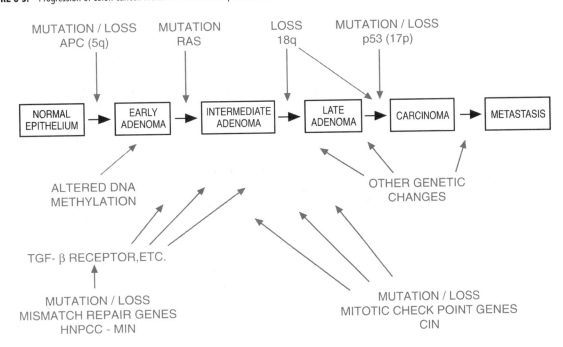

suspended in fresh medium, and transferred to a new dish; this constitutes a *secondary culture*. Normal cells in secondary cultures multiply until they begin to touch each other, at which point growth is inhibited (*contact inhibition*) until they are again diluted and *passed* to a new dish. Such cells would not, as a rule, grow in suspension in stirred medium; they have *anchorage dependence* and need to be attached to a surface, such as the bottom of a dish.

Cells in secondary cultures can be transferred in this way for many generations, but not indefinitely. After approximately 50 cell divisions the cultures begin to decline and die. It is thought that this *senescence* of secondary cultures may be related to the finite life span of humans. Cultures started from older individuals senesce earlier than those from younger people, as though the older cells had already used up some "passages."

Cultured cells may lose some of their properties, especially if they have been subjected to mutagens or certain viruses. A clone of cells may be observed in a dish where cells continue to grow and pile on top of each other after they reach confluence. In successive passes, subclones may originate that appear even more vigorous, less dependent on calf serum, and generally easier to grow. Cells that have undergone these types of changes are said to be *transformed*. (Note that this word is also used, in a completely different sense, to indicate the uptake of DNA by cells.) Transformed cells may also become anchorage independent (i.e., they become capable of growing in suspension) and immortal, giving rise to a *cell line*. Table 8-1 summarizes the main differences between secondary cultures and cell lines, although not all differences are found in all cases.

The properties of transformed cells are very similar to those of cancer cells. In fact, many of the immortal cell lines used in research were derived from samples of cancers in advanced stages.

ONCOGENES

The term *oncogene* was first applied in studies of the carcinogenic action of some viruses. The Rous sarcoma virus (RSV) is an avian RNA virus capable of inducing cancer in infected chickens or transforming cells in culture. Studies of this virus extend back to the beginning of the 20th century and it was discovered that some mutant strains, although unimpaired in their abilities to infect cells and grow, had lost their ability to cause cancer. This led to the postulate that there is a specific gene in the virus whose function it is to induce malignant cell growth: an oncogene. As this virus

TABLE 8-1. Characteristic differences between transformed and untransformed cells.

Untransformed Cells	Transformed Cells
Contact inhibition	Lack of contact inhibition
Anchorage dependence	Anchorage independence
Serum requirement	Reduced serum requirement
Senescence	Immortality

induces sarcomas, the putative gene was termed *src* (pronounced "sark"). Those strains that had lost the ability to cause cancer were thought to carry a mutation that inactivated the *src* gene or its product.

In the 1970s, when the first DNA cloning techniques became available, one of the earliest great successes was the cloning of *src*. Using oncogenic transformation of culture cells as an assay, the viral genome was fragmented and the various pieces tested for their ability to stimulate cell growth. Thus, viral fragments were introduced into cells and these were plated; those cells that received a functional *src* gene would divide more actively and form a distinct colony in a few days. (This is an example of expression cloning by functional assay as described in Chapter 4.) The viral *src* gene was then used as a molecular probe to test for its presence in virus-infected cells. The most unexpected result was obtained when it was found that the oncogene was present not only in infected but in noninfected cells as well.

If all cells carry *src*, why does it only cause cancer when coming from the virus? To differentiate the two sources, the viral gene was called v-*src* and the cellular one was called c-*src*. It was soon discovered that c-*src* is a normal constituent of the host genome and v-*src* is an allele that has acquired a dominant, gain-of-function mutation that makes it oncogenic. Thus, c-*src* was then termed a *proto-oncogene*. It was eventually discovered that proto-*src* functions in the regulation of cell division as a receptor-stimulated protein tyrosine kinase. The mutant allele codes for a protein that is constitutively active—that is, it cannot be regulated down. Note that with respect to cancer, the virus is totally incidental in this case; if a normal allele of c-*src* were to acquire a mutation similar to that found in v-*src*, it would also be carcinogenic. Proto-oncogenes become incorporated into the viral genome accidentally during replication of the virus. When they mutate to oncogenes, they give the virus a selective advantage and thus propagate preferentially in the viral population.

Using the transformation of mammalian cells in culture as an assay, many other oncogenes (such as *myc*, *ras*, and *raf*) were isolated from animal oncogenic viruses. Although almost none of those oncogenes was associated with virally transmitted cancers in humans (which are rare), most of them were found in their mutant (c-*onc*) form in human malignancies. Approximately 20 genes account for the vast majority of mutant oncogenes found in human malignancies (Table 8-2). In all cases the same general logic is followed: the genes code for proteins that stimulate cell division and are under tight control in the normal cell, but their protein products persist in an activated state, or are overproduced, in the mutant, oncogenic form. As a rule, the hyperactivating mutations are dominant at the cellular level and would be lethal as germline mutations. Given their critical role in stimulating cell division, germline null mutations in proto-oncogenes would be expected to be lethal in very early embryonic stages; this expectation has been confirmed in a few cases of knockout mutations in mice (Table 8-3).

Overexpression of cell-cycle control genes in cultured cells showed that many of them caused the cells to display some properties of transformed cells. Altogether, more than 60 genes have been described that have the potential to participate in oncogenic transformation (and these are sometimes included in the category of proto-oncogenes), although the majority of them have never been found associated with human malignancies. Because new searches have yielded very few unknown genes in recent years, it is estimated that that number accounts for most of the potential oncogenes in the human genome.

TABLE 8-2. Involvement of some oncogenes in human tumors.

Name of Oncogene	Tumor Associations	Mechanism of Activation	Properties of Gene Products
ERBB2	Mammary, ovarian, and stomach carcinomas	Amplification	Cell-surface growth-factor receptor
ERBB	Mammary carcinoma, glioblastoma	Amplification	Growth factor receptor
RAF	Stomach carcinoma	Rearrangement	Cytoplasmic serine/threonine kinase
HARAS	Bladder carcinoma	Point mutation	GDP/GTP binding
KIRAS	Lung and colon carcinomas	Point mutation	GDP/GTP binding
NRAS	Leukemias	Point mutation	GDP/GTP binding
MYC	Lymphomas, carcinomas	Amplification, chromosomal translocation	Nuclear transcription factor
NMYC	Neuroblastoma	Amplification	Nuclear transcription factor
LMYC	Small cell lung carcinoma	Amplification	Nuclear transcription factor
BCL2	Follicular and undifferentiated lymphomas	Chromosomal translocation	Cytoplasmic perhaps mitochondrial
GSP	Pituitary tumor	Point mutation	Cytoplasmic GDP/GTP signal transducer
HST	Stomach carcinoma	Rearrangement	Growth factor
RET	Thyroid carcinoma	Rearrangement	Growth factor receptor
TRK	Thyroid carcinoma	Rearrangement	Growth factor receptor

SOURCE: Based on Table 2 in RA Weinberg. Oncogenes and tumor suppressor genes. Cancer J Clin 44:160–170 (1994), with permission.

TABLE 8-3. Phenotypes of mutations in oncogenes. *Somatic mutations* refers to the effect of a mutation on somatic cells in an otherwise homozygous normal individual. *Germline mutations* refers to the phenotype of an individual who received the mutation through one of the gametes and is therefore wholly mutant. (Shading corresponds to those cases found in human cancers.)

Somatic	Loss of function	Unknown. Probably recessive cell-lethal by blocking cell division.
	Gain of function (overexpression)	Dominant, oncogenic.
Germline	Loss of function	Unknown in humans, recessive embryonic lethal in knockout mice.
	Gain of function (overexpression)	Unknown. No familial cancers due to oncogenes. Perhaps dominant lethal because of the production of tumors early in development?

In the following sections we will review briefly specific cases of genes that normally play a role in stimulating cell division and how they are altered to become oncogenes. Table 8-2 summarizes the effect of different types of proto-oncogene mutations.

Growth Factors and Their Receptors as Oncogenic Proteins

The genes for both growth factors and their receptors (see Fig. 8-3A) can act as oncogenes. The structural gene for the growth factor subunit PDGF B becomes an oncogene (v-*sis*) when it is carried by the genome of the simian sarcoma virus and therefore overexpressed, leading to oncogenic transformation in infected cells. Although this gene does not play an extensive role in human cancers, co-expression of the PDGF B chain and PDGF receptor in some glioblastomas (malignant brain tumors) suggests its participation in *autocrine* growth stimulation, in which growth factor is secreted by the same cells that are targets of the growth factor (Fig. 8-6).

Synchronous expression of another growth factor, EGF, and its receptor (a protein tyrosine kinase), again suggesting autocrine stimulation, was detected in a significant number of advanced (but not early) carcinomas. Overexpression of the EGF receptor

FIGURE 8-6. Autocrine stimulation. A cell produces the mitogen on which it depends for proliferation.

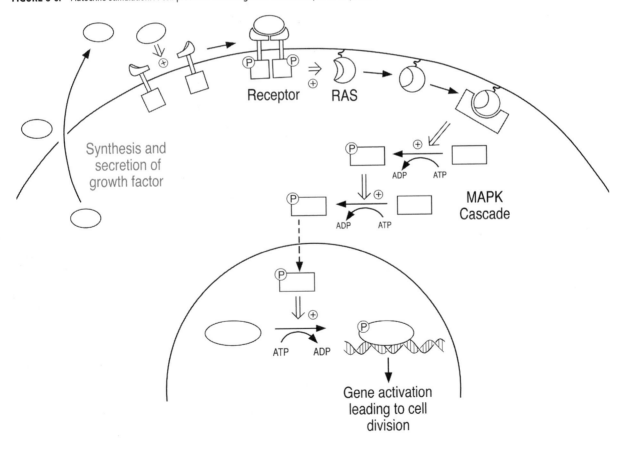

gene—either because of increased transcription or gene amplification—has been detected in numerous tumors. The EGF receptor gene belongs to the *ERBB* family of oncogenes.

As we saw above, *SRC* belongs to the receptor-stimulated protein tyrosine kinases. Overall, approximately one-third of a sample of solid tumors overexpressed protein tyrosine kinases or receptors.

RAS and the MAP Kinase Cascade

Immediately downstream of the receptors, or receptor-associated protein tyrosine kinases, are members of the Ras family (see Fig. 8-3B). The gene product, also known as p21 or p21^{RAS}, is a monomeric GTP-binding protein with GTPase activity. (Note that although the name is similar, p21^{RAS} is different from the protein p21^{CIP1}, an inhibitor of cyclin-dependent kinases we saw earlier.) GTP-Ras is the active form of the protein, which becomes inactive upon hydrolysis of GTP (Fig. 8-7). In addition to other targets, GTP-RAS acts on the MAP kinase pathway to stimulate cell proliferation. Amino acid substitutions at some positions in RAS (especially residues 12 and 13) result in a reduction of the GTPase activity; others (residues 116, 117, 119, and

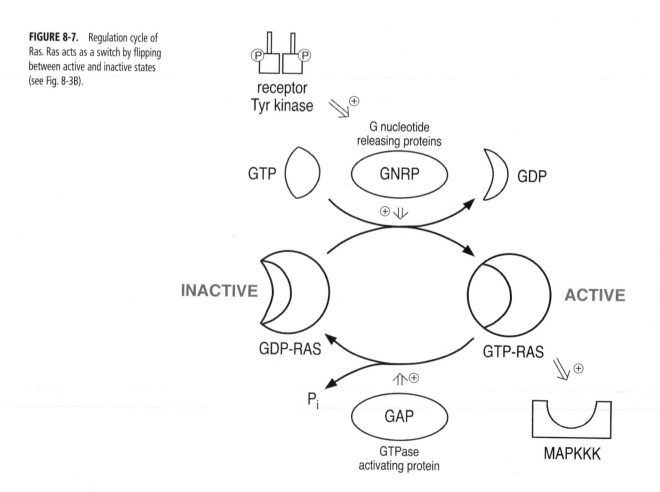

FIGURE 8-7. Regulation cycle of Ras. Ras acts as a switch by flipping between active and inactive states (see Fig. 8-3B).

146) increase the rate of exchange of GDP for GTP. In both cases, the direct consequence is an excessive accumulation of the active form in the cytoplasm. Such hyperactivated *RAS* oncogene products are found in approximately 30% of all cancers. (In Table 8-2 *RAS* is listed under the names of individual members of this family: *HARAS*, *KIRAS*, and *NRAS*.)

The equilibrium between active and inactive RAS depends on other proteins, which in turn respond to extracellular signals. The intrinsic GTPase activity of RAS is poor, but it is greatly stimulated by a *GTPase-activating protein* (GAP) (see Fig. 8-7). Mutant RAS protein, however, does not respond to GAP and tends to accumulate in the active form. The GAP family of proteins—one of whose members is the product of the neurofibromatosis 1 gene (*NF1*)—thus acts to suppress cellular proliferation in the absence of specific extracellular signals, and its members are tumor suppressors rather than oncogenes (see below).

Exchange of GDP for GTP also requires the action of other proteins, the *guanine-nucleotide–releasing proteins* (GNRPs). GNRPs are activated by receptor tyrosine kinases and seem to be the main regulatory system governing RAS activation by mitogens (see Figs. 8-3, color asterisk, and 8-7), although in some cell types extracellular stimulation may also result in repression of GAP activity.

The most notorious of the phosphorylation cascade proto-oncogenes encodes a MAPKKK (see Fig. 8-3C), and was first identified as *RAF*, an oncogenic form carried by certain mouse and chicken sarcoma viruses. Its mutation is prevalent in stomach cancer.

Nuclear Oncogenes

Genes for proteins that function farther down in the signaling pathway (see Fig. 8-3D) may also become oncogenic. The first genes whose transcription is stimulated by growth factor signals (early response) code for transcription factors that, in turn, activate the expression of other genes important for cell division. Three of these transcription factors are the proto-oncogenes *MYC*, *FOS*, and *JUN*, which become oncogenic by mutation or overexpression.

There are three proteins in the MYC family—MYC, NMYC, and LMYC—and they play a central role in cell proliferation and differentiation in various tissues. Thus, cells in which *MYC* is not expressed are unable to divide and cells in which *MYC* is expressed constitutively are not able to enter G0. Gene amplification (see Plate 6) is the most common cause of overexpression of *MYC* in human cancers.

Suppression of Apoptosis

Apoptosis, or programmed cell death, is a developmental mechanism used to remove cells in certain morphogenetic processes. Among blood cell precursors, apoptosis eliminates as many as 95% of some lymphocyte precursor cells due to improper rearrangement of immunoglobulin or cell surface receptor genes. Apoptosis also functions to do away with cells that have become damaged, or that have lost the ability to respond normally to environmental signals; in particular, cells that proliferate excessively are subject to apoptotic control. The products of certain genes, such as *BCL2*, act to prevent apoptosis except under the proper conditions. Overexpression of *BCL2*, however, leads to complete suppression of apoptosis and thus contributes to carcinogenesis.

TUMOR SUPPRESSOR GENES

We will now consider *tumor suppressor genes,* the genes whose products act to control rather than stimulate cell division. Contrary to oncogenes, they become carcinogenic when *inactivated* by a mutation, and are often associated with familial cancers (Table 8-4). We will review the function of a few of them.

Retinoblastoma

As can be expected from the central role the retinoblastoma protein, pRB, plays in the control of the cell cycle, a cell homozygous for amorphic mutations in this function would be unable to respond to signals directing it to pause in G1 or go to G0 (see Figs. 8-1 and 8-2). Such mutations are characteristic of the eye tumors that give this gene its name. As in all tumor suppressor genes, cancer-causing mutations lead to a loss of function and are recessive, because a single working copy of the gene is sufficient to provide a significant amount of normal function.

Tumor suppressor gene mutations are recessive at the cellular level; at the organismal level, however, they are dominant, as illustrated by the *RB* gene (Table 8-5): children carrying one copy of the mutant allele develop eye tumors with a penetrance of 85% and later on suffer other forms of cancer. This apparent contradiction was resolved by studying the genetic composition of cancer cells in these patients, where it was found that in addition to the original mutation inherited from a parent, the other allele is now also mutant or lost through chromosomal rearrangement or loss. These results had been accurately predicted by Knudsen, who in the early 1970s, long before the function of pRB was known, proposed a "two-hit" hypothesis of gene inactivation to explain the *RB* phenotype.

The dominant behavior of *RB* mutations in these familial cancers is witness to the unrelenting and ubiquitous nature of somatic mutations; the implication is that in the development of the majority of normal eyes there is at least one cell with a mutation that inactivates the *RB* gene. In most people, the chance that another mutation will knock out the second allele in the progeny of that heterozygous cell is very rare, and so is the incidence of eye tumors. In patients who inherit only one normal allele, however, that single mutation is sufficient to put a cell on the path to neoplasia.

It should be remembered that the *RB* gene is expressed in all tissues. It is not clear why in human heterozygotes RB^+/RB^- the retina is more sensitive to develop tumors than other tissues. In the mouse, strains with a mutation of this same *RB* gene suffer predominantly from pituitary tumors—mice homozygous RB^-/RB^- die after a few days of gestation.

Practically all human cancers seem to have mutations that affect the regulatory functions of the pRB pathway (p16, cyclin D1, CDK4, or pRB itself) (see Fig. 8-2), but it is interesting that different mutations predominate in the various tissue-specific cancers. Thus, in one study, 48 out of 55 small cell lung carcinomas lacked normal pRB, and six of the other seven lacked the inhibitor p16. On the other hand, in many esophageal, breast, and squamous cell carcinomas a cyclin D gene is amplified; in glial tumors the CDK4 gene is often amplified.

pRB is so important in cell proliferation because, in addition to its control of G1 to S transition, it seems to play a role in regulation of rRNA, tRNA, and other small RNAs synthesis—and, through them, in the control of cellular biosynthesis and cell growth.

TABLE 8-4. Tumor suppressor genes: a summary of selected inherited cancer syndromes.

Syndrome	Primary Tumor	Associated Cancers or Traits	Chromosome Location	Cloned Gene	Proposed Function of Gene Product
Dominant inheritance					
Familial retinoblastoma	Retinoblastoma	Osteosarcoma	13q14.3	RB1	Cell cycle and transcriptional regulation; E2F binding
Li-Fraumeni syndrome (LFS)	Sarcomas, breast cancer	Brain tumors, leukemia	17p13.1	p53 (TP53)	Transcription factor; response to DNA damage and stress; apoptosis
Familial adenomatous polyposis (FAP)	Colorectal cancer	Colorectal adenomas, duodenal and gastric tumors, CHRPE jaw osteomas and desmoid tumors (Gardner syndrome), medulloblastoma (Turcot syndrome)	5q21	APC	Regulation of β-catenin microtubule binding
Hereditary nonpolyposis colorectal cancer (HNPCC)	Colorectal cancer	Endometrial, ovarian, hepatobiliary and urinary tract cancer glioblastoma (Turcot syndrome)	2p16, 3p21 2q32, 7p22	MSH2, MLH1 PMS1, PMS2	DNA mismatch repair
Neurofibromatosis type 1 (NF1)	Neurofibromas	Neurofibrosarcoma, AML, brain tumors	17q11.2	NF1	GAP for p21 ras proteins; microtubule binding?
Neurofibromatosis type 2 (NF2)	Acoustic neuromas, meningiomas	Gliomas, ependymomas	22q12.2	NF2	Links membrane proteins to cytoskeleton?
Wilms' tumor	Wilms' tumor	WAGR (Wilms', aniridia, genitourinary abnormalities, mental retardation)	11p13	WT1	Transcriptional repressor
Wiedemann-Beckwith syndrome (WBS)	Wilms' tumor	Organomegaly, hemihypertrophy hepatoblastoma, adrenocortical cancer	11p15	?p57/KIP2 ?Others-contiguous gene disorder	Cell cycle regulator
Nevoid basal cell carcinoma syndrome (NBCCS)	Basal cell skin cancer	Jaw cysts, palmar and plantar pits, medulloblastomas, ovarian fibromas	9q22.3	PTCH	Transmembrane receptor for hedgehog signaling molecule
Familial breast cancer 1	Breast cancer	Ovarian cancer	17q21	BRCA1	Interacts with Rad51 protein; repair of double-strand breaks

TABLE 8-4. Tumor suppressor genes: a summary of selected inherited cancer syndromes. *(Continued)*

Syndrome	Primary Tumor	Associated Cancers or Traits	Chromosome Location	Cloned Gene	Proposed Function of Gene Product
Familial breast cancer 2	Breast cancer	Male breast cancer, pancreatic cancer, ?others (for example, ovarian)	13q12	BRCA2	Interacts with Rad51 protein; ?repair of double-strand breaks
von Hippel-Lindau (VHL) syndrome	Renal cancer (clear cell)	Pheochromocytomas, retinal angiomas, hemangioblastomas	3p25	VHL	?Regulates transcriptional elongation by RNA polymerase II
Hereditary papillary renal cancer (HPRC)	Renal cancer (papillary type)	?Other cancers	7q31	MET	Transmembrane receptor for HGF
Familial melanoma	Melanoma	Pancreatic cancer, dysplastic nevi, atypical moles	9p21	p16 (CDKN2)	Inhibitor of CDK4 and CDK6 cyclin-dependent kinases
			12q13, ?Others	CDK4	Cyclin-dependent kinase
Multiple endocrine neoplasia type 1 (MEN1)	Pancreatic islet cell	Parathyroid hyperplasia, pituitary adenomas	11q13	MEN1	Unknown
Multiple endocrine neoplasia type 2 (MEN2)	Medullary thyroid cancer	Type 2A pheochromocytoma, parathyroid hyperplasia, Type 2B pheochromocytoma, mucosal hamartoma, familial medullary thyroid cancer	10q11.2	RET	Transmembrane receptor tyrosine kinase for GDNF
Multiple exostoses	Exostoses (cartilaginous protuberances on bones)	Chondrosarcoma	8q24.1, 11p11-13 19p	EXT1, EXT2, EXT3	Unknown
Cowden disease	Breast cancer, thyroid cancer (follicular type)	Intestinal hamartomatous polyps, skin lesions	10q23	PTEN(MMAC1)	Dual-specificity phosphatase with similarity to tensin
Hereditary prostate cancer (HPC)	Prostate cancer	Unknown	1q25, ?Others	Unknown	Unknown
Palmoplantar keratoderma	Esophageal cancer	Leukoplakia	17q25	Unknown	Unknown

Recessive syndromes

Ataxia telangiectasia (AT)	Lymphoma	Cerebellar ataxia, immunodeficiency, ?breast cancer in heterozygotes	11q22	ATM	DNA repair; ?Induction of p53
Bloom's syndrome	Solid tumors	Immunodeficiency, small stature	15q26.1	BLM	?DNA helicase
Xeroderma pigmentosum	Skin cancer	Pigmentation abnormalities, hypogonadism	Multiple complementation group	XPB, XPD, XPA	DNA repair helicases, nucleotide excision repair
Fanconi's anemia	AML	Pancytopenia, skeletal abnormalities	9q22.3, 16q24.3, ?two others	FACC, FACA	?DNA repair

AML, acute myelogenous leukemia; CHRPE, congenital hypertrophy of the retinal pigment epithelium; GAP, GTPase-activating protein, a negative regulator of the p21 ras guanine nucleotide-binding proteins; contiguous gene disorder, alterations in several distinct genes in a particular chromosomal region account for the phenotype seen in patients with the disorder; hedgehog, a secreted factor that regulates cell fate determination via its binding to the PTCH protein; HGF, hepatocyte growth factor; GDNF, glial-derived neurotrophic factor

SOURCE: Based on Table 1 in ER Fearon, Human cancer syndromes: clues to the origin and nature of cancer. Science 278:1043-1050 (1997). Copyright 1997 American Association for the Advancement of Science. Used with permission.

TABLE 8-5. Loss-of-function mutations in tumor suppressor genes: phenotypes of mutations in tumor suppressor genes. Gain-of-function mutations in tumor suppressor genes are unknown. *Somatic mutations* refers to the effect of a mutation on somatic cells in an otherwise homozygous normal individual. *Germline mutations* refers to the phenotype of an individual who received the mutation through one of the gametes and is therefore wholly mutant. (Shading corresponds to those cases found in human cancers.)

Somatic	Recessive, oncogenic in homozygous cells.
Germline	Dominant, familial cancer predisposition, very high penetrance. Homozygous lethal in knockout mice.

An exception to the generalization in this table is provided by *BRCA1* mutations that appear to be no more severe in homozygous than heterozygous condition (see Fig. 2-7).

Isolation of the *RB* gene was done—as was the isolation of most of the tumor suppressor genes with familial syndromes—by positional cloning. Its cDNA codes for a protein of 928 amino acids. Most alterations in transformed cells are deletion, frameshift, nonsense, or splice-site mutations that grossly alter a region of the protein identified as a binding "pocket." Within this pocket, deletion mapping of binding sites and crystallographic studies defined two "boxes," A and B—from residue 393 to 572 and from 646 to 772, respectively—to which the transcription activator E2F binds (Fig. 8-8). Interestingly, to the same box B bind CDK-associated D cyclins, responsible for phosphorylative inactivation of pRB, and viral proteins that block pRB function. What cyclins and viral proteins have in common is the pentapeptide sequence LXCXE, in which the residues Leu, Cys, and Glu point toward a shallow, but tight-fitting, groove on the surface of pRB and interact directly with it.

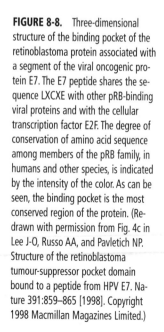

FIGURE 8-8. Three-dimensional structure of the binding pocket of the retinoblastoma protein associated with a segment of the viral oncogenic protein E7. The E7 peptide shares the sequence LXCXE with other pRB-binding viral proteins and with the cellular transcription factor E2F. The degree of conservation of amino acid sequence among members of the pRB family, in humans and other species, is indicated by the intensity of the color. As can be seen, the binding pocket is the most conserved region of the protein. (Redrawn with permission from Fig. 4c in Lee J-O, Russo AA, and Pavletich NP. Structure of the retinoblastoma tumour-suppressor pocket domain bound to a peptide from HPV E7. Nature 391:859–865 [1998]. Copyright 1998 Macmillan Magazines Limited.)

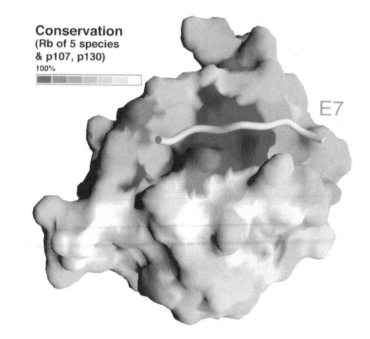

Conservation
(Rb of 5 species
& p107, p130)
100%

E7

p53

The protein p53 has multifaceted biological significance and plays a central role in the regulatory pathways that control cell fate. p53 is sensitive to the presence of DNA damage (Fig. 8-9), foreign DNA (such as viral infections), or hypoxia, and it responds to these conditions by blocking DNA synthesis, at least in part, by suppressing pRB phosphorylation (see Fig. 8-2). Under some conditions when it is unable to stop cells in G1 (in the absence of pRB product, for example), p53 activates genes that put the cell in a programmed death pathway.

p53 is a 393-amino-acid transcription factor with four domains (Fig. 8-10B):

1. A C-subterminal region (residues 324–355) through which four monomers bind into active tetramers.
2. A central DNA-binding region (residues 102–292, where most of the mutations that inactivate *p53* are found) that recognizes the binding site 5'PuPuPuC(A/T)3' in DNA. These binding sites occur in groups of four, one for each subunit of the p53 tetramers, in the regulatory region of genes controlled by p53.
3. An N-terminal transcriptional activation domain (residues 1–42) that interacts with basal transcription factors associated with promoters located in the vicinity of p53 binding sites.
4. A C-terminal regulatory region (residues 367–393). Newly synthesized p53 is inactive, and it is incapable of sequence-specific DNA binding unless it is sterically modified and activated through this regulatory region. Activation can be achieved by degradation or phosphorylation of the C-terminal domain or by its association with single-stranded DNA. The regulatory domain also binds preferentially to internal single-stranded DNA loops (as produced by slippage during synthesis) and to DNA ends.

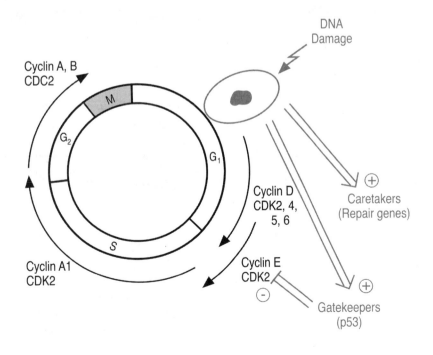

FIGURE 8-9. Regulatory effects of DNA damage. Induction of DNA repair processes and of p53. p53 in turn blocks cell transit through the late G1 restriction point.

FIGURE 8-10. **(A)** Frequency distribution of mutations along the p53 polypeptide sequence. The vertical axis indicates the number of times that mutations at each site have been found in various tumors. The horizontal axis indicates amino acid positions in the polypeptide chain. Some of the amino acids and their codons are shown. **(B)** Various functional regions along the p53 polypeptide sequence. The amino terminus is to the left, and the numbering of amino acids corresponds to part A. (Based on Cho et al., 1994.)

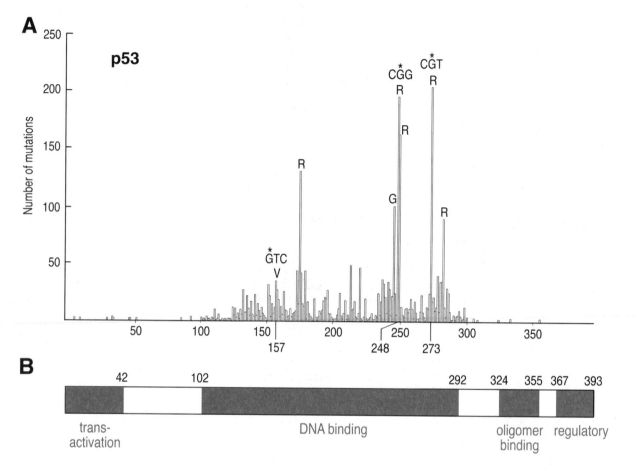

Among the genes that respond to p53 activation, under various conditions, are those that code for inhibitors of cyclin-dependent kinases (such as p21); GADD45, which binds to proliferating cell nuclear antigen (PCNA) and thus arrests the cell cycle in response to DNA damage; IGF-BP3, an insulin-like growth factor receptor binding protein, which blocks the receptor and thus prevents the cell from responding to the mitogen IGF; and *BAX*, a gene that promotes apoptosis.

Under normal conditions p53 has a very short half-life and it is present at low concentrations, mostly in its inactive form. Cells in distress have been shown to stimulate activation and increase the translation rate of p53. Conditions that increase active p53 levels include viral or damaged DNA (including repair and recombination intermediates), low nucleotide pools, and hypoxia.

In addition to the direct detection of DNA damage by p53, there must be other proteins that sense unfavorable conditions and act on p53—one such protein is the product of the *ataxia telangiectasia* (AT) gene, *ATM*. Cells deficient in *ATM* function fail to repair DNA damage, apparently because p53 levels remain low. For the same reason

ATM$^{-/-}$ cells (homozygous for the mutation) seem to be deficient in ionizing-radiation–induced apoptosis. (AT is a complex hereditary disease that includes neural and immune systems defects, cancer, and hypersensitivity to ionizing radiation.)

Thus, p53 plays a pivotal role in the cellular response to various unfavorable conditions. It ascertains the type and extent of damage suffered and it evaluates a host of intrinsic and environmental variables. Its output is the stimulation of one of two pathways: 1) cell cycle arrest or 2) apoptosis, programmed cell death (Fig. 8-11).

p53 acts as a tumor suppressor gene because, in its absence, cells enter S despite the presence of DNA damage, leading to a greatly increased mutation rate, especially aneuploidies; when aneuploidies involve loss of other tumor suppressor genes, neoplastic transformation results. Normal *p53* functions as a tumor suppressor also by committing to apoptosis those cells in which there is a loss of pRB function, or in which there are hyperactivated oncogenes. It also places cells on a programmed cell

FIGURE 8-11. Processes controlled by p53. Signals converging on and emanating from p53 (top half) and various conditions (bottom half) under which p53 function ultimately results in a delay of the cell cycle (right side) or programmed cell death (left side).

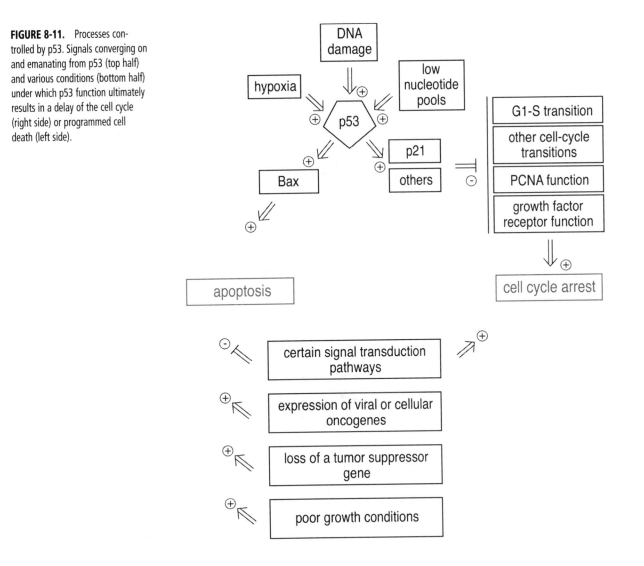

death pathway under conditions of low O_2, as is found in solid tumors before vascularization. It is therefore not surprising that loss of *p53* function is an early step in development of at least some cancers and that over 50% of all cancers have deleted or nonfunctional *p53* genes.

In some tumors (mostly sarcomas) p53 is not altered, but there is overproduction of the p53 inhibitor MDM2. Finally, in teratocarcinomas (tumors of undifferentiated embryonic cells) inactivation of *p53* is not selected for because the stem cells from which they are derived do not express *p53*—but at least two other genes related to *p53* have been identified, which may play a similar role in these cells.

Li-Fraumeni syndrome is a condition analogous to retinoblastoma. It is transmitted as an autosomal dominant trait, patients suffer cancers at an early age, and it is caused by heterozygosity for a loss-of-function mutation in a tumor suppressor gene, in this case *p53*. Given the rarity and severity of the heterozygous condition, homozygous *p53⁻/⁻* individuals are not known in humans, but such null genotypes have been produced in mice. Homozygous mutant mice develop relatively normally, but are more prone to cancer and have a higher frequency of malignancies when treated with ionizing radiation or mutagens. These mice do not show a higher incidence of point mutations in other genes.

The effectiveness of therapies in some cancers can be explained by the biological function of p53. Anticancer treatments usually work by causing DNA damage. Cancer cells, having lost the G1 restriction point, proceed to S with unrepaired damage and are shunted into apoptosis by p53. In normal cells, by contrast, the G1 to S transition is delayed until the damage is repaired, and they survive. In childhood acute lymphoblastic leukemias, there is very good correlation between functional p53 and a good response to radiation or chemotherapy. In cases of relapse, loss of treatment effectiveness is usually associated with loss of p53 function. Some other cancers (melanoma and lung cancers) are often associated with loss of p53 function; lacking access to p53-dependent apoptosis, they are more resistant to treatment. Many issues of basic cancer biology remain to be learned, however, because the correlation between p53 function and response to treatment is not applicable in all cases.

HNPCC Family of Mismatch Repair Genes

The impact of defects in DNA repair on cancer first came to light with the discovery that many syndromes characterized by a high risk of tumors, such as xeroderma pigmentosum (*XP*) and Bloom's syndrome (*BSM*), actually resulted from mutations in genes involved in repair. Significant progress in the field was made when it was discovered that a family of genes that is mutant in a rare form of familial cancer—*hereditary nonpolyposis colon cancer* (*HNPCC*)—is responsible for various steps in the mismatch repair process (see Chapter 7). These same genes were subsequently found mutant or missing in many sporadic cancers; for example, 15% of sporadic colorectal cancers have mutations in the *HNPCC* genes. As might be expected from the role of mismatch repair, such mutations are characterized by high levels of microsatellite DNA instability (MIN).

In hereditary nonpolyposis colon cancer families, the *HNPCC* mutations behave as dominant, just as retinoblastoma and Li-Fraumeni syndrome do. They could be considered mutations in tumor suppressor genes whose loss of function contributes to neoplastic transformation. At least part of the mechanism by which a deficiency in mismatch repair genes causes colorectal cancer has been elucidated, and it involves mutational inactivation of subunit II of the receptor for TGF-β, which acts to repress cell division in epithelial cells. The coding region of the gene for subunit II of the TGF-β receptor includes a segment with 10 adenines in a row. This is the kind of sequence prone to expansion or contraction, and therefore frameshifts, in cells deficient in mismatch repair. Cells with the resulting frameshift mutations would have a nonfunctional receptor, be unresponsive to TGF-β, and immediately acquire a replicative advantage.

BRCA1 and BRCA2

The *BRCA1* and *BRCA2* genes are mutant in families with dominant breast cancer predisposition—up to 80% of carriers develop breast cancer, usually at an early age (see Example 2.2, "Familial Breast Cancer," in Chapter 2). Females with mutations in *BRCA1* also have increased risk of ovarian cancer. But note that familial breast cancers represent only a small percentage of the overall incidence of breast cancer, as is the case with other forms of familial cancer predisposition.

Mutations in these genes seem to contribute to cancer indirectly, by interfering with DNA repair. The products of both genes have been found associated with RAD51, a protein involved in repair of double-strand breaks and in recombination. Mice homozygous for *BRCA* null mutations die during embryogenesis and cells from those embryos are very sensitive to ionizing radiation. Thus, *BRCA1* and *BRCA2* may contribute to cancer in a manner similar to *HNPCC* and *XP*.

Gatekeepers and Caretakers

Discovery of the role of DNA repair genes in cancer development led to the classification of tumor suppressor genes into two groups. *Gatekeepers* are tumor suppressor genes whose products have direct control of the cell cycle, such as *RB* or *p53*. *Caretakers* are genes whose products help maintain the integrity of other genes, such as the *BRCA* and *HNPCC*. In the case of caretaker genes, loss of function at two loci is required for a cell to increase proliferation. Mutation at the first locus, the caretaker gene, simply increases the probability that mutations in one or more gatekeeper genes will occur.

It has been suggested that tumor suppressor genes may not be completely recessive at the cellular level, but rather that cells heterozygous for a null mutation may have a phenotype intermediate between wild type and mutant. Thus, heterozygous cells would have increased mutability in the case of a caretaker gene, or increased replicative advantage in the case of a gatekeeper gene. In either case, the likelihood that mutation of both alleles occurs is greater in heterozygous cells than in homozygous normal cells. On the other hand, the "two-hit" view or carcinogenesis may not apply uniformly–one conflicting example may be provided by the *BRCA1* gene–according to one report, mutations appear to be equally carcinogenic in the homozygous or heterozygous condition (see Fig. 2-7).

MUTATIONS AND OTHER CHANGES THAT ALTER THE FUNCTION OF CANCER GENES

We will briefly review here the types of mutations found in neoplastic cells: those that inactivate tumor suppressor genes and those that lead to overexpression of oncogenes.

In addition to mutations, mention will be made of two other factors that affect gene expression: viruses, and epigenetic changes.

Epigenetic Changes—Altered Chromatin Methylation

Changes in the composition and organization of chromatin can alter the expression of single genes or entire chromosomes. Gene silencing accompanied by hypermethylation of DNA is stably transmitted from one cell generation to the next—the same as X-inactivation in females—and is functionally equivalent to a null mutation. In cancer cells, this type of modification is sometimes found in the gene for the inhibitor of cyclin-dependent kinases *p16* (see Fig. 8-2). Altered methylation levels have also been observed in early stage colorectal adenomas.

Point Mutations

Every kind of point mutation discussed in Chapter 7 is found in one or another of the mutant genes in cancer cells. Their distribution is, however, surprisingly nonrandom. In *p53* and *RAS* the mutations are mostly amino acid substitutions, but in *RB* and *APC* they are for the most part chain terminations.

Cancer-causing base substitutions in *p53* occur at many sites within the DNA-binding domain. The positions of the mutations are not uniformly distributed, as there are many distinctive hotspots (see Fig. 8-10A). Furthermore, the position of the hotspots is not entirely conserved from one type of cancer to another. Position 157, for example, is a prevalent hotspot in lung cancer, but not in other malignancies. In the case of lung cancer, the distribution of mutations can be explained in part by the preferential binding of the potent mutagen and carcinogen BPDE to guanines in positions 157, 248, and 273 of the non-template strand. BPDE is the metabolically activated dihydroxy, epoxy, form of benzo[a]pyrene, a ubiquitous component of cigarette smoke and a cause of G→T transversions. That mutated G is preferentially in the non-template strand is probably explained by its much slower repair rate compared to the template strand (see Chapter 7). Other such specific correlations can be found between dietary aflatoxin and G→T transversions at residue 249 in liver cancer, and between ultraviolet light and C→T transitions in skin cancer. On the other hand, prevalent *p53* mutations in colon cancer, often C→T transitions at CpG pairs, seem to be explained better by the intrinsic mutability of CpG pairs, rather than the specificity of carcinogens.

In 6% of the Ashkenazi Jewish population there is an inherited point mutation in the coding region of the *APC* gene that acts in an unusual way to give a twofold increase in the probability of colon cancer. In this mutation a T→A transversion converts the sequence GAA ATA AAA to GAA AAA AAA. This substitution, in itself (Ile1307Lys), does not affect protein function significantly; rather, in creating a string of eight As, it produces a hotspot for strand slippage and frameshift mutations. Such apparently neutral polymorphisms may play a similar role in other tumor suppressor genes.

Chromosomal Rearrangements

In addition to deletions—which result in null mutations of tumor suppressor genes—translocations play an important role in the creation of oncogenes. Translocations activate growth promoting genes in one of two ways:

1. By positioning a proto-oncogene in the vicinity of a strong enhancer, which leads to increased or misplaced transcription.

2. By fusing two genes, usually through an intron, in such a way that a chimeric protein is produced.

Specific translocations have been found in association with a large number of tumor types, such as t(12;16)(q13;p11) in myxoid liposarcoma, t(2;13)(q37;q14) in alveolar rhabdomyosarcoma, and t(11;22)(q24;q12) in Ewing sarcoma. These rearrangements have been very useful in the identification and positional cloning of oncogenes.

Activation of proto-oncogenes by translocations. The process of activating proto-oncogenes through translocation is found frequently in acute as well as chronic leukemias. The gene providing the activating enhancer is usually one of the antibody genes (immunoglobulin heavy or light chains) or the gene for the T-cell receptor. In the course of lymphocyte maturation, all of these genes are involved in intragenic rearrangements designed to generate antibody diversity. With certain regularity, however, the rearrangements engage DNA in other chromosomes, resulting in translocations. In most cases cells with such translocations undergo apoptosis, except when the translocation occurs near a proto-oncogene and oncogenic transformation results.

The proto-oncogenes involved are often transcription factors. The first molecular demonstration of the carcinogenic effect of a translocation was for Burkitt's lymphoma, in which the gene for the transcription activator cMYC was translocated into the vicinity of the immunoglobulin heavy chain (Fig. 8-12). Translocation of other transcription factor genes occurs in various T-cell acute leukemias.

Another type of proto-oncogene whose expression is enhanced by translocations to the vicinity of immunoglobulin genes is the suppressor of apoptosis *BCL2*. Such translocations are found in the chronic cancer follicular lymphoma.

FIGURE 8-12. Molecular analysis of a Burkitt's lymphoma cell line with the reciprocal translocation t(8;14)(q24;q32). Organization of the *MYC* and immunoglobulin heavy chain genes in the normal and translocated chromosomes. The breakpoint in chromosome 8 is approximately 1 kb upstream of the transcription initiation site of the *MYC* gene, whereas the breakpoint in chromosome 14 is in an intron between the J and C regions of the Igμ immunoglobulin gene. It is not entirely clear what part of the Ig gene acts as an enhancer.

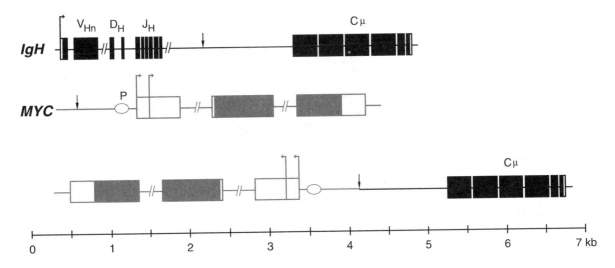

Chimeric proteins. The fusion of coding sequences to produce a *chimeric protein* is a more common mechanism of carcinogenesis than gene activation. The first documented case was provided by the *Philadelphia chromosome*, the shortened chromosome 22 that is one of the partners in the translocation t(9;22)(q34;q11) and is found in cells from 90% of chronic myeloid leukemia patients (Fig. 8-13). The translocation breaks chromosome 9 in the 200-kb-long first intron of the *ABL* proto-oncogene and breaks chromosome 22 in intron 7 of the *BCR* gene. The tip of 9q is added to 22q in such a way that a chimeric gene is formed with the 5' end of *BCR* and the 3' end of *ABL*. The ABL protein is a nonreceptor protein Tyr kinase that interacts with DNA, pRB, and the product of the ataxia telangiectasia gene. It seems to be involved in the detection of DNA damage and cell cycle checkpoints. A polypeptide segment encoded by the first exons of *BCR* interacts with the regulatory region of the ABL protein, thereby stimulating its kinase activity and promoting oncogenesis.

Other examples of protein fusions include PDGF-β receptor kinase and a transcription factor, which thus produces a DNA-binding protein with kinase activity (found in chronic myelomonocytic leukemia); and t(15;17)(q21;q11–22), which produces a chimeric protein between the retinoic acid receptor and another transcription factor (found in acute promyelocytic leukemia) (Plate 5). More than 30 cases of gene fusions have been described, in solid tumors as well as hematopoietic malignancies.

Gene Amplification

Increase in copy number of large DNA segments is characteristic of many tumor types and is another mechanism of overexpression of oncogenes. Typically it occurs in tumor cells displaying genomic instability—those that have already lost the ability to correct DNA damage or undergo apoptosis. DNA amplification manifests itself cytologically in the form of *homogeneously stained regions* (HSR), chromosomal segments

FIGURE 8-13. Gene fusion as a result of the translocation t(9;22)(q34;q11). The exon–intron organization of the *ABL* and *BCR* genes is shown, as well as the fusion gene present in the Philadelphia chromosome (Ph).

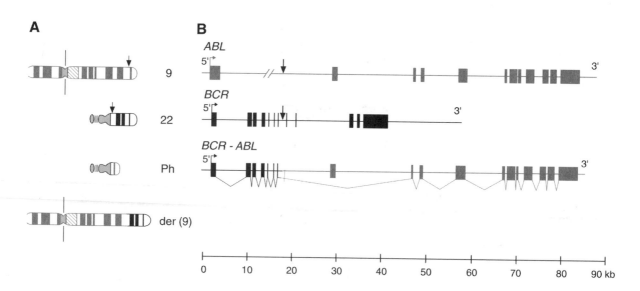

with an absence of Giemsa bands, and as *double minutes* (DM), small chromatin bodies with the appearance of very small chromosomes. DNA amplification has received particular attention in breast cancer, where amplification of the oncogenes *HER2* (*erb*B2), *MYC* (Plate 6), and H-*RAS* has been observed.

Viruses

As was mentioned earlier, the oncogenic role of viruses has been more prevalent in animal systems than in humans. It is now well established, however, that under certain conditions viral oncogenesis occurs in humans. Molecular analyses have shown that viruses act, as mutations do, to interfere with the regulatory network of tumor suppressors and oncogenes. And just as any mutation, viruses do not cause cancer by themselves—in cancer cells, viral infection is accompanied by changes in cellular genes.

We will consider separately the action of *small DNA tumor viruses* and *retroviruses*. Two other viruses of the *herpes* family are responsible for specific cancers: Epstein-Barr virus connected with some forms of Burkitt's lymphoma, and Kaposi-sarcoma–associated herpesvirus.

Small DNA tumor viruses. The small DNA tumor viruses include three main groups: *adenoviruses, papilloma viruses,* and *polyoma viruses,* all with genomes between 5 and 35 kb. They are common and benign parasites in all human populations. Infections are chronic, and for the most part are asymptomatic, except for the occurrence of limited cellular hyperproliferation. For example, common warts are caused by some forms of papilloma virus.

Although a minority of papilloma virus infections result in cancers, some types of cancers are often associated with viral infections. Over 90% of cervical and other anogenital carcinomas are associated with papilloma virus infection. The oncogenicity of these viruses is due to two viral proteins interfering with the tumor suppressor proteins pRB and p53. Thus, inactivation of pRB, as is prevalent in cervical cancer, is carried out by the binding of the viral oncoprotein E7 to the same pocket in pRB through which pRB binds and inhibits the transcription factor E2F (see Figs. 8-2 and 8-8). The oncogenic viral proteins E1A of adenovirus and large-T antigen of polyoma virus also attach to the binding site of pRB. E6, another papilloma virus oncogenic protein, binds p53 and accelerates its degradation, thus reducing the effectiveness of one of the primary tumor suppressor products in the cell.

Retroviruses. *Retroviruses* are so called because the RNA genome carried by infecting particles is copied into DNA within the host cell and then integrated into the cellular genome. The viral infection is not cytotoxic, and the host cell may carry the infecting virus indefinitely. These are the viruses that led to the identification of so many oncogenes in the last two decades.

There are two general mechanisms of oncogenesis. The first, characteristic of *nonacute* retroviruses, consists on the insertion of the viral genome in the proximity of a proto-oncogene in such a fashion that the latter falls under the control of strong viral enhancers, or is in some other way overactivated by the virus. The second mechanism, characteristic of *acute* retroviruses and leading to much more potent oncogenic activity, consists of the incorporation of an oncogenic fragment of a proto-oncogene into the genome of the retrovirus. This is how the first oncogene, *src*, was discovered in association with the avian Rous sarcoma virus. Most acute retroviruses result in mesodermal sarcomas, leukemias, and lymphomas rather than carcinomas.

HTLV (human T-cell leukemia virus), the only oncogenic human retrovirus, operates by a mechanism slightly different from the two mentioned previously. Rather than the virus working as a *cis*-acting activator of an oncogene (by inserting next to it), one of the proteins coded by the virus is a transcription regulator that is capable of stimulating transcription of various proto-oncogenes. Thus, HTLV acts by *transactivation* of oncogenes.

GENETIC INSTABILITY

Genetic instability is a general characteristic of cancer cells. This instability results from defects either in DNA repair or chromosome segregation—although in a few cases both forms may be present. Genetic instability contributes to the production of new mutations, thus fueling the selective process that gives rise to increasingly aggressive and drug-resistant cells.

Defects in repair include mutations in nucleotide excision repair (NER) genes, which lead to inability to repair pyrimidine dimers in skin in patients with xeroderma pigmentosum (see Chapter 7). This NER-deficient instability is sometimes called NIN. Mutations in mismatch repair genes (MMR) can be easily detected because they result in microsatellite DNA instability (MIN) and were discussed earlier in connection with *HNPCC* (see Chapter 7). In sporadic cancers, MMR defects are found in 10% to 15% of colorectal, gastric, and endometrial cancers, and are very rare in other types of cancer.

There is another interesting difference between NER and MMR mutations and their effect on cancer: *XP* mutations are recessive, whereas *HNPCC* mutations are dominant (as are mutations in most tumor suppressor genes). This difference could be due to the fact that for cancer to develop in individuals who are heterozygous for an *XP* mutation would require 1) a mutation in the second allele of *XP*, and 2) ultraviolet mutagenesis of other cancer-causing genes. In individuals who are heterozygous for an *HNPCC* mutation, a mutation of the second allele (knocking out MMR) would itself be mutagenic without the action of external agents. The difference in dominance between NER and MMR mutations could also be due to intrinsic differences in the mode of cancer development in the various tissues.

Chromosomal instability (CIN), the other form of genetic instability, results in highly aneuploid cells. It is much more frequent than instability due to defects in DNA repair, being found in practically all tumors that do not display MMR defects. It has been estimated that in colorectal cancers displaying CIN, the chromosomal mutation rate can be as high as 1% per chromosome per cell division. Homozygous loss of *p53* function (and concomitant loss of apoptosis) has been held responsible for the survival of aneuploid cells, but it appears that there must also be dominant mutations in other gene(s) involved in the increased frequency of disjunction failures.

In the progression of benign tumor to carcinoma there seems to be a transition at which chromosomal losses and duplications begin to accumulate at high rate, and in some cases defects in the mitotic checkpoint were observed. In normal cells, the mitotic checkpoint ensures that anaphase does not start until all kinetochores are stably associated with the mitotic spindle, thus ensuring normal disjunction and euploidy.

Chromosomal losses late in neoplastic transformation are mainly responsible for "uncovering" mutant alleles in tumor suppressor genes in a process known as *loss of heterozygosity* (LOH) (Figs. 8-14 and 4-10). Thus, a mutation may inactivate one of the two *p53* alleles in an early stage adenoma, producing heterozygous cells. Loss of

FIGURE 8-14. Detection of loss of heterozygosity (LOH) by PCR analysis. PCR products (dashes) generated by primers flanking a (TG)$_n$ microsatellite in intron 4 of the cardiac muscle actin gene in chromosome 15q (several artifactual echo bands are also visible). The sample on lane 1 was normal tissue and shows that this individual was heterozygous for the number of TG repeats at this locus. The sample on lane 2 was from breast cancer tissue of the same individual and shows that one of the two alleles was lost. This observation was part of a large screen of genetic alterations in breast cancer cells; this case of LOH may or may not be associated with a specific tumor suppressor gene. (Courtesy of Dr. Kathleen Conway, Molecular Epidemiology Laboratory, Department of Epidemiology, University of North Carolina.)

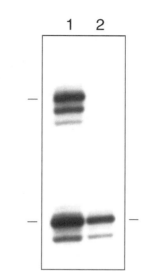

the chromosome 17 with the normal allele, later on, would then lead to complete loss of function for this gene. The resulting cells would actually be *hemizygous* for *p53*, although later nondisjunction may lead to a duplication of the mutant chromosome and homozygosity for the mutant allele. In a study of LOH for markers dispersed throughout the genome, among 50 colorectal tumors, half had LOH in 5% to 25% of alleles, and half had LOH in 25% to 70% of alleles.

LOH in cancer development can be studied using microsatellite loci linked to a tumor suppressor gene. Because of the very high level of polymorphism in microsatellites (see Chapter 4), normal cells are usually heterozygous at the microsatellite locus. A point mutation inactivating the tumor suppressor gene does not change the status of the microsatellite, but later loss of the chromosome with the normal allele can be detected by the loss of heterozygosity in the microsatellite.

LOH has been used successfully to identify tumor suppressor genes. One approach is to prepare karyotypes of many tumor samples for a given type of cancer to identify chromosomes or chromosomal arms that are lost in a high proportion of samples. In addition, DNA from the tumor samples is tested for LOH of any member of a panel of microsatellites selected to represent a significant fraction of the genome. LOH for any given microsatellite, in multiple samples, points to the presence of a putative tumor suppressor gene in that vicinity; such tumor suppressor gene can be later isolated by positional cloning.

Studies of this type led to the identification of chromosomes 5, 17, and 18 as chromosomes in which LOH is commonly associated with colorectal cancers. The gene in chromosome 17 was quickly identified as *p53* because this was already known to be a cancer gene. The search in chromosome 5 led to the identification of *APC* (*adenomatous polyposis coli*), which, when it is inherited as a mutant allele, is responsible for familial adenomatous polyposis (see Table 8-4).

The most difficult step in these searches is usually to demonstrate that a particular candidate gene is in fact responsible for the neoplasia. The ideal confirmation is to introduce a functional copy of the candidate gene into a line of cancer cells and observe a reversion of their proliferative capacity. This is not always possible though, and there have been a few cases of misidentified putative tumor suppressor genes.

INVASIVENESS

The mechanisms by which cancer cells gain the ability to detach from their neighbors, move across the basal membrane, and become invasive seem to involve fewer gene products than the mechanisms that alter the regulation of cell division, although less is known about them. To become invasive, cancer cells must sever the junctions that intimately bind together all epithelial cells.

There is evidence that mutations in the Ca-dependent, cell adhesion protein *E-cadherin* play a role, at least in some cases. Thus, linkage analysis of gastric cancer in a large kindred from New Zealand revealed tight linkage between the disease and molecular markers flanking the gene for E-cadherin. Direct sequencing of the E-cadherin gene in this family, and other families with familial gastric cancer, showed heterozygosity for mutations that lead to protein truncation.

Somatic mutation and LOH in the E-cadherin gene is common in sporadic carcinomas of various tissues, and its inactivation by hypermethylation of CpG sites has also been observed. In animal models of cancer cells, invasiveness was reduced significantly by introduction into those cells of expression vectors carrying E-cadherin cDNA. Conversely, invasiveness was regained by treatment with E-cadherin antibodies.

Introduction of dominant negative E-cadherin mutations into otherwise normal mouse enteric epithelial cells results in increased cell motility and programmed cell death. Thus, loss of E-cadherin function in cancer cells is probably preceded by other mutations that disable apoptosis. Given the behavior of E-cadherin mutations as dominant in familial studies, but recessive at the cellular level, its gene could be considered a tumor suppressor. Because of its cellular function, the term *invasion suppressor* has been used.

CELLULAR IMMORTALITY AND TELOMERES

As mentioned earlier, cell cultures established from healthy tissues eventually die out, whereas cultures from cancers are immortal. We do not know the full spectrum of cellular functions that determine whether cells will senesce and die, or continue to divide indefinitely. It appears clear, however, that the maintenance of telomeres is one cellular function that is centrally involved.

Telomeres, the ends of chromosomes, were recognized as special structures by early cytogeneticists. Thus, while the free ends of broken chromosomal arms are highly reactive and tend to fuse with each other, natural chromosomal termini never do. Telomeres have also been observed associated with the nuclear membrane, and in meiotic prophase they may facilitate synapsis of homologous chromosomes. Molecular analyses have shown that human chromosomes end in several kilobases of sequences derived from repeats of the basic unit 5'TTAGGG3', to which the *telomere repeat binding proteins* (TRF1 and TRF2) bind (Fig. 8-15).

The ends of chromosomes present special difficulties for DNA replication. The strand with a 5 at the telomere cannot be synthesized all the way to the end, because the 5'-most RNA primer cannot be replaced. With every round of replication there is a slight shortening of the DNA molecule (Fig. 8-16). Some shortening can be tolerated because of the telomeric repeats and because of the presence of subtelomeric repeated sequences with no coding function. Eventually, however, shortening results in

FIGURE 8-15.   Telomeric loop. The 3' overhang at each chromosomal end seems to tuck itself into double-stranded DNA, thus avoiding degradation or interaction with other ends. **(A)** Proposed model of loop formation. In native chromatin the DNA is associated with proteins, including the telomeric repeat binding proteins TRF1 and TRF2, that would collapse the loop into a DNA-protein complex. Base pairing is possible because of the repeating nature of telomeric DNA. **(B)** Electron micrograph of telomeric DNA from human HeLa cells. In this particular end the loop measures approximately 20 kb. To visualize the loop, proteins are removed so that DNA can expand freely as it floats on a water surface. (A, Based on Griffith et al., 1999; B, Courtesy of Dr. Jack Griffiths, Lineberger Cancer Center, University of North Carolina.)

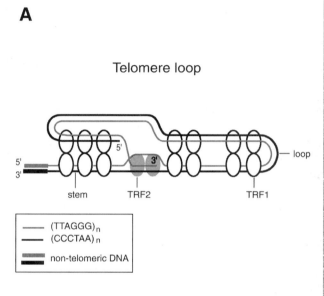

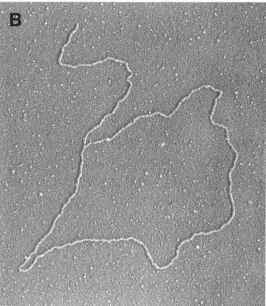

FIGURE 8-16.   Telomeres. Shortening of the lagging strand at its 5' end in each round of replication. Wavy line represents the RNA primer. The chromosomal end is shown here extended rather than in a telomere loop as in Figure 8-15.

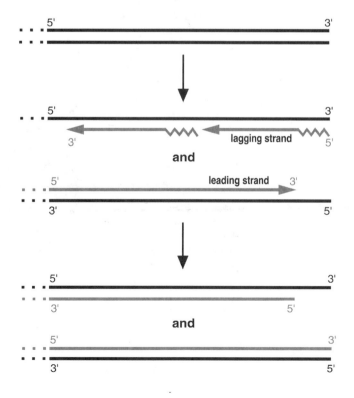

the deletion of essential sequences and the consequent targeting of cells for apoptosis. This process seems to be responsible for the finite replicative capacity of cells in cultures and may be an element contributing to the aging process (Fig. 8-17).

Telomere shortening is counteracted by the action of the enzyme *telomerase*, which has as a cofactor an 11-nucleotide RNA molecule complementary to the sequence of almost two consecutive telomeric repeat units. Using this RNA as a template, telomerase acts as a reverse transcriptase to elongate the 3' overhang and thus compensate for the shortening of its complementary strand (Fig. 8-18). Telomerase is active in germline cells so that each new individual starts out with telomeres of a predetermined length, but not in normal somatic cells.

CONCLUSION

Oncogenic transformation is a process of natural selection for the various cell populations that constitute the organism. Cells that acquire the ability to proliferate beyond the limits imposed by normal development, because of their increased numbers and relative advantage, have an enhanced chance of accumulating new mutations, which makes them even more proliferative. This selection at the cellular level is contained within the life span of an individual, and begins anew with each generation. Its impact on the evolution of the species, however, resides in providing an advantage to those individuals who have more efficient built-in mechanisms to prevent this uncontrolled cell division. Thus, genotypes more adept at using cellular control mechanisms (such as apoptosis) or whole-organism systems (such as the immune system) to suppress or delay malignancies until the individual is beyond the reproductive years would have been favorably selected in the evolution of multicellular organisms.

The struggle to control cell proliferation is probably as old as life itself, which helps explain the many layers of regulatory mechanisms. It also explains why so many mutations are necessary to accomplish a malignant transformation.

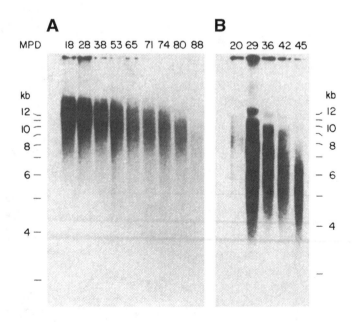

FIGURE 8-17. Telomere shortening in cultured cells. Southern analysis of genomic DNA from cultured fibroblasts derived either from **(A)** a fetal cell line, or **(B)** an elderly adult. DNA was digested with the restriction enzyme *Msp*I, which recognizes the sequence CCGG, and the hybridization probe was derived from the telomeric repeat sequence TTAGGG. Because this telomeric repeat sequence is found in large numbers only in telomeres, significant hybridization occurs only to terminal chromosomal fragments. By using an enzyme such as *Msp*I with a restriction site that is found every few-hundred bases in unique sequence DNA but not at all in the repeats, it was ensured that the terminal fragment released is made up mainly of telomeric repeats. The cultures were analyzed after the indicated mean population doublings (MPD). We see that in all cases the size distribution of telomeres is very broad (extending over several thousands of kilobases), that cells from the older donor have shorter telomeres, and that telomeres shorten with cell doublings. (Redrawn with permission from Fig.1a and 1g in Harley CB, Futcher AB, and Greider CW. Telomeres shorten during ageing of human fibroblasts. Nature 345:458–460 [1990].Copyright 1998 Macmillan Magazines Limited.)

FIGURE 8-18. Action of telomerase to extend the 3'overhang and thus compensate for the shortening of the complementary 5' end. **(A)** A shortened 5' end, as in Figure 8-16, showing telomeric repeats. (Note that in fact there are thousands of these repeats.) **(B)** Elongation of the 3' end by telomerase, which slides downstream in stepwise fashion. After a number of iterations, an RNA primer (wavy line) allows synthesis of a bottom strand that is at least as long as the original bottom strand (this illustration shows it a little longer).

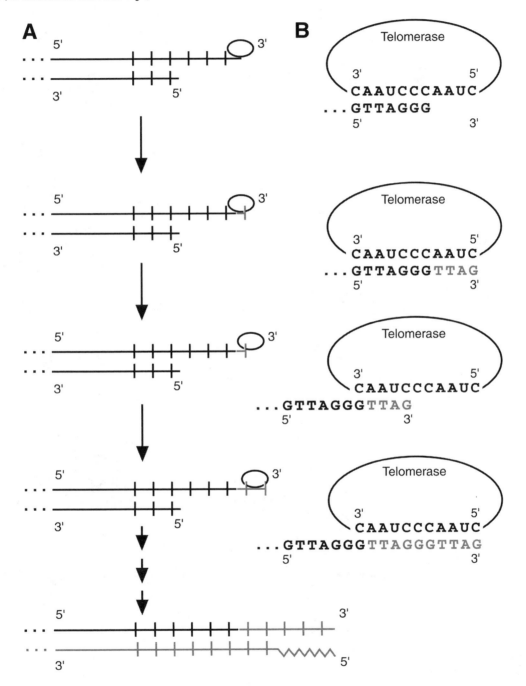

---

www *Box 8.1* *Internet Sites*

The Cancer Genome Anatomy Project (CGAP) is an effort to catalog all the genes differentially expressed in cancer cells. Using established and developing methods to detect gene expression, the long-term goal is to have a database of all types of tumor cells at various stages of cancer development and the pattern of gene expression in those cells. It may thus be possible to develop a *tumor gene index* so that a patient's diagnosis and treatment will be based on the pattern of gene expression in that particular tumor.

This program is being carried out under the auspices of several agencies of the U.S. government (the National Center for Biotechnology Information, the National Cancer Institute, and the Department of Energy) and several pharmaceutical companies:

http://www.ncbi.nlm.nih.gov/CGAP/

The Genetics of Cancer, a site supported by the Robert H. Lurie Comprehensive Cancer Center of Northwestern University, has general information on cancer as well as a number of interesting case studies. It can be found at:

http://www.cancergenetics.org/home.htm

The Telomere Database (TelDB), the GenLink Multimedia Telomere Resource at Washington University in St. Louis, is a rich source of information on chromosome ends:

http://www.genlink.wustl.edu/teldb/index.html

The examples and studies presented in this chapter were selected so as to provide a coherent picture of the genetic and cellular interactions that contribute to cancer. It should be emphasized that such a "neat" description remains highly incomplete, even for the best-known cancers. It is entirely possible that less well-characterized malignancies will reveal oncogenic mechanisms that are outside our current understanding of this phenomenon.

Exercises

8-1. How would you design an experiment using X-inactivation in females to test that all cancer cells in a patient are clonally derived from a single ancestral cell?

8-2. Study the various components in Figures 8-2, 8-3, and 8-7, and classify them according to whether each is a product of a tumor suppressor gene or a proto-oncogene.

8-3. The proteins activated by kinases in Figure 8-3 are inactivated by phosphatases that remove the pertinent phosphate groups. How would you classify the genes for these phosphatases in terms of their potential effect on oncogenesis (oncogenes, tumor suppressors, etc.). Not many phosphatase mutations have been found that participate in oncogenesis—how can you explain this?

8-4. In cases of familial retinoblastoma, the disease is usually bilateral, but in sporadic cases the tumor affects only one eye. Explain these observations on the basis of the "two-hit" hypothesis.

8-5. In addition to the functions discussed in the text, p53 also induces transcription of a gene, *MDM2*, whose product binds to the transactivator region of p53, thus inhibiting it. This circuit serves as a safety device so that p53 does not interfere with cell proliferation unless the proper distress signals overcome the *MDM2* block. Make a copy of Figures 8-2 and 8-11 and draw the arrows necessary to symbolize those processes.

8-6. Explain how a translocation that brings closer the immunoglobulin heavy chain gene and the suppressor of apoptosis *BCL2* can be a step toward carcinogenesis.

8-7. Describe, by means of a diagram, the location of the $(TG)_n$ microsatellite shown in Figure 8-14, the position of the primers used, and the genotypes of the healthy and cancer cells.

8-8. How would you separate telomeric DNA, such as that seen in Figure 8-15B, from the bulk of chromosomal DNA?

REFERENCES

General references

Brown MA. (1997) Tumor suppressor genes and human cancer. Adv Genet 36:45–135.

Fearon ER. (1997) Human cancer syndromes: clues to the origin and nature of cancer. Science 278: 1043–1050.

Fearon ER, and Vogelstein B. (1990) A genetic model for colorectal tumorigenesis. Cell 61:759–767.

Kinzler KW, and Vogelstein B. (1997) Cancer-susceptibility genes: gatekeepers and caretakers. Nature 386:761–763.

Lengauer C, Kinzler KW, and Vogelstein B. (1998) Genetics instabilities in human cancers. Nature 396:643–649.

Levine AJ. (1997) p53, the cellular gatekeeper for growth and division. Cell 88:323–331.

Parker PJ, and Pawson T. (1996) Cell signaling. In: Cancer surveys, vol 27. Tooze J, ed. New York: Cold Spring Harbor Laboratory Press.

Nathanson N. Viral pathogenesis. (1997) Philadelphia: Lippincott-Raven.

Popescu NC, and Zimonjic DB. (1997) Molecular cytogenetic characterization of cancer cell alterations. Cancer Genet Cytogenet 93:10–21.

Rabbitts TH. (1994) Chromosomal translocations in human cancer. Nature 372:143–149.

Royle NJ. (1996) Telomeres, subterminal sequences, variation and turnover. In: Human genome evolution. Jackson M, Strachan T, and Dover G, eds. Oxford: BIOS, Oxford, 147–169.

Sherr CJ. (1996) Cancer cell cycles. Science 274:1672–1677.

Weinberg RA. (1994) Oncogenes and tumor suppressor genes. Cancer J Clin 44:160–170.

Weinberg RA. (1995) The retinoblastoma protein and cell cycle control. Cell 81:323–330.

Specific references

Aaltonen LA, Peltomaki P, Leach FS, et al. (1993) Clues to the pathogenesis of familial colorectal cancer. Science 260:812–816.

Cho Y, Gorina S, Jeffrey PD, and Pavletich NP. (1994) Crystal structure of a p53 tumor suppressor-DNA complex: understanding tumorigenic mutations. Science 265:346–355.

Denissenko MF, Pao A, Tang MS, and Pfeifer GP. (1996) Preferential formation of benzo[a]pyrene adducts at lung cancer mutational hotspots in *p53*. Science 274:430–432.

Fearon ER, and Vogelstein B. (1997) Tumor suppressor and DNA repair gene defects in human cancer. In: Cancer Medicine, 4th ed. Holland JF, Frei E, Bast RC, Kufe DW, Morton DL, Weichselbaum RR, eds. Baltimore, MD: Williams & Wilkins, 97–117.

Griffith JD, Comeau L, Rosenfield S, Stansel RM, Bianchi A, Moss H, and de Lange T. (1999) Mammalian telomeres end in a large duplex loop. Cell 97:503–514.

Guilford P, Hopkins J, Harraway J, McLeod M, McLeod N, Harawira P, Taite H, Scoular R, Miller A, and Reeve AE. (1998) E-cadherin germline mutations in familial gastric cancer. Nature 392:402–405.

Harley CB, Futcher AB, and Greider CW. (1990) Telomeres shorten during ageing of human fibroblasts. Nature 345:458–460.

Laken SJ, Petersen GM, Gruber SB, Oddoux C, Ostrer H, Giardello FM, Hamilton SR, Hampel H, Markowitz A, Klimstra D, Jhanwar S, Winawer S, Offit K, Luce MC, Kinzler KW, and Vogelstein B. (1997) Familial colorectal cancer in Ashkenazim due to a hypermutable tract in *APC*. Nat Genet 17:79–83.

Lee J-O, Russo AA, and Pavletich NP. (1998) Structure of the retinoblastoma tumour-suppressor pocket domain bound to a peptide from HPV E7. Nature 391:859–865.

Lengauer C, Kinzler KW, and Vogelstein B. (1997a) Genetic instability in colorectal cancers. Nature 386:623–627.

Lengauer C, Kinzler KW, and Vogelstein B. (1997b) DNA methylation and genetic instability in colorectal cancer cells. Proc Natl Acad Sci USA 94:2545–2550.

Look AT. (1997) Oncogenic transcription factors in the human acute leukemias. Science 278:1059–1064.

Marx J. (1993) New colon cancer gene discovered. Science 260:751–752.

Peltomaki P, Aaltonen LA, Sistonen P, et al. (1993) Genetic mapping of a locus predisposing to human colorectal cancer. Science 260:810–812.

Thibodeau SN, Bren G, and Schaid D. (1993) Microsatellite instability in cancer of the proximal colon. Science 260:816–819.

9 | *Genetic Counseling*

Classical genetics and an understanding of the mode of inheritance of genetic diseases was applied for many years in what has become a unique area of medicine: genetic counseling. With knowledge of Mendelian genetics alone, the consultations used to revolve around Mendelian ratios and the probability of having affected children. Practically the only impact was on a person's decision to have any children, given the odds of having affected children (1 in 2, 1 in 4, etc.). Advances in human molecular genetics have had a revolutionary impact on the field by permitting the detection of specific alleles through *predictive testing*.

It should be stressed that although human disease was presented in this book in the form of anonymous and "sanitized" examples for the sake of illustrating scientific principles, the reality is that in each and every case there is human suffering associated. Physicians, nurses, and counselors regularly deal with both the scientific and the human sides of the question. Genetic counseling is not presented here in any systematic fashion, but the few issues reviewed should be enough to show that this is a very complex area that requires a good deal of thoughtful and specialized training. It is not just a question of applying Mendel's laws. The reader is encouraged to read more extended accounts of real cases, some of which can be found in the references at the end of this chapter.

PRENATAL, NEONATAL, CHILDHOOD, AND ADULT GENETIC TESTING

Prenatal Genetic Testing

There are two main areas of prenatal genetic testing and diagnosis, and they correspond to the analysis of either chromosomes or DNA sequences.

The study of chromosomes has been exploited since the middle of the century, and even today the most frequent reason for prenatal diagnosis is to test for Down syndrome (trisomy 21) in mothers at risk because of their age. Prenatal diagnosis for other chromosomal abnormalities is carried out in high-risk pregnancies, for instance, when one of the parents has a balanced translocation. These balanced, asymptomatic translocations are in turn often discovered when chromosome studies were indicated as a consequence of previous pregnancies (either from the same parents or close relatives of one of them) that terminated in spontaneous abortions or a newborn with serious abnormalities.

Prenatal diagnosis of DNA mutations lagged until the 1980s and the revolution in human molecular genetics. Before then, practically the only progress in prenatal diagnosis of genetic diseases caused by point mutations was in sex-linked diseases such as hemophilia, in which case for a pregnancy at risk (a heterozygous mother), chromosomal studies could establish whether the fetus was male (and at risk) or female and excluded.

With the development of DNA probes for markers linked to recessive mutations such as Tay-Sachs disease, Lesch-Nyhan syndrome, sickle cell anemia, hemophilia, and cystic fibrosis, prenatal diagnosis became possible for many genetic diseases—although for a few inborn errors of metabolism, biochemical assays that permitted prenatal diagnosis had been developed in the 1960s and 1970s. In these cases, heterozygosity of the parents determines the at-risk nature of the pregnancy.

There are no cures or even treatments for most of the conditions mentioned above, and the most common reason for these prenatal diagnoses is to terminate pregnancies of affected fetuses. Thus, in most of those cases, prenatal diagnosis is of limited value to couples who will not consider an abortion. This is, naturally, a very difficult question, and people respond in a variety of ways. Thus, beyond those who would not consider an abortion under any circumstance, there are individuals who would terminate a pregnancy in the case of a certain and severe disease in which the infant would not survive more than a few months, but would not consider it in case of a milder condition or in case of uncertainty as to the genotype of the fetus. At the other end of the spectrum, there are those who would not consider bringing to term a pregnancy if there is a significant chance that the child will be afflicted for the rest of his or her life (see Box 9.1, Case 1).

| Box 9.1 | *Case 1: Knowing too much* |

How do individuals who do not want to forego children deal with the risk of passing a mutant allele to their offspring? If parents are willing to undergo abortions for the sake of avoiding the possibility of passing HD, there is a method to do this and at the same time maintain the parent's lack of knowledge about at-risk status: it is called prenatal *exclusion testing* or *nondisclosing prenatal testing*.

Exclusion testing is illustrated in Figure 9-1.

Individual II-2 wants to avoid transmitting HD, present in his or her father, but without knowing his or her own status as a carrier. This can be done by terminating any pregnancy in which the fetus received a chromosome from the grandfather (I-1) rather than the grandmother (I-2). Use is made of a highly polymorphic marker closely linked to *HD*, as the one we saw in Figure 1-16. We can see that II-2 received allele *B* from his or her father and allele *A* from the mother. Unfortu-

FIGURE 9-1. Partial pedigree of a family with Huntington's disease. Note that a predictive test for II-1 would not have been possible without information from I-3 and her offspring, which helped establish that allele *A* is the one associated with HD in this family.

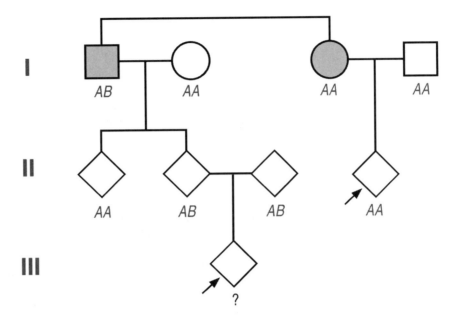

nately the spouse (II-3) also carries an allele *B* (in this case most likely not associated with the *HD* mutation), and the only way to exclude the fetus from carrying the *B* allele of the affected grandfather is to terminate all pregnancies except those that are *A/A*.

Note that, from studies of the offspring of I-1, we cannot know whether *B* is associated with the mutation or not. In this family there was another twist that complicated matters further. Unknown to II-2, a cousin (II-4) requested predictive testing in the same clinic. The study of the cousin's mother established that it was the *A* allele that was associated with the mutation in generation I, and II-2 could be given a modified risk close to 0%. Should II-2 be informed of the situation?

The decision was not to inform II-2, and to proceed with the prenatal test, advising that an abortion was not indicated no matter what the result. The rationale was based on the fact that if II-2 were told that he or she was not at risk despite the expressed instruction not to be told, this would create a false expectation on any other member of the family who came in with a similar request. Such persons might assume that if they were not told anything it meant that they were carriers, because they would falsely expect to be told if they were not carriers, just as it happened in the case of II-2.

As we can see there is a conflict here between respecting the instructions not to know of II-2 and possibly other members of the family on one hand, and providing both a psychological benefit to him or her, and avoiding a small but certain risk of spontaneous abortion as a consequence of the prenatal test. Also, the course taken might involve deceiving II-2 about whether the fetus has received a chromosome from the grandfather by telling the proband that an abortion is not necessary.

Another solution might have been to keep the studies of II-2 and II-4 totally separate, perhaps by having different teams carry them out. In this way the counselor working with II-2 would truly not know the patient's status. The price of this course of action would be a very high probability of unnecessary abortions.

Neonatal and Childhood Genetic Testing

In addition to prenatal diagnosis, the genetic testing of children and adults is sometimes indicated, although this is a hotly debated area. There is consensus that the testing of newborns is advisable in the few cases for which there are effective treatments, such as the control of hypothyroidism and phenylketonuria, in which treatment must be initiated early to avoid irreversible damage. Beyond these relatively simple cases, there is no consensus. There are some, probably a minority, who would use genetic testing whenever practical and inform parents of all results. At the other extreme there are those who would withhold information from parents, even when the genetic counselor is in possession of the information. An example of this last case is the decision by some counselors not to tell parents of the detection of sickle cell trait (heterozygotes β globin S/A) in an infant being tested for sickle cell anemia (homozygous S/S). The argument presented is that it is not in the best interest of the patient (the child) for the parents to know, because their attitude toward the child may be tainted by the information.

Adult Genetic Testing

Mainly as a consequence of the ability that scientists developed in the 1980s and 1990s to determine whether a mutant allele is present in an asymptomatic proband (and not just a simple probability of being affected by a genetic disease), genetic testing of adults became a reality. This is of particular importance in the case of late onset, dominant diseases such as Huntington's disease (HD), where the individual may live half a lifetime before the symptoms present themselves. Although geneticists have benefited enormously from very large HD pedigrees, most cases occur sporadically or in families with just one or two affected members. Often young people in their teens, 20s, and 30s learn for the first time that they may be at risk for HD when one of their parents begins to develop symptoms; sometimes they may already have children themselves. Before the gene was cloned, these individuals could

---

EXAMPLE 9.1

Phenylketonuria

Early detection of phenylketonuria and institution of a strict, low phenylalanine diet a few days after birth are keys to the satisfactory development of individuals who are homozygous for phenylalanine hydroxylase (*PAH*) mutations. The diet does not need to be continued into adulthood as it appears that it is only the developing brain that is sensitive to elevated levels of phenylalanine catabolites.

Dietary treatment of PKU infants started in the 1950s, and when the first treated PKU females reached their reproductive years, most of their offspring were severely retarded, although they were, as a rule, heterozygous for the *PAH*

mutation. These parents were individuals who had abandoned their diet as adults, and the high levels of ketonic bodies interfered with the normal development of the fetus. Once this problem became apparent, the solution presented itself readily: pregnant PKU women must return to monitoring phenylalanine levels and dietary restrictions. Simple as it is, the avoidance of unnecessary cases of mental retardation among the children of PKU mothers requires a robust information system that educates PKU patients about a liability that may appear many years after they are finished with any treatments needed because of their own health.

Box 9.2	*Case 2: Disagreement between spouses*

J is 6 weeks pregnant and contacts a genetics counseling center because her husband, T, has a 50% risk of having inherited HD. T, however, is opposed to learning his own status and because his parents are deceased, there is no way to do a prenatal exclusion test. He wants to continue the pregnancy in the expectation that in the next 20 or 30 years treatments will be available. Some possible conflicting claims in this case could be:

1. Claim to life for the fetus.
2. Claim for the fetus to avoid the suffering of HD, or even the uncertainty of the possibility of HD.

3. Claim for J to have the right to terminate her pregnancy.
4. Claim for J to have the right to know the genotype of the fetus so that she can terminate her pregnancy only if the fetus carries the mutation.
5. Claim for T to have the right to remain ignorant of his modified risk.
6. Claim for T to have the right to veto the abortion.

Some of these are legal issues that have been decided, but all of them have an ethical component that needs to be weighed by J, T, and the counselor. The decision was in favor of performing the test.

only know that they had a 50% chance of one day being affected themselves, and that there was thus a 25% probability that any one of their children would receive the mutant gene.

When linkage of *HD* to molecular markers was detected, even before the *HD* gene and mutation were identified, individual risk in affected families sometimes could be reassessed by identifying the linked markers (see Figs. 1-16 and 1-18), and predictive testing became possible. Given this possibility, individuals with HD relatives could choose to ascertain their risk by predictive testing. This is called having the *risk modified,* because after the test the individual goes from having a 50% risk to either having close to 0% or close to 100% risk. (The risk does not go to exactly 0% or 100% because of the possibility of crossing-over or of new mutations.) The modified risk can then be used in an individual's decision to have children. Their offspring also have modified risks, either 0% or 50%, as a consequence of the test.

The knowledge derived from predictive testing is not welcome by all. Some individuals who have a 50% probability of having received the mutant gene find that the possibility of testing positive would be so destructive to the remaining years of health they have left that they prefer *not* to submit to predictive testing and have their risk modified. Thus, they choose to live with a 50% risk of being affected but also a 50% chance of having dodged it. One of the ethical tenets of genetic counseling that has developed as a consequence of the testing of adults is that a patient has not only the *right to know* but also the *right not to know* (Box 9.2, Case 2).

A situation where the right not to know may become an issue is with identical twins. If one of them requests a genetic test, the results would apply equally to the sib, and almost certainly the sib would discover his or her own status whether he or she wants to or not. Conflict may also arise between an adult child who requests predictive test-

| Box 9.3 | *Case 3: Sharing information with adult offspring* |

R's mother died many years ago of the autosomal dominant form of polycystic kidney disease (*ADPKD*), and when an accurate test for the mutation became available, he requested a DNA test to evaluate his status. The results indicated that he was carrying the mutant allele. R has two young adult daughters, but he declines to inform them of his condition on the grounds that the health insurance difficulties they would face if they knew their condition would be a more serious problem for them than possible health risks during pregnancy or the inability to make decisions concerning the transmission of this mutation to their offspring if they remained ignorant.

The consensus of a panel is that the daughters would be better off learning that they are at risk. The question is, should the counselor breach confidentiality with R and inform the daughters? To decide, we may want to consider some critical information that was left out of the report of this case. It is not clear whether the daughters know that their grandmother carried this mutation. If they do know, the new information about their father simply changes the risk from 25% to 50%, and one can argue that knowing they are at 25% risk, they could seek testing themselves if they want the information. Although others will say that lacking direct information that the daughters have received genetic counseling and acquired a clear understanding of their situation, they should be informed.

If the daughters do not know about their grandmother, it is somewhat easier to decide to inform them even if that means breaking the confidentiality agreement with R. The experience of counselors indicates that before breaching R's confidentiality, he should have the benefit of several counseling sessions. It has been observed that it takes time for probands to assimilate the news of their changed status, and to think through all the consequences of their actions. Usually probands need to resolve the implications of this type of news for their own lives before they proceed to appreciate the implications for their relatives. Given some time, R would probably agree to inform his daughters.

In a 1990 survey of U.S. geneticists, 58% indicated that under certain conditions they would breach confidentiality and disclose genetic information concerning risk of Huntington's disease to unsuspecting relatives over a patient's objections.

---

ing and a parent, because the carrier status of the child immediately defines that of the parent. Genetic counseling clinics try to get such relatives to agree on their requests; in the absence of agreement, the prevalent opinion seems to favor disclosing to the patient who requests the information rather than granting a third party a right to veto.

Cloning of the gene, and being able to identify the mutant allele directly rather than by linkage to a marker, has the advantage that it obviates the necessity for testing relatives to ascertain the transmission of the mutation in the family in question (see Box 9.1, Case 1).

Some rare familial forms of Alzheimer's disease present issues similar to HD. Other dominant, autosomal diseases of adults include polycystic kidney disease, and familial cancers such as familial adenomatous polyposis (colon cancer, *APC* gene) and familial breast cancer (*BRCA1* and *BRCA2* genes). This latter group presents significant differences in counseling situations because as opposed to HD, they are treatable conditions and predictive genetic testing can have a very real impact on the course of the disease (see Boxes 9.5, Case 4, and 9.6, Case 5). Also, some of these diseases have early onset forms and testing of underage children may be indicated.

We have touched on the right of people who know that they are at some risk, to have that risk altered by a predictive test. What about people who are totally oblivious to the presence of a condition in the family? Should they be informed when the test of a relative indicates they may be at risk? Does the geneticist have a duty to inform? Does the geneticist have a *right* to inform (even with the permission of the proband)? Considering Case 3 (Box 9.3) from the point of view of R's daughters, how do we know that they would prefer to know?

Without doubt the answers to the questions above depend on specific diseases and circumstances. Is treatment available? Does early detection affect the course of the disease? Is the unsuspecting relative planning to have children?

The power of predictive testing we have for a few diseases will become more accurate and extend to more afflictions. It will take yet many years of practice, thought, and discussion to develop reliable guidelines on how to use this power.

ACTUAL RISK AND PERCEIVED RISK

One of the serious difficulties faced by genetic counselors is the innate inability of humans to perceive risk accurately, especially to accept the independence of events. Let us imagine how individual IV-3 in Figure 7-20 or the youngest sister in Case 4 might feel: "Everybody in my family gets it, I am doomed." Or it could be the opposite: "If everybody else got it, that makes it less likely that I will." In the absence of extensive counseling, lay people without particular training in statistics probably will not accept coldly that both outcomes are equally likely. It appears that the less a person can be

---

Box 9.4   *Case 4: When ignorance is not bliss*

Tissue samples were taken from four sisters in search for *BRCA1* mutations. The three elder sisters suffered from ovarian or breast cancer in their 40s; the youngest is not quite 40 yet and she agreed to donate a sample, but insisted on remaining ignorant about the results. The results came back that she was not a carrier of the mutation, and naturally the counselor should not reveal her status to her.

The woman in question has a 12-year-old daughter, who, being a minor and many years from any risk, was not tested. The counselor might discuss the responsibility of the mother toward her daughter, who, knowing the family history, would live under a cloud unnecessarily if the mother were to test negative for the mutation.

On a second interview, the woman reveals that she is planning to undergo prophylactic mastectomy and oophorectomy (surgical removal of breasts and ovaries). Many counselors would still not break their pledge not to disclose, but note that in this case it is possible to make a strong argument that, if she is willing to have such serious prophylactic surgery carried out to reduce the risk of disease, it is in her benefit to find out what her status is (i.e., whether she even needs the surgery). The difference between this case and a similar one dealing with HD is that in the case of *BRCA1* there is something that can be done, even if prophylaxis has its own risks and costs.

A strong argument that it is in the patient's benefit to discover her true risk before engaging in prophylactic surgery applies even if she does carry the mutation, in this case to validate her choice and remove doubts later on (i.e., "Perhaps I shouldn't have done it").

indifferent about the outcome, the more difficult it is to be objective about the *chances* of being afflicted.

Other irrelevant issues that come into play are guilt, merit, and sense of worth, especially between sibs: "If it has to be one of us, it should be me," "It is OK that I got it so now my brother/sister will be spared," and so on.

There is also the question of how individuals who have been living with a certain assigned a priori risk deal with the modified risk. The ones who receive the bad news that they indeed carry a mutation must, of course, deal now with a new certainty. In many cases there is relief in the certainty, even if the prospect is not good, because plans can be made. Unfortunately, recipients of good news also face difficulties. Feelings of guilt when one is spared and others are not can create psychological problems. In a few cases individuals with a 50% risk are so sure that they would face premature death that they adopted "live for today" lifestyles and are totally unprepared, emotionally and economically, to receive assurance that they are expected to have a normal life span.

Perhaps it is in part because of our inability to comprehend accurately risk assessments that some studies seem to show that individuals who learn about their modified risks (i.e., who are told with a degree of certainty either that they are carriers, or that they are not) tend to cope as a whole better than those who remain ignorant.

Risk assessment also determines what course of action one takes after being informed of a modified risk. In diseases where some preventive measures can be taken, as with *BRCA* mutations, the decision as to what measures to take depends very much on each individual. Some may opt for less invasive surveillance regimes in cases where others would choose more radical prophylactic surgery (see Box 9.5, Case 5).

| **Box 9.5** | *Case 5: Similar situations, different courses* |

A 25-year-old woman goes for genetic testing because of a family history of breast cancer: her mother, a maternal aunt, maternal cousin, and grandmother have all been diagnosed with breast cancer at ages between 29 and 40. Two of these four relatives died of the disease. She has already decided that if she is a mutation carrier she will undergo prophylactic mastectomy and oophorectomy.

In another case, a 27-year-old woman requests testing because of a similar family history: her mother and maternal aunt and grandmother were diagnosed with breast cancer between ages 29 and 41. Contrary to the first case, this woman has decided that she will not undergo prophylactic surgery, regardless of the test outcome, but will instead choose increased surveillance, if necessary.

It is difficult to justify the difference in choices based on the facts of the cases, and both choices are equally rational. In addition to possible differences in personality and priorities, experience may have colored each patient's perception. Thus, seeing two of her close relatives die of the disease probably weakened the confidence of the 25-year-old in medicine's ability to control breast cancer; the 27-year-old, on the other hand, is probably more confident after having witnessed three relatives "beat" the disease.

EXERCISES

9-1. Given what you know about the example presented in Figure 9-1, what is the prognosis for II-1 and II-4? What information should be given to them?

9-2. See Box 9.2, Case 2. Not knowing what the results of the test actually were, do you think that the *perceived* risk to herself of the 12-year-old daughter would be affected by her mother's surgery?

9-3. See Box 9.2, Case 2. Suppose the woman approaches a surgeon who is unrelated to the genetics counseling clinic for prophylactic surgery, citing her family history. What do you think should be the proper response of the surgeon?

REFERENCES

Applebaum EG, and Firestein SK. (1983) A genetic counseling casebook. New York: The Free Press.

Gert B, Berger EM, Cahill GF Jr, Clouoser KD, Culver CM, Moeschler JB, and Singer GHS. (1996) Morality and the new genetics. Boston: Jones and Bartlett.

Editorial. (1995) When is genetic screening permissible? Nature 377:273.

Marshall E. (1993) A tough line on genetic screening. Science 262:984–985.

Offit K. (1998) Clinical cancer genetics: risk counseling and management. New York: Wiley-Liss.

Smith DH, Quaid KA, Dworkin RB, Gramelspacher GP, Granbois JA, and Vance GH. (1998) Early warning: cases and ethical guidance for presymptomatic testing in genetic diseases. Bloomington: Indiana University Press.

Appendix

ONE-FACTOR CROSSES

FIGURE A-1. Mendelian ratios in the first and second generation of a monohybrid cross for a trait with complete dominance.

Parents

$\male$ $\qquad$ $\female$

Genotypes (2n) $\qquad \dfrac{M}{M} \ \times \ \dfrac{m}{m}$

Phenotypes $\qquad$ "M" $\qquad$ "m"

Gametes $\qquad \underline{M} \qquad \underline{m}$

First Generation (F1)

Genotype $\qquad \dfrac{M}{m}$

Phenotype $\qquad$ "M"

Mating   F1 x F1

$\male \qquad \female$

Genotypes $\qquad \dfrac{M}{m} \ \times \ \dfrac{M}{m}$

$\female$ gametes / $\male$ gametes	$\frac{1}{2}$ M	$\frac{1}{2}$ m
$\frac{1}{2}$ M	$\frac{1}{4} \dfrac{M}{M}$	$\frac{1}{4} \dfrac{M}{m}$
$\frac{1}{2}$ m	$\frac{1}{4} \dfrac{m}{M}$	$\frac{1}{4} \dfrac{m}{m}$

Second Generation (F2)

Genotypes $\quad$ 25% $\dfrac{M}{M}$: 50% $\dfrac{M}{m}$: 25% $\dfrac{m}{m}$

Phenotypes $\qquad$ 75% "M" : 25% "m"

INDEPENDENT ASSORTMENT IN TWO-FACTOR CROSSES

The independent assortment is determined by the randomness of the orientation of the various chromosome pairs in the equatorial plate of the first meiotic division (Fig. A-2). If the colored chromosomes were received from one parent (the mother) and the black ones from another (the father), during meiosis it is equally likely that the two maternal chromosomes will go together to one pole and the two paternal ones to the other (A), or that one maternal and one paternal will go to one pole while the other maternal will go to the other pole with the second paternal chromosome (B). Thus, the frequency of *AB* and *ab* gametes will be equal to the frequency of *Ab* and *aB* gametes.

FIGURE A-2. Chromosomal constitution of an individual that is a double heterozygote (F1) for genes on different chromosomes.

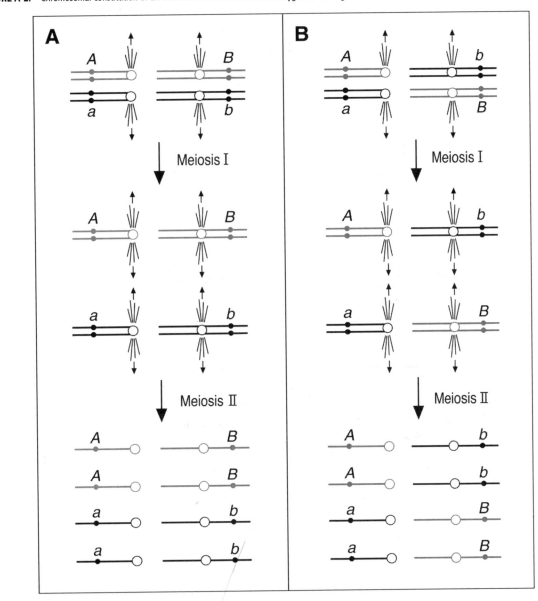

LINKAGE IN TWO-FACTOR CROSSES

If it were not for crossing-over, alleles on the same chromosome would always by transmitted together showing complete linkage. Crossing-over allows for a certain amount of assortment. But if the genes are relatively close together, the probability of crossing-over in the segment separating the genes (Fig. A-3A) is smaller than the probability of crossing-over elsewhere (Fig. A-3B). Thus, the parental allelic combination (in this case, *CD* and *cd*) are found more frequently than the recombinant combinations (*Cd* and *cD*).

FIGURE A-3. Linkage of genes on the same chromosome.

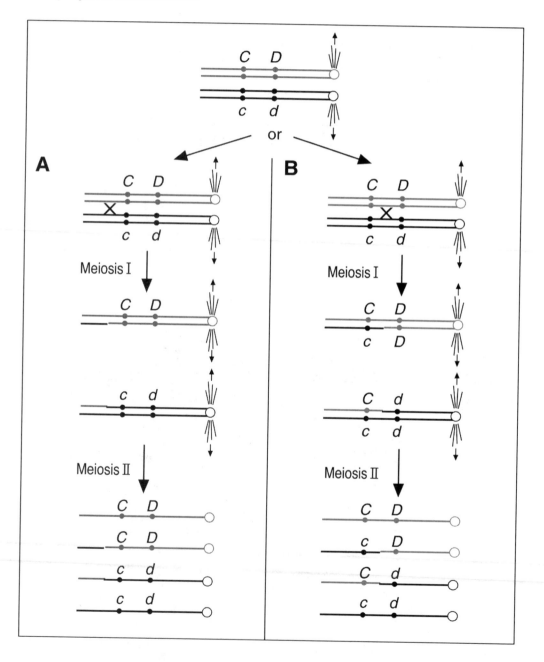

SYNTHESIS AND MATURATION OF MESSENGER RNA

Black lines and boxes, in Figure A-4, represent double-stranded DNA, and colored lines and boxes represent RNA. Boxes indicate sequences that are found in the mature mRNA. Shaded areas indicate the coding regions which, with the assistance of ribosomes, tRNAs, and numerous other factors, are translated into the amino acid sequence of a polypeptide.

FIGURE A-4. Synthesis and maturation of messenger RNA.

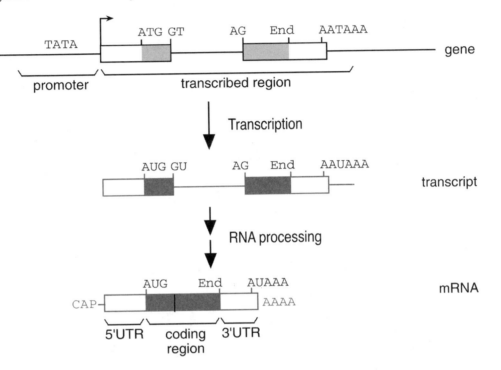

TABLE OF CODONS AND THE AMINO ACIDS

FIGURE A-5. Table of codons. AUG (Met) is the translation initiation signal. UAA, UAG, and UGA are codons that indicate the end of translation.

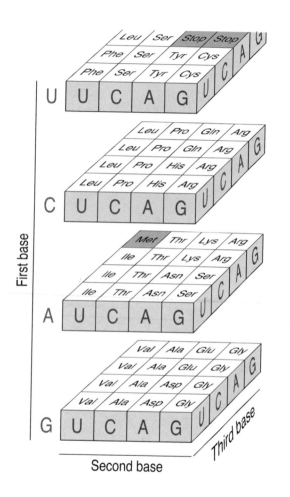

First base

Second base

Third base

FIGURE A-6. The amino acids, grouped according to the properties of the side chains.

The Amino Acids

Smallest Amino Acid (R+H)		
Glycine	Gly	G

Hydrophobic Amino Acids:
Aliphatic

Alanine	Ala	A
Valine	Val	V
Isoleucine	Ile	I
Leucine	Leu	L
Proline	Pro	P
Methionine	Met	M

Aromatic

Phenylalanine	Phe	F

Polar Amino Acids:
Aliphatic

Serine	Ser	S
Threonine	Thr	T
Cysteine	Cys	C
Asparagine	Asn	N
Glutamine	Gln	Q

Aromatic

Tyrosine	Tyr	Y
Histidine	His	H
Tryptophan	Trp	W

Charged Amino Acids:
Acidic

Aspartic acid	Asp	D
Glutamic acid	Glu	E

Basic

Lysine	Lys	K
Arginine	Arg	R

FIGURE A-7. Restriction enzymes and gel electrophoresis: Fractionating DNA into pieces of reproducible size.

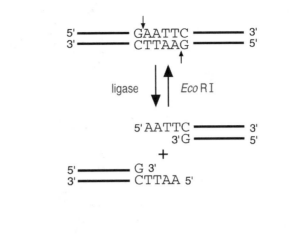

A

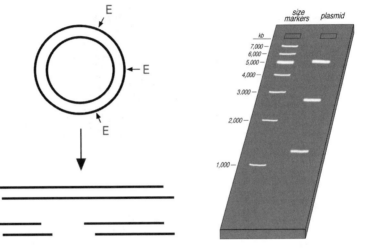

B

ANALYZING DNA WITH RESTRICTION ENZYMES AND GEL ELECTROPHORESIS

Endonucleases are enzymes capable of digesting the phosphate (ester) bond that hold neighboring nucleotides together and maintain the integrity of the DNA molecule. A special kind (called *restriction endonucleases*) are capable of digesting, or cutting, the DNA only where certain specific sequences occur. These sequences are known as *restriction sites*. In many cases, restriction sites have the sequence of an *inverted repeat*, for example, 5'GAATTC3'. In this example, GAA is an inverted repeat of TTC because the complement of TTC, read in the 5' to 3' direction is GAA. This property can be seen more easily in the double-stranded sequence, which shows that the entire six-base sequence is an inverted repeat of itself, or palindrome. Restriction enzymes have themselves a symmetry, such that when they cut, they digest the bond between the G and the A on both strands, leaving two half-molecules held together by the hydrogen bonds that connect the four bases on one strand to the four on the

other. These hydrogen break spontaneously, fragmenting the DNA (Fig. A-7A).

The most commonly used restriction enzymes recognize target sites of either four, six, or eight base pairs. In laboratory jargon they are called, respectively, four-cutters, six-cutters, and eight-cutters. The size of the fragments generated by a particular enzyme depends on the frequency with which the restriction sites occur in a given DNA preparation. In an idealized random DNA sequence of base composition [A] = [T] = [G] = [C], four-cutters have restriction sites every 256 nucleotides $(1/4^4)$, six-cutters every 4096 $(1/4^6)$ nucleotides, and eight-cutters every 65536 $(1/4^8)$. Often, these frequencies are not realized in practice, because DNA sequences depart significantly from the idealized random one.

Digestion of small plasmid or viral DNA usually produces one or a few fragments (see Fig. A-7B), which can be fractionated by size by agarose gel electrophoresis. A sample of the digested DNA is placed within a notch in a gel matrix and an electric field is applied. The DNA

FIGURE A-7.   (Continued)

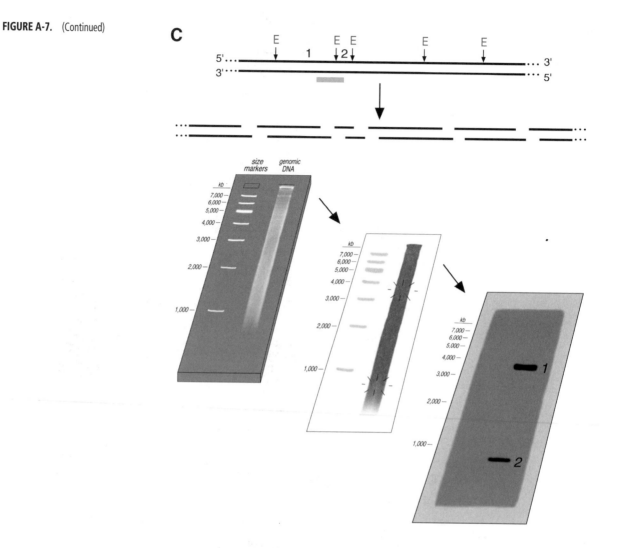

fragments migrate toward the positive pole at a speed that is inversely related to their size: smaller fragments find their way more easily in the gel matrix and move faster. Each gel is calibrated by running, next to the sample of digested DNA, a sample of DNA fragments of known size (called *size markers* or *ladder*). When electrophoresis is finished, the gel is stained with ethidium bromide, a dye that binds to DNA and produces a yellow fluorescence when illuminated with ultraviolet light.

Genomic DNA, by contrast to plasmids, is of practically unlimited size, so digestion with a restriction enzyme produces not a few, but an enormous number of fragments of all conceivable sizes. When the digested DNA is run on a gel, individual fragments cannot be resolved from one another; upon staining with ethidium bromide, the DNA appears as a smear. To identify any one fragment in particular, it is necessary to resort to the specificity of DNA hybridization. Figure A-7C shows a

portion of genomic DNA and the distribution of restriction sites for an enzyme. Let us suppose that a small piece of DNA is available to us that corresponds to a segment in this region, and that we can make this DNA radioactive (the probe). After running digested DNA on a gel, the DNA can be transferred onto a nylon or nitrocellulose membrane by blotting, and then denatured in situ. When the membrane bearing the denatured DNA fragments is immersed in a solution that contains the radioactive probe (also denatured), this forms a double-stranded helix with DNA fragments on the membrane that are complementary to the probe. The membrane is then washed of excess probe and pressed against a photographic film. The radioactivity on the membrane makes an impression on the film, which upon development appears as black bands in the positions on the membrane that correspond to where the DNA fragments had migrated (Fragments 1 and 2 in Fig. A-7C).

cDNA LIBRARIES

cDNA stands for *copy DNA*, or DNA copied off RNA by the enzyme reverse transcriptase. mRNA is usually purified by separating poly(A)-containing RNA from the rest (mostly ribosomal RNA and tRNA) by means of an oligo(dT) resin to which poly(A)$^+$ RNA binds.

To make cDNA from mRNA, the most common method takes advantage of the poly(A) tail in most mRNAs. A synthetic poly(dT) oligonucleotide is used as a universal primer; this binds to all mRNA containing a poly(A) tail and provides a 3'OH for elongation. Elongation is catalyzed by viral reverse transcriptase. After this first strand is synthesized, subsequent rounds of synthesis are carried out by a subunit of *Escherichia coli* DNA polymerase (Klenow fragment), supplied with all four deoxyribonucleotide triphosphate (one of them radioactive, if one needs a radioactive probe) in the proper buffer conditions.

mRNAs derived from more active genes are present at higher concentrations, represented here by the two copies of mRNA 2.

Once cDNA is available, linkers are attached to both of their ends. These linkers include a restriction site (in this case *Bam*H1). In this way, a plasmid *vector* can be digested with the same enzyme and then ligated to the cDNA fragments to produce clonable recombinant molecules. *E. coli* cells are then transformed with these recombinant plasmids so that each cell receives only one plasmid. When *E. coli* is grown in agar plates, each independent colony will harbor a different clone, and the aggregate of all these colonies is called a *cDNA library*. With an appropriate probe the library can then be screened for the clone of interest. Figure 4-9 shows the screening of a genomic library; the screening procedure is the same for both types of libraries.

FIGURE A-8. Preparing a cDNA library.

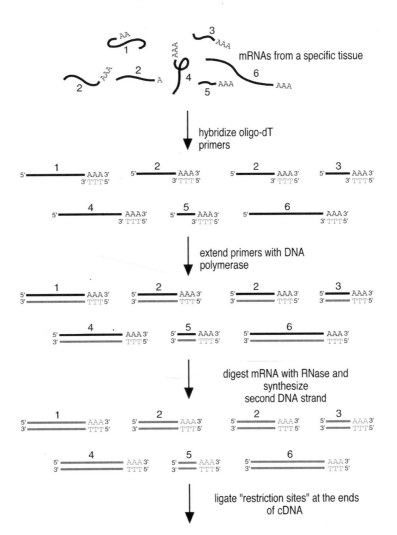

mRNAs from a specific tissue

hybridize oligo-dT primers

extend primers with DNA polymerase

digest mRNA with RNase and synthesize second DNA strand

ligate "restriction sites" at the ends of cDNA

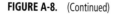

FIGURE A-8. (Continued)

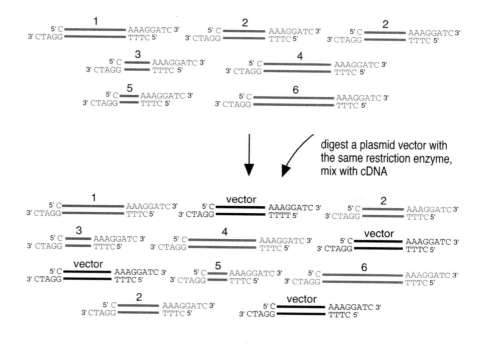

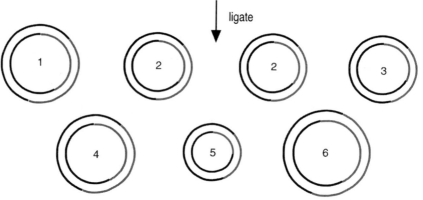

GENOMIC LIBRARIES

The top two lines in Figure A-9 represent two copies of a segment of chromosomal DNA that includes two exons of a particular gene for which we have a probe (perhaps a cDNA that hybridizes to both exons, but not the introns). The two copies shown here are meant to represent the millions of copies that exist in a preparation of DNA used to prepare a genomic library. The position of all restriction sites for a particular enzyme are indicated with tick marks.

Below each line, arrows point to the subset of those restriction sites that are actually cut in a *partial* digestion with the enzyme in question; the particular subset of sites digested will be different from one molecule to another. Partial digestion allows the generation of longer fragments to be cloned.

Each fragment from the two copies is identified with a number. Note that the fragments are partially overlapping. From one molecule or another, all regions of the molecule will be cloned, and all fragments are repre-

FIGURE A-9. Preparing and screening a genomic library.

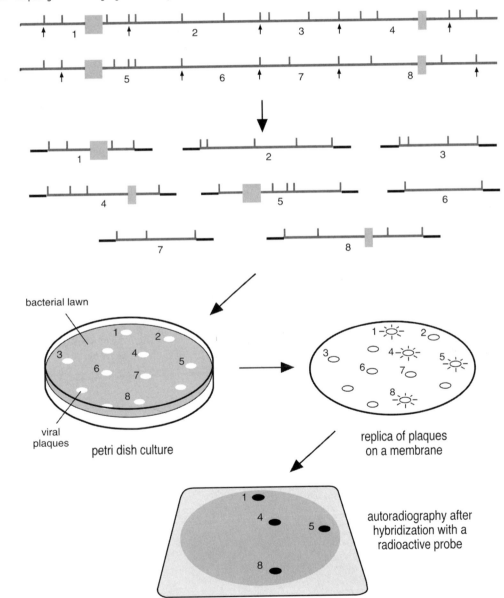

sented more or less equally (although there are certain sequences that make some fragments difficult to clone).

Digesting viral DNA with the same restriction enzyme forms *sticky ends* that allow the ligation of the human DNA to the viral DNA. The virus thus becomes a vector for the cloning of human DNA. Each recombinant virus is unique because it carries a unique piece of human DNA. The viruses are allowed to grow on a *lawn* of bacteria where they form *plaques*. Each plaque is derived from a single viral particle, so all the viruses in a plaque are identical.

If the viral culture is overlaid gently with a nylon or nitrocellulose membrane, a sample of material from each plaque is transferred to the membrane, preserving the overall arrangement of the plaques on the surface. These membranes can be processed to expose the DNA in each phage and allow hybridization to a radioactive probe. After washing away the unbound probe, the filter is pressed against a photographic film for a few hours to allow the radioactivity of the labeled plaques to make an impression on the film.

In Figure A-9, we would expect clones number 1, 4, 5, and 8 to hybridize to the probe. In practice, of course, one needs to test tens of thousands of plaques to find one or a few that hybridize. (These *screens* are often run by testing several dozen Petri dishes each with 1000 to 2000 plaques.) Once a radioactive spot is detected on the film, its location is identified on the Petri dish and a sample of the corresponding plaque is retested and eventually turned into a pure culture of the recombinant virus.

THE POLYMERASE CHAIN REACTION

The polymerase chain reaction (PCR) depends on the ability of DNA polymerase to synthesize DNA only when there is a single-stranded molecule to serve as a template and when a 3' end is available as a primer. If the DNA sequence is known for a segment of DNA, short primers (20 to 30 nucleotides long) can be synthe-

sized that are complementary to sequences spaced a few hundred bases apart on the DNA molecule (Fig. A-10). The native DNA molecules are denatured and mixed with a great excess of primers, such that upon renaturation, the primers associate to the complementary sequences (a) very quickly, before the complementary single-stranded DNA molecules have a chance to reassociate. DNA polymerase then extends the primers to the extent of a few thousand nucleotides until it "falls off" the DNA (b). After a short interval of synthesis, the DNA is denatured again and the cycle repeated.

For simplicity's sake in (c) and (d), we only show extension of primers associated with the newly synthesized molecules (in color) to demonstrate that short fragments are produced. Their size represents exactly the distance between the positions of the primer sequences. As the process is repeated, the short fragments (called *PCR products*) increase in amount in geometric proportion, while the native DNA remains a small fraction of the total. Thus, in 30 to 35 cycles of *amplification* from a trace amount of DNA so small that it would be undetectable, enough PCR product is produced that it can be run on a gel and visualized (when stained with a fluorescent dye) as a sharp band. The position of the band on the gel indicates the size of the fragment and is predicted by the distance between the primers.

In practice, PCR is made extraordinarily simple by the use of DNA polymerases from bacteria adapted to life in very hot thermal springs. Such polymerases are resistant to the temperatures needed to denature DNA (close to 100°C). All the ingredients for the reaction (DNA, primers, the four deoxyribonucleotide triphosphates, DNA polymerase, etc.) are mixed in a small tube and placed in a special bath whose temperature cycles every few minutes between the denaturation temperature and the synthesis temperature. Thus, the repeated reaction cycles are carried out simply by changing the temperature of the reaction tube.

FIGURE A-10. Polymerase chain reaction (PCR).

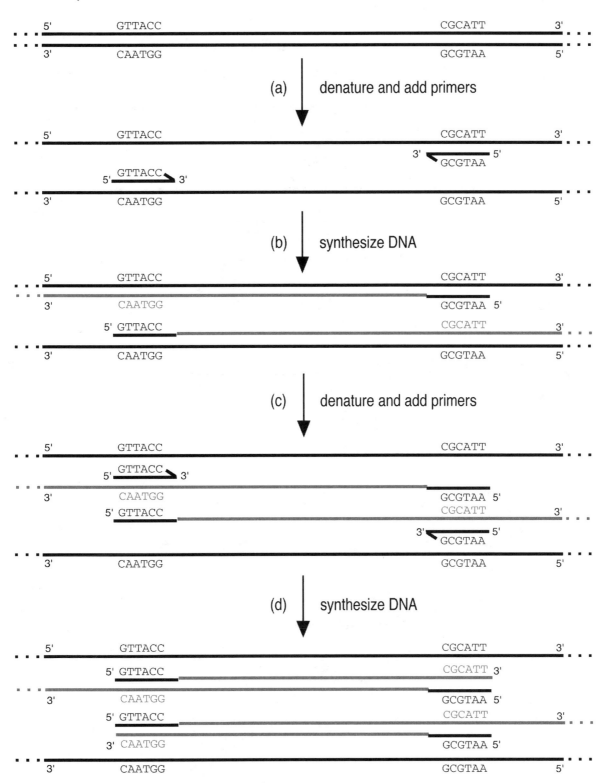

FIGURE A-10. (Continued)

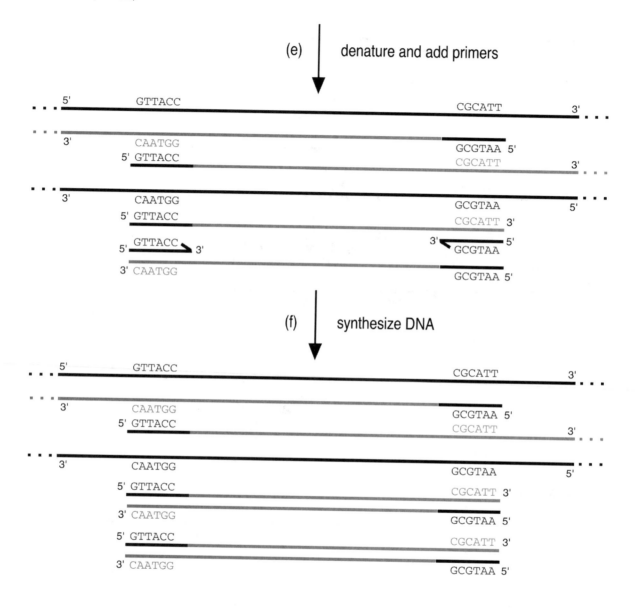

Index